AF560238

THE TIME KEEPERS OF THE VEDAS

The Time Keepers of the Vedas

History of the Calendar of the Vedic Period
(from *Ṛgveda* to *Vedāṅga Jyotiṣa*)

PRABHAKAR GONDHALEKAR

MANOHAR
2013

First Published 2013

ISBN 978-81-7304-969-9

Published by
Ajay Jain *for*
Manohar Publishers & Distributors
4753/23 Ansari Road, Daryaganj
New Delhi 110002

Printed by
Salasar Imaging Systems
Delhi 110007

Contents

Acknowledgements		9
Abbreviations		11
Introduction		13
I.	Bricks and Books	25
1	Indus Civilization	26
2	Ārya and the Vedic Culture	32
3	The Vedic Corpus	37
	3.1. *Ṛgveda*	40
	3.2. *Yajurveda*	44
	3.3. *Sāmaveda*	45
	3.4. *Atharvaveda*	45
	3.5. *Brāhmaṇas*	46
	3.6. *Śrauta Sūtras*	49
	3.7. *Gṛhya Sūtras*	50
4	The Geography of the Vedic Texts	50
5	The Vedic Age	52
	Notes	57
II.	Seasons	61
1	Seasons in the *Ṛgveda Saṃhitā*	63
	1.1. Ṛgvedic climate and paleoclimatic data	70
2	Seasons in the post-Ṛgvedic Saṃhitās and Brāhmaṇas	72
3	Seasonal New Year	81
	Notes	81

III. Days, Months and Years 83
1 The Vedic View of the Sky 89
2 The Vedic Calendar 91
2.1. Day – *aha* 92
2.2. Week – *ṣaḍaha* 97
2.3. Month – *māsa* 98
2.4. Half-Year – *ayana* 101
2.5. Year – *Śarad* and *Saṃvatsara* 103
2.5.1. Vedic New Year 107
2.6. Yuga 108
2.7. Intercalation 110
2.8. Piling of the Fire-altar – *agnicayana* 125
2.8.1. *Mahā Cayana* 126
2.8.2. *Sāvitra Cayana* 129
2.8.3. *Nāciketa Cayana* 131
2.8.4. *Cāturhotra Cayana* 132
Notes 133

IV. *Nakṣatra*s 137
1 *Nakṣatra*s in the Vedic texts 139
2 *Nakṣatra*s – stars and asterisms 145
3 *Nakṣatra*s – lunar mansions 155
3.1. The *jāvādi* arrangement of *nakṣatras* 155
3.2. *Nakṣatra*s – ecliptic and equatorial coordinates of lunar mansions 162
4 Vedic New Year – revisited 167
5 Seasons – revisited 169
6 Vedic Months – revisited 174
7 Epoch of the *Nakṣatras* 181
Notes 190

V. Vedāṅga Jyotiṣa 195
1 Preamble 197
2 The Calendar of *Vedāṅga Jyotiṣa* 197
2.1. Measurement of time 202
2.2. Days 204
2.3. Seasons and the Year 205
2,4. Intercalation – synodic and sidereal 206

2.5. Yuga 216
2.6. *Nakṣatra*s 221
2.6.1. Solar *Nakṣatra*-sectors 222
2.6.2. Lunar *Nakṣatra*-sectors 225
3 Conclusions 227
4 The Time and Place of *Vedāṅga Jyotiṣa* 230
Notes 235

Epilogue 241

Appendix 247

Glossary 273

Bibliography 279

Index of the passages from the Vedic texts 289

Index 295

Acknowledgements

I would like to thank Elizabeth Tucker for her encouragement and help in my studies of the Vedic texts. James Benson is thanked for introducing me to the pleasures and pains of Sanskrit language. Annelie Gondhalekar and Daphne Gondhalekar are thanked for translating German texts into English. B.N. Narahari Achar and Balachandra Rao are thanked for their encouragement. A number of topics in this book have been published in the *Indian Journal of History of Science*. These topics are presented here, in slightly revised and sometimes enlarged form. I would like to thank S.R. Sarma and Christopher Minkowski for their comments and suggestions on the original articles.

Patrick Wallace (Rutherford Appleton Laboratory) is thanked for providing the SLALIB positional-astronomy software and advising me on its use.

I would like to thank J. Kenoyer and Omar Khan for providing Figure I.1.

Awinash Gondhalekar carefully read drafts of this book His numerous suggestions to make the contents of this book comprehensible to a reader not familiar with the subject are greatly appreciated. Needless to say, any shortcomings in the final text are solely my responsibility.

Lastly, I would like to thank my wife, Jane, for her patience and understanding while I laboured over this book.

This research has made use of the SIMBAD database, operated at Centre de Données astronomiques de Strasbourg, France.

This book has made use of material from the TITUS (Thesaurus Indogermanischer Text- und Sprachmaterialien) archive.

The diagrams in this book have been produced with the software GRAPH. I would like to thank Ivan Johansen for making this software freely available.

PRABHAKAR GONDHALEKAR

Abbreviations

AB : *Aitareya Brāhmaṇa*
RJ : *Ārca(Ṛk)-Jyotiṣa*
ĀGR : *Āśvalāyana Gṛhya Sūtra*
AV : *Atharvaveda Saṃhitā*
JB : *Jaiminīya Brāhmaṇa*
KB : *Kauṣītaki Brāhmaṇa*
KpKS : *Kapiṣṭhala-Kaṭha Saṃhitā*
KŚS : *Kātyāyana Śrauta Sūtra*
KS : *Kāṭhaka Saṃhitā*
LŚS : *Lāṭyāyana Śrauta Sūtra*
MS : *Maitrāyaṇī Saṃhitā*
PB : *Pañcaviṃśa Brāhmaṇa*
PGR : *Pāraskara Gṛhya Sūtra*
ṚV : *Ṛgveda Saṃhitā*
ŚB : *Śatapatha Brāhmaṇa*
SGS : *Sāṅkhāyana Gṛhya Sūtra*
TB : *Taittirīya Brāhmaṇa*
TS : *Taittirīya Saṃhitā*
VS : *Vājasaneyī Saṃhitā*
VS(M) : *Vājasaneyī Saṃhitā* (Mādhyaṃdina recension)
VS(K) : *Vājasaneyī Saṃhitā* (Kāṇva recension)
VJ : *Vedāṅga Jyotiṣa*
YJ : *Yājuṣa-Jyotiṣa*
YV : *Yajurveda Saṃhitā*

References to passages in the Appendix: A.n.nn ('n' indicates the chapter number and 'nn' the serial number in the Appendix).

Reference to a passage in the Notes: – N.x.xx ('x' indicates the chapter number and 'xx' the serial number in the Notes).

Introduction

An obsession with time is one of the distinguishing characteristic of human beings. This obsession may have emerged with the realization of human mortality – all life starts at birth and ends with death, and the sensation of passage of time is ever present during this period. As early humans became aware of the regularity of the rising and setting of the sun, the regularity of the phases of the moon, and the regularity of the seasons, they probably realized that it was possible to think of passage of time as an ordered process. This realization may have sparked numeration, mathematics and astronomy. The recognition of this ordered passage of time may have led early humans to identify, classify, and correlate the naturally occurring units of time, namely the day, month and the season. The fundamental unit of time is the day – the period of rotation of the earth about its axis, or as common sense and observations dictate, the time taken by the sun to complete its journey and return to the 'same' point in the sky. Except in the polar regions, the alternating periods of light (day) and dark (night) mark the passage of a day. The month is the regularity of the waxing and waning of the moon. Similarly, the year is a complete cycle of seasons, the time taken by the earth to revolve around the sun in its orbit (or for the sun to move in a full circle against the backdrop of stars). The starting point of a day is well-defined; it is either sunrise or sunset. Similarly, the starting point of a month is also well-defined; it is either the new moon or the full moon. In ancient times, the beginning of the year was not so well defined: a year can be measured from one season to the same season the following year, but seasons do not start on the same day each year. The recurring starting point of the year was determined after careful observation of the apparent motion of the sun across the sky and the identification of the two solstices and the two equinoxes. A very long period of careful observation and record-keeping must have preceded the recognition of these four cardinal days in a year.

A calendar is a system of organizing units of time for social, religious, agricultural, commercial, and administrative purposes. To achieve this it is necessary to identify the 'clocks of nature', classify and label the units of time defined by these clocks and synchronize these clocks. Most (but not all) calendars replicate astronomical cycles according to some fixed rules. The smallest calendric unit is the day (the measurement of fractions of a day is time-keeping). All calendars are based on the succession of days and nights, the waxing and waning of the moon and the rhythm of the seasons. The cycle of seasons controls the availability of food (a primary requirement for survival) for hunter-gatherers, pastoralists, and agriculturists. It is highly likely that the earliest calendar was the recognition and measurement of the duration of the seasons. The ability to count (by the number of days) the length of a season would have given the early man the ability to predict the time of migration of game or migration to pastures and for sowing and planting. The beginning and end of a season is not well defined as it depends on the local weather pattern and that can change in a random manner. However, a delay of a day or two in the start of migration, sowing or planting will not adversely affect a tribe or a village. The early man could thus have counted the length of seasons in integer number of days. It is impossible to guess the next step in the evolution of the calendar. It seems most likely that the early man attempted to group the number of days in a season in a unit of time larger than a day. For this, he turned to the second natural clock, the regular waxing and waning of the moon or the month (Box 1). At this stage, our ancestors would have encountered their first problem with the calendar –

Box 1

He appointed the moon for seasons; the sun knoweth his going down.
– Old Testament, Psalm 104:19.

It was He who gave the sun his brightness and the moon her light, ordaining her phases that you may learn to compute the seasons and the year. – al-Qur'an the number of days in a lunar (synodic) month can sometimes be 29 and sometimes 30. When our ancestors measured the length of a year from a well-defined starting-point like the solstice or equinox or the heliacal rising of a star, as was done by the Egyptians and the Babylonians, they would have recognized the myriad of problems that plague the calendar. The number of days in a year is sometimes 365 and sometimes 366; there can be 12 new moons in a year but sometimes there can be 13. Twelve synodic (lunar) months account for approximately 354 days of the year, leaving a disconcerting deficit of about 11 or 12 days. It would have been apparent that there was not an integer number of days in a month, nor an integer number of days or months in a year. The history of calendars is man's frustrated attempts to reconcile these three naturally occurring units of time or to synchronize these three naturally occurring clocks (Box 2).

Box 2

The calendar is intolerable to all wisdom, the horror of all astronomers, and a laughing-stock from a mathematician's point of view.

– Roger Bacon, CE 1214-94

The calendar has been an indispensable requirement of human life for an awful long time; it has a pivotal role in formulating and regulating the religious, civic, agricultural and economic life of a society. A well-constructed calendar can also be a record of the chronology of a culture. Since the construction of a calendar requires the use of mathematics and astronomy and its use involves a level of sophistication, the calendar and its history provides an insight into the development of a culture. In India, varieties of calendars are in use today. The Calendar Reform Committee, appointed in 1952, identified more than thirty well-developed calendars (Saha and Lahiri 1992). The situation in Afghanistan and Pakistan is somewhat simpler as the Islamic calendar is the principal calendar in these countries. Owing to the continuity of South Asian culture and the diversity of cultural influences in the region, the history of the calendar in South Asia is an exceptionally complex subject. Absolutely nothing is known about the calendar of the Indus Valley Civilization that flourished in this part of the world from about 3000 to about 1300 BCE. The calendar of the Vedic culture, that dominated the northern parts of South Asia from about 1500 to about 500 BCE, forms the subject of this book. The calendar of this period culminated in the calendar of *Vedāṅga Jyotiṣa*, the earliest extant codified calendar of South Asia. From 300 BCE onwards, there were increasing incursions into South Asia by people and cultures from the west and north. These contacts profoundly modified South Asian arts, sciences, sculpture, state-craft, and of course, the calendar. The calendar of *Vedāṅga Jyotiṣa* was gradually modified and eventually replaced. By about CE 400 it had disappeared from all parts of South Asia, leaving only traces in the new calendars that came into use.

In South Asia, an awareness of the order and periodicity of celestial phenomena is found in the oldest available Sanskrit text, namely *Ṛgveda Saṃhitā*. This awareness of the cosmic order is carried over into later Saṃhitās and their associated Brāhmaṇas. In the *Ṛgveda Saṃhitā* the notion of a cosmic order is encapsulated in the concept of *ṛta*; one of the great conceptions of early South Asian culture. This word, derived from the root *ṛ* – to go (and with the suffix *ta*) means 'the correct order of going' or the settled order; it encompasses the concept of both moral and physical world-order. Implicit in this is the idea of a dominant world-order of what is and what must be. The *ṛta* is conceived as absolute and eternal, impersonal in itself and entirely objective and mechanistic. It cannot be influenced in any way and both humans and gods have mandatory duties and functions under it. Events in nature that occur regularly are obedient to *ṛta* or their regular occurrence is the *ṛta*. For example, dawn (*uṣas*) does not violate the heavenly ordinance/order (*daivyāni vratāni*) for each day she returns to her appointed place. The Vedic ritualists saw the calendar as a part or an aspect of *ṛta* and in their ceremonies and observances, they attempted to emulate the cosmic order they saw in the sky. For example, they made offerings to gods from left to right in emulation of the path of the sun (as seen from locations north of Tropic of Cancer); this practice is seen in a number of present-day Hindu rituals. In the post-Ṛgvedic texts, the year is synonymous with both Prajāpati (Creator God) and *yajña* (sacrifice). The Vedic people identified the Creator God with phenomenal time and considered the year to have come into existence with Prajāpati or through his creative activity—a numerical congruence is often noted; there are four syllables in both Prajāpati and *saṃvatsara* (year). Outside the year, there is nothing;

in the beginning there was nothing, before the first creative act there was no year. Without the year, there are no natural processes. Like Prajāpati, the year implies totality and completeness. The cosmic or the divine order was reflected in the rituals and ceremonies of the Vedic people; offerings were made everyday at sunrise and sunset, at every full and new moon, at the beginning of (at least three) seasons and the commencement of the northern and southern course of the sun, i.e. the solstices. The Vedic texts indicate that the daily rituals were (at least in part) performed to maintain a calendar, for example, the *sattra* (ceremony) of *gavam ayana* was performed for 360 days, the length of the Vedic ritual year.

The calendar of the Vedic period came under the scrutiny of the Western Sanskrit scholars from the time they became acquainted with Sanskrit texts. These scholars considered *Vedāṅga Jyotiṣa*, the earliest extant codified calendric text in South Asia (believed to have been composed around the middle of the first millennium BCE), to be the starting point of the calendar of this period. However, the earlier Vedic texts describe and justify the performance of *yajña*s (the ritual sacrifices). In South Asia, astronomy, geometry and mathematics were, at least partly formulated and developed to cater for the requirements of these *yajña*s. These rituals were performed at prescribed times and their performances presuppose a calendar. Therefore, a Vedic calendar must have existed long before *Vedāṅga Jyotiṣa*. In a seminal study in late nineteenth century, Dīkṣita (1896) identified all elements of the calendar of *Vedāṅga Jyotiṣa* in these earlier ritual and sacerdotal texts. He also identified following fundamental elements of a calendar in the *Ṛgveda Saṃhitā*.

- *ṛtu* – season
- *saṃvatsara* – year; that is, tropical or the seasonal year and the ritual year
- *māsa* – month.
- *adhimāsa* – intercalary month.
- *pakṣa* (or *ardhamāsa*) – half-month or a fortnight.
- *aha* – day and *ahorātra* that is day and night.

In addition, there are tantalizing clues to a stellar frame of reference (*nakṣatra*s) to define the position of the moon. The final redaction of *Ṛgveda Saṃhitā* has been preserved with exceptional fidelity. The calendric parameters identified in this Saṃhitā are, therefore, original and authentic to this text. There is not enough information in *Ṛgveda Saṃhitā* to reconstruct a calendar. However, it is possible to deduce that this is a lunisolar calendar incorporating various approximations. It is not clear if these approximations reflect the available measuring technologies, were introduced for ease of computations, or mirror the beliefs of the Vedic people. However, it is abundantly clear that the three 'clocks of nature' were identified for this calendar and a scheme was developed to synchronize these clocks. It is not possible to trace the origins of this calendar from the available Vedic texts but the sophisticated calendar of *Ṛgveda Saṃhitā*, requiring astronomical observations over long periods, record-keeping, and computations, is more likely to have been developed in an urban culture rather than in a pastoral or semi-nomadic society. This conclusion is supported by the identification in this text of the quasi-synodic or 'ideal' year of 360 days and the intercalation of five days to synchronize this year with the tropical/seasonal year. This scheme of synchronization has been identified in the calendars of a number of ancient cultures, separated widely in space and time (N.III.9), and these are Bronze Age urban cultures and not nomadic or pastoral cultures.

Although the fundamental elements of a calendar can be identified in the *Ṛgveda Saṃhitā*, this Saṃhitā does not explicitly refer to the use of a calendar. The days and times at which rituals or ceremonies were performed are not given. This ambiguity is removed in the later texts in which the chronological markers like the position of the sun (that is, a season or a month) and the position and the phase of the moon among the stars, for performance of rituals, are unambiguously identified. The calendar of these texts is a development of the calendar of the Ṛgvedic Saṃhitā with the addition of following refinements:

- A fiducial season and month are defined for start of the tropical/seasonal year and the ritual year.
- The 'length' of a season is defined and the 'date' of the performance of the three seasonal sacrifices is identified by the coincidence of the full moon and predefined asterisms or *nakṣatra*s.
- Months are labelled by their seasonal characteristics and by asterisms or *nakṣatra*s that conjoin twelve full moons in a year.
- Algorithms are given for intercalation.

- A greater degree of precision is introduced in the timing of the rituals and ceremonies by identifying the stars with which a phase of the moon should coincide for the performance of the ceremonies.
- The heliacal rising (and probably setting) of prescribed stars is used to determine the appropriate time for the performance of some rites.
- Time-keeping is explicitly introduced by dividing a day (sunrise to sunrise) into smaller time units.

In the post-Ṛgvedic texts, astronomers (*nakṣatra darśa*, *TB*.3.4.4) and mathematicians (*gaṇaka*) are mentioned, and a sacrificer is advised that in order to determine the correct time of a ceremony 'he should either study astronomy, or ascertain the correct time from those who know it'. These texts describe both an arithmetic calendar and an astronomical calendar. In an arithmetic calendar, the synodic (lunar) year is synchronized with the seasons by simple but rigid arithmetic rules. The astronomical calendar, on the other hand, synchronizes the synodic (lunar) year with seasons by astronomical observations of the solstice, the new moon, the heliacal rising of specific stars, etc. The calendar that can be formulated from these texts is substantially the calendar described in *Vedāṅga Jyotiṣa*. Not surprisingly, this text also describes an arithmetic and an astronomical calendar but the astronomical calendar of this text is different from the astronomical calendar of the Saṃhitās and the Brāhmaṇas. Moreover, *Vedāṅga Jyotiṣa* introduces a number of new and in particular mathematical concepts. Significantly, in this text the *nakṣatra*s are not stars and asterisms, but are redefined as sectors of the ecliptic. These sectors may have been identified by *nakṣatra*s (stars and asterisms) of the Saṃhitās and the Brāhmaṇas but the absolute position (on the ecliptic) and width of the sectors are independent of *nakṣatra*s. For calendric purposes, *Vedāṅga Jyotiṣa* defines the position of the sun and the moon within these sectors. The position of the sun with respect to the *nakṣatra*s is not mentioned in the Saṃhitās and the Brāhmaṇas and is original to *Vedāṅga Jyotiṣa*. This text also introduces mechanical means (clocks) for time-keeping.

A number of publications have explored the Vedic texts to show the link between the Vedic calendar and calendar of the *Vedāṅga*

Jyotiṣa. Others have asserted that the calendar of *Vedāṅga Jyotiṣa* has been almost entirely imported from sources outside South Asia and is a totally new development with no links to the South Asian past. In this book it is shown that there is an unbroken thread of development of the Vedic calendar from the *Ṛgveda Saṃhitā* through the later Saṃhitās and their associated Brāhmaṇas and Sūtras, culminating in the calendar of *Vedāṅga Jyotiṣa*. The history of this calendar is punctuated with a number of novel features, original and unique to South Asia. All civilizations and cultures are influenced by their neighbours and this would be true of the Vedic culture as well, but the analysis presented in this book demonstrates that there is no *a priori* reason to believe that, in its totality, the Vedic calendar was not a product of the cultures that flourished in South Asia.

The chronology of the Vedic culture has been a matter of contention since the time when Western scholars became aware of Sanskrit and in particular the Vedic texts. It has also been recognized, since this time, that astronomical references in these texts hold the promise of independent corroboration of chronology based on philological analysis. In the last decade of the nineteenth century and the first of the twentieth century, the astronomical significance of various passages of the Vedic texts and the dates that could be inferred from these passages, were subjects of intense speculation and debate. It is not the intention of this book to reopen this debate, but the analysis of the Vedic texts and the calendar derived from this analysis does lead to independent conclusions regarding the chronology of this period. The following four independent considerations suggest a date of 1400±300 BCE:

- The visibility throughout the year of all seven stars of *sapta ṛṣis* (constellation Ursa Major or Big Bear) from (mid) South Asian locations.
- The time of performance of the seasonal sacrifices defined by the coincidence of the season, the full moon and the prescribed *nakṣatra*s.
- The time of start of a *yuga* defined by the heliacal rising and setting of a prescribed *nakṣatra* at the winter solstice.
- Cross correlation of the *nakṣatra*-sectors and the *yogatārā*s of *nakṣatra*s.

It is highly likely that around this time the New Year's Day or the start of a *yuga* was moved from spring to the winter solstice. This analysis challenges the assumption and belief of (mostly) Western scholars and commentators that the Vedic people had only a crude calendar; that they were passive recipients of stellar astronomy (*nakṣatras*) and mathematical astronomy from West Asian sources, which they did not understand and were incapable of refining for their location. This analysis also does not support the high antiquity of 'Indian astronomy' favoured by some Indian scholars and commentators.

The calendar of the Vedic people, described in this book, is based entirely on the religious, liturgical and sacerdotal texts left to us by the Vedic culture. To put the Vedic calendar in a historic context, the complexities of the early history of South Asia are briefly described in Chapter I. The contentious 'Āryan question' is very briefly touched upon in this chapter. In addition, an opportunity is taken to briefly describe the Vedic texts from which the Vedic calendar is derived. A brief description of the geography and the current thinking on the chronology of these texts are given in this chapter.

The identification of the seasons is the foundational element of all calendars, and in Chapter II, the references to seasons in the Vedic texts are analysed to show that the Vedic people had identified and labelled the weather-system of the lands they occupied before the composition of the *ṛcā*s (verses) of *Ṛgveda Saṃhitā*. They moulded their classification to the requirements of their belief-system. Discussions over the last 150 years, of the seasons in the Vedic texts, are based on the belief that the climate of South Asia has been static for last 4,000 years and that the climate during the Vedic period was not particularly different from the present climate. Not a great deal of paleoclimatic data from South Asia is available at present, but enough is known to show that the assumption of a static climate of South Asia is false, and paleoclimatic data should be included in any future studies of the climate of the Vedic period. Chapter III describes the Vedic view of the heavens and how the 'clocks of nature' were identified and codified to establish the fundamental parameters of a calendar. The vexed question of intercalation in the Vedic calendar is addressed. It is shown that far from 'no systematic intercalation scheme can be attributed to this (Vedic) period' (Pingree 1970-80), the Vedic people had 'experimented' with a number of intercalation schemes. Mathematical interpretation of some passages of the Vedic texts suggests that a 'perfect' intercalation scheme may have been formulated. The importance of a calendar to the Vedic people is demonstrated by the fidelity with which they mapped their calendar onto the altars on which the ritual fire, which was fundamental to their belief-system, was lit.

The stellar frame of reference to chart the path of the moon (and later the sun) in the firmament reaches its full maturity in the post-Ṛgvedic Saṃhitās and the Brāhmaṇas. This frame of reference, called the *nakṣatras*, is the subject of Chapter IV. It has been known for some time that, in the Vedic texts, the *nakṣatras* can mean both the stars and sectors of the ecliptic. Nevertheless, almost all discussion has been about *nakṣatras* as stars; *nakṣatras* as sectors of the ecliptic, independent of the stars, have received no attention. It is shown in this chapter that a passage in *Vedāṅga Jyotiṣa* enables absolute (and invariant) coordinates of *nakṣatra*-sectors to be determined without reference to the stars. The division, in *Vedāṅga Jyotiṣa*, of the ecliptic into twenty-seven equal sectors to define the position of the moon (and the sun), is unique to South Asia.

An invariant coordinate system is essential to interpret unambiguously the algorithms of *Vedāṅga Jyotiṣa*. The algorithms of *Vedāṅga Jyotiṣa* are presented and discussed in Chapter V. These algorithms are mathematical expressions of the calendric concepts developed in the Saṃhitās and the Brāhmaṇas.

The analysis presented in Chapters II to V suggests that the calendar of the Vedic Age can be divided into three broad overlapping periods; these are the *ṚV*-period, the post-*ṚV*-period and the *VJ*-period. Each period builds on the developments of the previous period but there were also unique developments in every period. The developments in each period and links to the following period are presented in Chapter VI.

The discussions in this book are based on the Vedic texts. The distance (in time) from the Vedic culture, the complexities of this culture and Vedic Sanskrit and the lack of understanding of these complexities makes interpretation of passages of Vedic text both difficult and not wholly unambiguous. The interpretation of the passages of the post-Ṛgvedic texts is marginally simple as the purpose of these texts is well defined. This is not so of *Ṛgveda Saṃhitā*; the true purpose of this text has not been identified, as its contents are a complex of liturgical formulae, mythological tales, magic spells, and philosophical speculations. The verses of *Ṛgveda Saṃhitā* that reinforce the discussions in this book are given in the Appendix. The available English translations and interpretations are also given in the Appendix. To allow for evolution in the understanding of Vedic Sanskrit and the Vedic texts, only translations made in the twentieth century are given. A number of verses referred to in this book have no published translations and translations of these verses in the Appendix are my translations. Some passages from the post-Ṛgvedic texts that are also deemed to be significant, are given in the Appendix. The reference to a passage in the Appendix is identified by 'A' followed by the chapter number and a serial number. Topics of relevance but not essential to the subject of this book are presented in the Notes. The notes relevant to the topic of a chapter are given at the end of the chapter. A Note is identified by 'N' followed by the chapter number and a serial number. A brief description of ritual and technical terms is given in the Glossary.

The primary purpose of a calendar is to identify days of civil, agricultural, religious or social significance. Thus, a calendar by its very nature should have predictive capability. In the past, passages in Vedic texts that have calendric significance were analysed in isolation and their predictive potential was rarely explored. Similarly, the astronomical observations implied in the calendric passages were not interpreted in terms of concepts with which the Vedic calendar-makers would have been familiar. These explorations are only possible mathematically and these are presented in this book. It is hoped that this will not deter the lay reader, as every effort has been made to make this book comprehensible to the general reader and the simple description and the supplementary notes should guide a reader through unfamiliar terrain.

The people whose calendar is described in this book called themselves Ārya. The word *ārya* has been much abused in the West and has racial connotations. In South Asia, *ārya* used as an adjective means 'noble, honourable' and *ārya* can also mean 'those who maintain the world order by means of sacrifice and gifts' (to the gods). In South Asia, the word *ārya* has never been used to describe or define a race, and this tradition is followed here. Ārya is not meant to indicate a distinct ethnic group but a group of people with common religious practices and a common spoken language (archaic Sanskrit). The rather cumbersome phrase South Asia(n) is used throughout this book because the usual practice of identifying the Vedic culture with India is geographically restrictive and incorrect. In the long history of South Asia; Afghanistan, India and Pakistan are recent constructs. The Indus and Vedic Civilizations are as much part of the heritage of the people of Afghanistan and Pakistan as they are of the people of India.

I

Bricks and Books

Our knowledge of ancient South Asia (modern-day Afghanistan, India and Pakistan) rests on twin pillars of philology (or linguistics) and archaeology. The Indo-European language family was discovered when Western scholars became acquainted with Sanskrit language and its texts. Sanskrit was brought to the attention of Europeans through a letter written in 1544 (Amaladass 1992) by the Jesuit St. Francis Xavier (1506-52), he described Sanskrit as equivalent to Latin in the West that is, a scholarly language used for both ecclesiastic and secular writings by speakers of different languages of India. In 1660, the German Jesuit Fr. Heinrich Roth wrote the first Sanskrit grammar in Latin (Amaladass 1992). In 1783 an East India Company official, Sir William Jones (1746-94) arrived in Calcutta as the new Supreme Court judge. William Jones was an established oriental scholar and knew Hebrew, Arabic and Persian and he rapidly learned Sanskrit. On his initiative and with him as the first president, the Royal Asiatic Society was founded in 1784. In his now famous address to the Society in 1786, William Jones identified the strong affinity both in 'root of verbs and the forms of grammar' between Sanskrit, Greek, Latin, Persian and also Gothic and Celtic. William Jones and William Marsden (another East India Company employee) also demonstrated that the Romani language of the Rom or the Gypsies of Europe was Indo-European and that these people came from South Asia a long time ago. These discoveries, towards the end of eighteenth century, marked the start of the study of comparative philology of the Indo-European languages and ultimately the science of linguistics.

In 1816 Francis Whyte Ellis, a colonial civil servant in Madras and his Indian collaborators Pattabhiraman Shastri and Shankara Shastri (at the College of Fort St. George, Madras) demonstrated that the south Indian languages, Tamil, Telugu and Kannada shared a common stock of roots and forms of grammar and that these were different from those of Sanskrit. Later this analysis was extended to other south Indian languages. These languages form the Dravidian language family. It has since been established that there are three principal language families in South Asia. The Indo-European family (that is Sanskrit and the languages descendant from it in the north and the western parts of India), the Dravidian family (dominating south India) and the Munda or Austro-Asiatic language family (represented in parts of east and central India and the north-eastern parts of India and Bangladesh). These language families have co-existed in South Asia for millennia. The discovery of these language families on the subcontinent demonstrated that the early civilization of South Asia was a fusion of cultures of different linguistic groups. It also undermined the traditional view among many learned Hindus that Sanskrit was eternal and a source of all other languages.

Early European visitors to South Asia took a keen interest in its ruins, monuments and ancient history generally. In 1853, Sir Alexander Cunningham (appointed director-general of the newly-established Archaeological Survey of India (ASI) in 1871) initiated a programme of archaeological surveys to identify ancient Buddhist cities, visited by the Chinese Buddhist pilgrim Hsüang Tsang in the seventh century. At Harappa (a collection of mounds in the Punjab), Cunningham discovered seals inscribed in an unknown script. He assumed that these were a foreign import and believed that this was a site of one of the cities visited by the Chinese pilgrim. Unknown to Cunningham, remains of a Bronze Age city lay buried at this site.

1. INDUS CIVILIZATION

In the winter of 1911-12, Devadatta Ramkrishna Bhandarkar, the Superintending Archaeologist of the ASI, while visiting the province of Sind (in Pakistan) noted one large mount and six small ones at Mohenjo-daro (the Mount of the Dead Men). In his report Bhandarkar dismissed the site as 'not representing the remains of . . .

any ancient monument'(Keay 2000), words that must have haunted Bhandarkar in later years. Ignoring his report, archaeologist Rakhal Das Banerji undertook exploration at Mohenjo-daro in 1922-3 and immediately discovered seals similar to those found at Harappa by Cunningham. At about this time, Rai Bahadur Daya Ram Sahni who had returned to the site at Harappa discovered more seals similar to those found by Cunningham during his excavations at Harappa in 1853. In 1924, Sir John Marshall, the then Director General of ASI published these finds in the *Illustrated London News*. Within weeks, the Indus seals were recognized to be similar to those found at Mesopotamian sites of third millennium BCE; the wraps of millennia had come off the Indus Civilization and it was revealed, to challenge both archaeologists and historians.

Following these discoveries in the first quarter of the twentieth century, a host of archaeologists from India, Pakistan and many other countries have carried out excavations at these sites and other sites in the Indus Valley and the surrounding regions. Possehl (1991) has published an account of excavations at Harappa from 1826 to 1990. Harappan sites have been discovered as far as the Iranian frontier in Baluchistan, in the North-West Frontier Province of Pakistan, Uttar Pradesh and Gujarat in India (Figure I.1). The north-south extent of this civilization, that is, from Lothal, a site in Gujarat, to Shortughai (a site in the Amu Darya (Oxus River) Valley in north-east Afghanistan), is over 1,600 km. The east-west extent, from Alamgirpur on the upper Ganges to Sutkagen-dor on the Makran coast, is similar. Guided by their colonial prejudices, pioneers like Marshall looked to the West to explain the Indus Civilization and suggested that the vast and sophisticated civilization was an offshoot or a colony of the Mesopotamian Civilizations and the Indus Valley Civilization was initially dubbed as the Indo-Sumerian Civilization (Possehl 2002). This conjecture has now been conclusively refuted (Kenoyer 1998, Possehl 2002, McIntosh 2008). At numerous sites in Baluchistan, Afghanistan, in the Indus Valley itself and along the now dry Sarasvati or Ghaggar-Hakra River, evidence of thousands of years of continued occupation have been found at pre-Harappan and Early Harappan sites. Evidence of refinements in agriculture, animal husbandry and other technologies have established a local progression from the hunter-gatherer to urban dwellers with all the

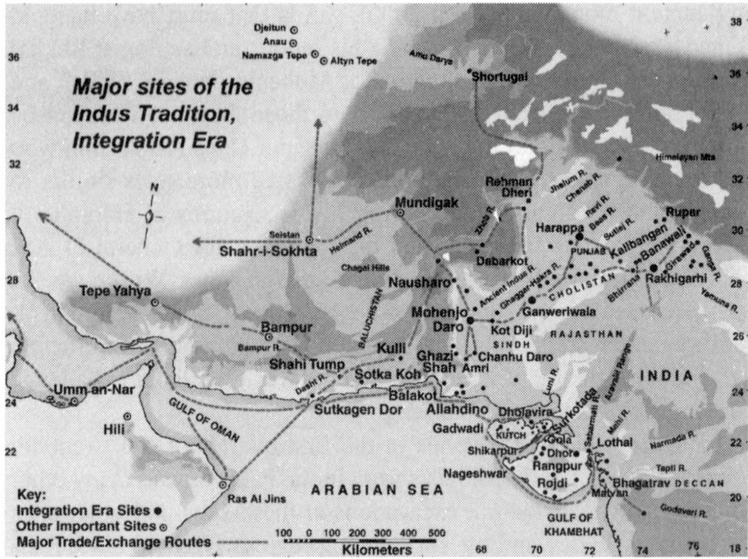

Figure I.1: The currently known sites of the Indus Valley Civilization. Note the distribution of sites along the course of both the Indus and Ghaggar-Hakra (the dried-up bed of Sarasvatī) rivers.

Copyright J.M. Kenoyer/Harappa.com, Courtesy Dept. of Archaeology and Museums, Govt. of Pakistan.

intermediate stages of pastoral culture, agricultural settlements and technological developments. Excavations have continued throughout the twentieth century but the basic questions about the Indus Valley people who created this complex culture(s) remain unanswered. Only a brief outline of this culture(s) is known at present and an even briefer outline can be given here.

The early civilization in South Asia is called the Indus Civilization because the urban centres along the Indus River were discovered first. It has since been recognized that another ancient river flowed to the east and parallel to the Indus. Today it is generally called Ghaggar in India and its lower stretch in Pakistan, mostly dry, is the Hakra. According to some scholars, linguistic, archaeological and historical data show that Sarasvatī of the Vedas is modern Ghaggar-Hakra (Possehl 2002, Gupta 2005). Numerous surveys in the deserts of present-day Cholistan and Rajasthan have revealed a large number of settlements along the banks of this river system. Urban centres like Ganweriwala and Rakhigarhi were comparable in size to Mohenjo-daro and Harappa on the Indus. The Indus and Sarasvatī rivers provided rich alluvial soil that made the region extremely fertile for farming. The rivers also provided easy routes for river-barge commerce, the hallmark of Indus Valley Culture. Settlements along the two rivers flourished from about the fourth millennium BCE. However, due to various geological factors the Sarasvatī River started to dry up towards the beginning of the second millennium BCE (Kenoyer 1998, Gupta 2005, McIntosh 2008) and the settlements along this river system were gradually abandoned. Some scholars would like to rechristen the early civilization of South Asia as Indus-Sarasvatī Civilization (Gupta 2005). However, in the following discussion, the traditional usage, i.e. Indus Civilization, is followed but the civilization along the Indus and Sarasvatī rivers is meant.

It is now recognized (Kenoyer 1998, McIntosh 2008) that the Indus and Sarasvatī River systems and the surrounding region was home to the largest of the four urban civilizations of Egypt, Mesopotamia, South Asia and China. In Pakistan and north-west India, over 1,000 settlements of this civilization have been identified (Gupta 2005). It is estimated that at its peak the total population was about a million. The principle strata of this civilization are shown in Table I.1. The Indus Civilization developed out of pastoral and farming communities on the western fringes of the Indus plain. The earliest evidence for agricultural and pastoral communities in South Asia is found at Mehrgarh, dated to around 6500 BCE. Here farmers cultivated barley and wheat and had herds of cattle, sheep and goats. They lived in mud-brick houses that were sub-divided into rooms and had elaborate bead ornaments made of shells and coloured stones. These ornaments appear to have been traded with distant lands. Towards the end of this period, the first coarse chaff-tempered pottery begins to appear. A

Table I.1: Chronology of Indus and Post-Indus Civilizations in South Asia (Kenoyer 1998)

Post-Indus or Indo-Gangetic Tradition	1300 to 300 BCE
Localization Era (Late Harappan Phase)	1900 to 1300 BCE
Integration Era (Harappan Phase)	2600 to 1900 BCE
Regionalization Era (Early Harappan Phase)	5000 to 2600 BCE
Early Food Production	6500 to 5000 BCE

period of development started at around 5000 BCE and for next 2,500 years different pottery designs, ornaments, architectural styles and methods of animal husbandry evolved. In this era, wheel-thrown pottery, copper tools, stone beads and carved seals were manufactured. Writing in the form of graffiti on pottery appeared in this period. This was a dynamic era with extensive trade links along the major river routes and through mountain passes. Raw materials and manufactured goods moved between communities and settlements scattered over a large area. This is known as the Early Harappan Period. Towards the end of this period, initial steps towards urbanization were taken. This period is characterized by specific type of painted pottery and ornaments, the appearance of seals with rudimentary writing and an expanded trade network (Kenoyer 1998, McIntosh 2008). A synthesis of various cultural groups occurred from about 2600 BCE to 1900 BCE (the Harappa Phase). Communities along the Indus and Sarasvatī (or Ghaggar-Hakra) rivers and the adjacent lands were organized into cities and towns. The geographic spread of this period was vast, almost twice as large as the Mesopotamian or Egyptian communities at the same time in history. The cities and towns were built in a grid pattern with major and minor streets. The architecture of the cities and the houses suggests a well-organized civic authority. Unifying symbols appear on painted pottery, ornaments and ritual objects and weights and measures were standardized. The cities traded with both the local rural agricultural communities and with distant sources of raw materials. The archaeological records offer evidence of long-standing commercial contacts with Central Asia, the Middle East Gulf, the Mesopotamian cities and Syria. A legacy of this commercial ability is echoed in the combined binary and decimal system of weights and measure that have been handed down to the modern world. These city communities developed a distinctive form of writing, the most prolific examples of which are found in the seals probably used to identify an individual or an organization (may be a trade guild). Only short inscriptions have been found at present and the writing system has not been deciphered nor the underlying language identified. Nothing definite can be said about the belief system of the Indus Valley people although clay figurines of men, women and animals and narrative scenes on the seals suggest rituals and ceremonies. Uniquely among all civilizations, the Indus Civilization appears to have developed and thrived without (excessive) internal or external conflict. There is also no evidence of an overarching monarchy and no palaces or temples have been discovered and very few structures can be identified as having had religious function. The discovery of a public bath and the architecture of this bath suggest rituals in which water was used to either 'initiate' and/or 'purify' the worshippers. The people buried their dead in wooden coffins along with grave goods that included few simple ornaments. No royal burials have been found.

After about 1900 BCE, there was a gradual weakening of the central economic (and probably political and religious) authority of Indus Valley cities. The Indus-Sarasvatī region began to fragment into smaller economic and political units (Ratnagar 2005). Commercial links with the distant lands of Mesopotamia and the Middle East Gulf were severed, and the focus of attention shifted to the lands east of the two river systems. The causes of decline of the Indus Valley cities are many and varied. Geological and hydrological factors seem to have played a major part. The Indus and Sarasvatī River systems were 'life-lines' for the cities. The waters of these rivers (and their tributaries) provided bountiful harvests and were communication arteries to surrounding towns, resource centres and coastal regions that were linked to distant lands. This civilization and in particular the core cities were vulnerable to shifts and changes in the course of the two river systems. Due to sedimentation and tectonic movement, the water that had flowed into the Sarasvatī was diverted to the west into river Sutlej, a part of the Indus River system, and eastwards into river Yamunā (Gupta 2005). The settlements along the Sarasvatī were thus deprived of water and had to be abandoned. Due to the increased water-load the Indus shifted its course to the east, flooding many settlements and burying them in silt. The cities on higher ground and those protected by flood-walls survived but repeated flooding and shifts in the course of the rivers damaged the agricultural foundation of the societies and destroyed the economic infrastructure.

Although the changes in the river systems were devastating, there was never a total collapse of this civilization (Ratnagar 2005). Some people moved, adapted, survived and ultimately flourished. These people adapted their agricultural practices to the monsoon dominated weather systems of the Gaṅgā-Yamunā Doab and Gujarat and new crops like rice, millet, etc., were cultivated. The observed changes in the ceramic styles, design of the seals and burial customs have enabled identification of three phases of the Late Harappa Period. These phases are named after the sites where specific pottery styles were first discovered. The Punjab Phase represents the northern cultural region that encompassed Harappa and sites to the east in northern India. The Jhukar Phase, named after a site near Mohenjo-daro in the southern Indus Valley refers to sites in Sind and parts of Baluchistan. The Rangpur Phase refers to the entire Kutch region, Saurashtra and Gujarat (Kenoyer 1998).

Excavations at Harappa and other sites have shown that the transition from the Harappa Phase to the Punjab Phase was gradual and extended to the Gaṅgā-Yamunā Doab to the east and throughout northern Pakistan and as far as Swat Valley where it mixed with the distinctive local tradition. However, the trade contacts with the western highlands and the coast were lost. But this was also a period of innovation; earliest glass (in South Asia), dated approximately to 1730 BCE is found during this phase, a deep azure faience, made probably in imitation of the unavailable *lapis lazuli*, is encountered for the first time and new techniques were developed to drill small beads (Kenoyer 1998). Similarly, the Jhukar Phase gradually emerges but shows strong continuity with the preceding Harappa Phase. Major differences are observed in the pottery style, the absence of script, and the increased use of circular seals with geometric designs. Changes and evolution were also taking place further south on the islands of Kutch and in Saurashtra and Gujarat. Some Harappa Phase artefacts disappeared and new ones appeared and began to dominate. However, some manufacturing techniques invented during the Harappa Phase continued. The major difference is in the almost total cessation of trade with the northern regions and lands overseas. This trade did not recover until around 600 BCE. Excavations and discoveries in and around Indus and (ancient) Sarasvatī River systems, the Gaṅgā-Yamunā Doab and Gujarat demonstrate that the period following the decline of the Indus Valley cities was not one of abandonment and desolation as portrayed in the earlier literature.

2. ĀRYA AND THE VEDIC CULTURE

The late nineteenth-century Western 'theory' of Indian civilization was that the Indian civilization is a product of the invasion of South

Asia by Sanskrit-speaking Āryans (or *ārya* in their own language) and the subjugation of the dark skinned, indigenous savages. The twentieth-century discovery of the Indus Civilization has led to two minor modification of the 'invasion theory'. One version is that the Indus Civilization had collapsed before the invasion of the Āryas. Another version is that the invading Āryas destroyed the Indus Civilization. The celebrated British archaeologist Sir Mortimer Wheeler declared, 'on circumstantial evidence, Indra (a Āryan god) stands accused' of destroying the Indus Valley cities (Wheeler 1947). This conclusion was based on uncritical and inaccurate understanding of *Ṛgveda Saṃhitā* [written in archaic Sanskrit (N.I.4), it is the earliest surviving collection of poems in any Indo-European language]. This view was, however, in accord with the nineteenth- and early twentieth-century European militarism (that is, conquest and subjection were the engines of cultural change) and preconceptions about the Āryas. The wide distribution of the Indo-European languages has brought to fore, in last few decades, theories of migration as the engine of change and development (McIntosh 2008). In response to this, the Aryan 'invasion theory' has evolved into the Āryan 'migration theory'. According to this theory, small bands or tribes of Indo-Aryan people entered South Asia via the western and/or north-western mountain passes. These people interacted and integrated with the indigenous population and some among the indigenous population and some among the immigrant population may have been bilingual (Witzel 2001).

The home of the Ārya and their immigration to South Asia are still contested topics. That there was migration(s) of Indo-European language speakers (Vedic Sanskrit belongs to this language family) possibly in waves is indicated by archaeological and epigraphic evidence from West Asia. Mesopotamia witnessed the arrival of the Kassites, who had Indo-European names, in about 1760 BCE. They introduced the horse and the chariot to this region. A Hittitte-Mitanni treaty lists four well-known Vedic divine names (Indra, Mitra, Nāsatya, and Varuṇa). This treaty, by the Mitanni king Matiwāza, can be independently dated to about 1350 BCE. Clay tablets dated to about 1400 BCE written at el-Amarna in Egypt in Akkadian cuneiform, mention princes with Indo-European names (Dumont 1947, Thieme 1960, Burrow 2001). Nearer South Asia, a striking correspondence is found between the Iranian religious text, the *Avesta*, and the *Ṛgveda*.

This suggests that the Indo-Iranians were settled over the lands stretching from the Iranian Plateau to north-western parts of South Asia. However, the evidence of the presence of Indo-European in West Asia tells us nothing about the time of arrival of Āryas on the Iranian Plateau or in South Asia as the movement of different groups of people need not be coordinated nor does it have to be sequential.

In central Asia and Afghanistan the Bactria-Margiana Archaeological Complex (BMAC), represents a diverse mix of nomadic and settled communities. The artefacts of these people are scattered over a large region, from deserts of Central Asia to southern Baluchistan and from the edge of the Indus Valley to Iran. The nomadic herders and traders of these communities were in contact with the Indus Valley before and during the Harappa and the Late Harappa Period (Ratnagar 2005). There would have been a continual exchange of goods, ideas and people (through marriage, trade alliances and wars) during this time. It has been suggested (Parpola 2004) that at about 2100-1900 BCE, Bactria-Margiana was taken over by 'Āryan-speaking' steppe nomads. These people, augmented with immigrants from southern Urals, steppes of Eastern Europe and Iran expanded into northern South Asia in two waves to form the core of the Ṛgvedic Āryas. It has been claimed that the *śrauta* ceremonies of *Agniṣṭoma* and *Agnicayana* mirror this eastward march of the Āryas or of the Vedic Culture (Staal 2004, N.I.1). This model of Āryan migration/ invasion, like all previous models, depends on major presuppositions. The model assumes that the archaeological complexes and artefacts discovered in southern Central Asia, Afghanistan and the Swat Valley 'belong' to the (pre-)Vedic Āryas. However, there is no corroboration for this assumption. Moreover, the Vedic texts do not encourage association of their rituals and hymns with any archaeological discoveries in Central Asia or the Indus Valley.

Migrations and invasions are quite common in the history of nations and it is unlikely that South Asia was immune from such movement of people. Migrants and invaders usually leave foot-prints in the language, art, architecture, religion, governance, etc., of the occupied lands. These foot-prints can be used to trace the emigrants to their 'homeland'. However, from the end of the Harappa Phase at about 1900 BCE to the beginning of the Early Historic Period around 600 BCE, no archaeological, epigraphic or biological foot-prints have been found to support the invasion or mass migration into the Indus Valley (Kenoyer 1998, Lal 2005)). The *Ṛgveda* is silent on the early history of Āryan migration. The world of *Ṛgveda* is the land of Sapta Sindhu (seven rivers – Indus and its five tributaries plus Sarasvatī, Section 4), i.e. the land stretching from the Kabul River to the Gaṅgā-Yamunā Doab. Lands outside this region are not mentioned. However, a number of non-Sanskrit names in the *Ṛgveda Saṃhitā* suggest that the Ārya were in close contact with people whose language was not Sanskrit. Moreover, there are some grammatical features of Sanskrit that are not found in any other Indo-European languages but are found in some languages of South Asia. This suggests that the contact between the non-Sanskrit-speaking people and the Sanskrit-speaking people occurred in South Asia, most probably in the Indus Valley and in the lands to the north of this valley.

The current assumption is that the (Proto-Indo-)Aryans arrived in South Asia around 1900 BCE (i.e. after the desiccation of the Sarasvatī River, but see N.I.2) and the early verses (*ṛcās*) of *Ṛgveda* were composed around 1700-1600 BCE (see Table I.3). These dates arise from the claim that the '*Ṛgveda*, (which) no longer knows of the Indus cities but only mentions ruins (*armaká* and [*mahā*]-*vailasthāna*)' (Jamison and Witzel 1992, Witzel 1995). This follows from the analysis of the word *árma*- and [*mahā*]*vailasthāna* by Burrow (1963, N.I.3). These words occur in only one hymn (A.I.1, 2) of the *Ṛgveda Saṃhitā* and Burrow's analysis suggests that the Āryas encountered both ruins and cities which they laid waste. Archaeological research, however, indicates that the Indus Valley was never a 'waste-land' (Ratnagar 2005). The metropolitan cities and the trade networks collapsed in the Late Harappan Phase (1900 to 1300 BCE). There were changes in the distribution and settlement pattern of the population. Lands in the Gaṅgā-Yamunā Doab and the Malwa Plateau were populated and new craft centres developed and prospered (Possehl 2002, Kenoyer 2006). However, small cities and towns continued to dominate the landscape of the Punjab and parts of Sindh (Kenoyer 2006). The drying up of Sarasvatī River system was a devastating blow for the people who relied on this river, but the Indus and its tributaries did not dry up and people continues to live along its banks – the Indus Valley was never abandoned. Moreover, recent analysis of *Ṛgveda Saṃhitā* (Stuhrmann 2008) suggests that the Āryas were familiar with towns/cities (*pur*) in South Asia. Sometimes they made alliance with the town/city dwellers, and sometimes they fought them.

The alternate view; that the Ārya were indigenous to South Asia and were the people of the Indus Valley Civilization, has also been around since the discovery of Indus Valley sites. The hypothesis that the peoples of the Indus Valley culture and Vedic culture may have co-existed is not without merit. Lal (2005) has shown that the disparity between the Vedic and Indus cultures is not particularly pronounced and the considerations of geography and chronology do not rule out an early Āryan presence in South Asia. In India, this so-called 'Āryan debate' has acquired a political dimension from about the last decade of the twentieth century and this has sprouted acrimonious newspaper articles and combative websites. This has done a disservice to this debate as it has provided the opponents of this hypothesis an excuse to denigrate and dismiss it. The proponents of the 'indigenous Ārya' hypothesis are dismissed as 'Indian nationalists' and 'Hindu chauvinists' although a number of proponents of this hypothesis are neither Indian nor Hindu. Trautmann (2005) has collated the 'reasoned and sober' arguments from both sides of the 'Āryan debate' but no conclusion has been reached and this debate is likely to rumble on.

The *Ṛgveda Saṃhitā* does not shed much light on the social and cultural milieu within which this text evolved. The Ārya of *Ṛgveda Saṃhitā* are believed to be pastoral nomadic and occasional farmers. Although pasture (*gavyūti* – *Ṛ*V.I.25.16; *Ṛ*V.X.14.2) is mentioned in the Saṃhitā, there is no evidence of seasonal movement of people nor can words for 'winter pasture' or 'summer pasture' be identified in the text. The large number of 'prayers' and 'invocations' for water both as rain and in rivers, words for a plough (*vṛka* – *Ṛ*V.I.117.21 and *sīra* – in post-*Ṛgveda* text), furrow (*sītā*) and barley (*yava* – *Ṛ*V.I.23.15; *Ṛ*V.I.117.21; *Ṛ*V.V.85.3; *Ṛ*V.VIII.2.3; *Ṛ*V.X.131.2; *Ṛ*V.X.27.8) and verses that describe farm activity (e.g. A.I.3, 4, 5; A.III.27) suggest people with stable ties to the land. It should also be pointed out that in the Saṃhitā, a number of words related to farming are not Sanskrit words, for example, *tilvila* (fertile – *Ṛ*V.V.62.7), *khala* (threshing-floor or granary – *Ṛ*V.X.48.7) and *lāṅgala* (plough – *Ṛ*V.IV.57.4). These words have been borrowed by the Ārya from the co-located non-Sanskrit-speaking people (Witzel 2001), most probably the

Harappans. The Ārya society is also familiar with commerce as suggested by the occurrence of the word for a merchant (*vaṇij* – *Ṛ*V.I.112.11). The principle measure of wealth in this society was cattle

The Saṃhitā suggests a society governed or held together by common observances, ritual practices and a common spoken language (archaic Sanskrit, N.I.4). Although this Saṃhitā and the texts that follow suggest a culture rich in rituals and ceremonies, no permanent structure for performance of Vedic rituals is mentioned. There are no temples. A location was chosen anew for each ritual or ceremony and there were well-defined rules for the choice of the location and the preparation of the ceremonial ground. A temporary structure of bamboo and thatch was constructed over the sacred space, which was burnt at the termination of the ritual. There is also no evidence for icons or images representing gods or their attributes. This was a society stratified into classes; nobility (kṣatriya) and 'the people' (*viś* or vaiśya) plus the brahmins (*brahman, ṛṣi*) and the lowest class (śūdra). This Saṃhitā leaves an impression of a prosperous society that had surplus capital to support a leisure class of poet/priests and organize ceremonies that involved elaborate and expensive rituals. In many ways, the post-Ṛgvedic society appears to be similar to the Ṛgvedic society. However, the ceremony for royal consecration (*Rājasūya*) suggests a consolidation of political, social and military authority into larger units. Unlike the Indus Valley society the Āryas were neither united nor peace-loving. There is ample evidence in the Vedic texts that the Āryan tribes were often in conflict with each other and with non-Āryan tribes. At times some Āryan and non-Āryan tribes united against other Āryan tribes.

Although no archaeological evidence of Āryas has been found, they have left the second source of history of (the northern part of) South Asia. This is in the form of liturgical, religious and sacerdotal texts that these people composed and which have been transmitted orally down the ages.

3. THE VEDIC CORPUS

The oldest texts in South Asia are those of the Vedic corpus. These were written in Vedic (Old or Archaic) Sanskrit (N.I.4). Extensive analysis of the Vedic texts has been carried out over last 100 years (Wallis 1887, Oldenberg 2005, Arnold 1905, Winternitz 1981, Griswold 1971, Gonda 1975, Jamison and Witzel 1992, Witzel 1997, Witzel 1999a). The gods and mythologies of the texts have also been categorized and discussed in detail (Hillebrandt 1980, Macdonell 2000). The Vedic corpus can be divided into five chronologically distinct periods. (The Vedic texts, Table I.2, are described in detail below.)

Table I.2: The Vedic Corpus

	Śākhā	Saṃhitā	Brāhmaṇa	Śrauta Sūtra
Ṛgveda	Śākalya	Ṛk	Aitareya	Āśvalāyana
			Kauṣītaki	Śāṅkhyāyana
Yajurveda	Kṛṣṇa			
	Taittirīya	Taittirīya	Taittirīya	Baudhāyana
				Vādhūla
				Bhāradvāja
				Āpastamba
				Hiraṇyakeśin
	Kāṭhaka	Kāṭhaka	Kaṭha	Vaikhānasa
		Kapiṣṭhala-		Kāṭhaka
		Kaṭha		Mānava
	Maitrāyaṇī	Maitrāyaṇī		Vārāha
	Śukla			
	Mādhyaṃdina	Vājasaneyi	Śatapatha	Kātyāyana
	Kāṇva			
Sāmaveda	Kauthuma		Pañcaviṃśa	Lāṭyāyana
			Ṣaḍviṃśa	Drāhyāyaṇa
	Jaiminīya	Sāma	Jaiminīya	Jaiminīya
	or			
	Talavakāra			
Atharvaveda	Śaunaka	Atharva	Gopatha	Vaitāna
	Paippalāda			

- The Ṛgvedic verses (*ṛc*, *ṛcā*s in plural) are the most archaic of the Vedic texts and have many linguistic elements in common with Indo-Iranian. These *ṛcā*s have sometimes been considered to be the precursors of Vedic literature, but it would be more appropriate to consider them the last stage of a long period of Āryan poetry. Many words that occur in these *ṛcā*s have cognates in Indo-Iranian poetry (*Avesta*). The *Ṛgveda* does not refer to any other text but all other Vedic texts presuppose the *Ṛgveda*. The *mantra* (sacred verse) period includes both the *mantra* and prose language of *Atharvaveda*, *Ṛgveda Khilani*, *Sāmaveda Saṃhitā* and the *mantras* of the *Yajurveda*. These texts, described below, are largely derived from the *Ṛgveda* but have undergone certain changes, both linguistic and of interpretation.

- The Saṃhitā period marks the beginning of the collection and codification of the Vedic corpus. The commentaries of the *Kṛṣṇa* (Black) *Yajurveda* (see below) belong to this period.
- The Brāhmaṇa period was the time of production of the explanations and the rationale of the rituals of the Vedas (*Ṛgveda*, *Yajurveda*, and *Sāmaveda*). The older Upaniṣads also belong to this period.
- The Sūtra period is the last layer of Vedic corpus and the bulk of the Śrauta Sūtras (texts on public or corporate ceremonies) and Gṛhya Sūtras (texts on domestic rituals) were produced during this period.

To this must be added the more esoteric Upaniṣads and Āraṇyakas. This chronological division is based on the existing redactions of the texts. It is extremely unlikely that these texts were composed in these strict chronological slots. Each text probably has a long history of foundation, discussion, debate and refinement. It is very likely that versions of texts overlapped. These earlier versions are now lost and what we have received is the final redaction. This final redaction of each text probably occurred at different times and the order of texts above probably reflects the chronological order of the final redactions. It is, however, possible that these final redactions include material and concepts from earlier versions of the texts. The chronology of these texts is discussed in Section 5.

The Saṃhitās and the Brāhmaṇas (also the Āraṇyakas and Upaniṣads) are considered to have been revealed to the *ṛṣis* (seers), i.e. these are *śruti* (heard) texts. The Sūtras, on the other hand, are considered to have been produced by human authors, i.e. they are *smṛti* (remembered) texts. The Vedic corpus is one of the most original and interesting product of human endeavour. It comprises texts composed over many centuries that deal with a variety of subjects. The texts were composed orally and still are an oral literature. For centuries, they were transmitted orally, without change of a single word. The earliest Vedic manuscript (written without accent marks) is from Nepal, this was produced around CE 1040, over 2,000 years after the composition of the original *mantras*.

The term Veda also denotes the four Saṃhitās (the fundamental collections) namely the *Ṛgveda* (the laudatory stanzas), *Yajurveda* (the sacrificial formulae), *Sāmaveda* (the chants to be sung to certain fixed melodies) and *Atharvaveda* (dealing with mundane matters like white and black magic, healing spells, marriage and death ceremonies, and also material discussed in other Saṃhitās). The last text was only later recognized as part of the Vedic corpus. The four Vedas were composed and transmitted in independent schools or *śākhā*s. A *śākhā* represents a particular community or school of priests which may have been localized and been unique to a tribe or a kingdom. A *śākhā* possessed a recension of a Saṃhitā and its own exposition of the rituals (Brāhmaṇa), ritual handbook (Sūtra), and other texts. The known *śākhā*s of the four Veda and their associated texts are given in Table I.2.

3.1. ṚGVEDA

The oldest Saṃhitā (methodically arranged collection of verses in praise of deities) is that of *Ṛgveda*. Study of the language of *Ṛgveda* suggests that it is an anthology of poems that were composed over a period of many centuries. It is difficult to determine the epoch of these poems but according to some scholars, some hymns may date back to the beginning of the second millennium BCE or earlier. There were probably a number of recensions of this Veda and five have been identified, but only that of Śākalya has survived. This *Ṛgveda Saṃhitā* consists of 1,028 hymns (average ten verses per hymn) or *sūkta* (well-recited, good recitation) arranged in ten chapters or *maṇḍala*. Each *sūkta* contains several verses or *ṛcā*s. *Maṇḍala*s II to VII, the 'family books', are traditionally regarded as the core of the collection. The hymns of each book have been composed by a single named family (which may have included students) of *ṛṣi*s, covering several generations and have been handed down as a family heritage. These books are arranged on a uniform plan, that is the sequence of deities and the number of verses in the hymns are similar. The plan of these books is different from the plan of the other books. *Maṇḍala*s I, VIII and X are not composed by a unique family of *ṛṣi*s. Of *maṇḍala* I, *sūkta* #51-191 are believed to be the earliest addition to the core formed by the family books. The first fifty *sūkta* appear to have been added later to this *maṇḍala* and show considerable affinity to *maṇḍala* VIII. The *maṇḍala* IX is addressed to the preparation, straining and flow of Soma (*soma pavamāna*). The *ṛṣi*s of the *sūkta*s of this *maṇḍala* are of the families of the *ṛṣi*s of the core *maṇḍalas*. There are only six Soma *sūkta*s in the family *maṇḍala*s and it is reasonable to assume that in *maṇḍala* IX all *soma pavamāna* hymns (some of which may originally have been in the family books) have been gathered for liturgical reasons. The last addition to the Saṃhitā was *maṇḍala* X. The supplementary character of this *maṇḍala* has been accepted for some time, although it is very likely that it contains some old verses. The contents of this *maṇḍala* are, however, significantly different from those of the other nine *maṇḍala*s. Speculations on the origins and the mysteries of the universe, on the ultimate principle, verses on family rites and gods not encountered in the other *maṇḍala*s differentiate this book from all other books. The uniform organization of the family books suggests the activity of redactor(s) rather then the actions of a particular family. The hand of the redactor can also be discerned in the inclusion in the same Saṃhitā the books of two mutually antagonistic families (Vasiṣtha and Viśvāmitra), the equal number of *sūkta*s in books I and X, and the dedication of the last *sūkta* of *maṇḍala* X to harmony (*saṃjñāna*). Although the Saṃhitā available to us contains the original *ṛcā*s of the composing *ṛṣi*s the arrangement that these *ṛṣi*s may have intended is now lost and in the Saṃhitā the old and new verses have been mixed up (Deshpande 1995).

Ṛgveda Saṃhitā is verbally and poetically sophisticated and often obscure. This text is not the crude expression of bards of primitive tribes as was once assumed. The Vedic Āryas went to extraordinary lengths to preserve the integrity of the *Ṛgveda Saṃhitā* (and other Saṃhitās). Supplementary texts were composed to preserve the fidelity of the texts. Śākalya's *Padapāṭha* (word-text) lists each word of the *Ṛgveda* separately and separates the compounds into constituent words. Śaunaka's *Prātiśākhya* (thesis on the euphonic combination and pronunciation when compounds are formed) shows how the *pada* texts can be transformed into the Saṃhitā. The *Anukramaṇī* (index) of Śaunaka provide a list of Ṛgvedic hymns, their metre, deities and composer. Kātyāyana's *Sarvānukramaṇī* (a complete index) gives in the *sūtra* form the first few words of each hymn, the number of verses in the hymn, the deities and the metre of each hymn. It also gives the names and the family of the composer. Śaunaka's *Bṛhaddevatā* gives in metrical form the gods worshipped in various hymns and also the associated myths and legends. These supplementary texts have helped ensure that over three millennia there have been only the most insignificant of changes in the text and the pronunciation of the words.

A large number of Ṛgvedic *sūktas* are invocations to various deities for water (in rain and in rivers), wealth, cattle, victory and sons (never daughters). However, this Saṃhitā does not have a single unified or unifying theme (unlike other Saṃhitās). It is a collection of verses on different topics. These topics range from profound to mundane and only a sample can be given here; the interested reader is invited to see O'Flaherty (1981) and others.

- *Creation.* There are many different theories of creation in the *Ṛgveda*, such as creation from a golden egg (*hiraṇyagarbha*), by sacrifice of the primordial man (*Puruṣa*) or unmotivated separation of heaven and earth. In many ways, the most intriguing and provocative is the hymn that seeks the origins of existence itself (*ṚV*.X.129). The foot-print of this 'riddle hymn' appears in the cosmogonic discourses of later Indian texts and the hymn has thus come under the scrutiny of several scholars. The hymn expresses radical doubts on the issue of creation: 'What was there in the beginning', 'Who really knows', 'Whence was it produced'. Brereton (1999) has given a recent reappraisal of this hymn.
- *History.* It is difficult to reconstruct the history of the Āryas from the hymns of the *Ṛgveda*. However, a battle of some importance is described in *ṚV*.VII.18.5-10; 19.2-4; 33.3-6; and 83.6-8. This 'battle of ten kings' was between the Bharata king/chief Sudās and his ally Tṛtsus against a confederation (led by the Pūru king/chief) of ten kings/chiefs of whom five were Āryan and five were non-Āryan. The battle was fought on the banks of river Puruṣṇī

(modern Rāvī). The odds were heavily against Sudās (this is the victor's version!) but due to timely intervention of Indra invoked by Vasiṣṭha (*ṛṣi* of Sudās) the enemy were killed or drowned and Sudās was victorious.

- *Weather*. A large number of hymns in the *Ṛgveda* are addressed to Indra, and many of these are prayers for water in rain or rivers. In *ṚV*.V.83 (see Chapter II) there is a powerful description of a storm or the onset of the monsoon. The deity (*devatā*) of the hymn is Parjanya, which is Indra as the 'sender of rain'. He drives clouds before him like a charioteer urging his horses and makes the wind blow and lightning flash. He makes plants grow, cattle thrive, rivers flow, and deserts bloom.
- *Unrequited Love*. The sad end of a story about an unhappy wife and a pining husband is narrated in the dialogue of *ṚV*.X.95. King Purūravas (mortal) is married to the (immortal) water-nymph Urvaśī. The husband was very happy with his wife and he thought she was happy with him. However, the wife was not and her friends contrived to take her away from him. In this dialogue, Purūravas urges Urvaśī to return. She refuses stating that she had been unhappy and implies that they were united only to produce a son and there was no need for her to return (Purūravas appears to have been unaware of this child until this moment). Urvaśī also chides Purūravas that he was neglecting his duties (as a king) in his pursuit of her.
- *Motherhood*. A woman's desire to have a child and an unworldly man's inability to notice the obvious are contemporary themes and the Vedic times appear to have been no different. In *ṚV*.I.179. Lopāmudrā the long-suffering wife of Agastya reminds him that she was 'getting-on' and soon will be past child-bearing age. Agastya is pursuing asceticism to achieve immortality and so has taken a vow of chastity. Down-to-earth Lopāmudrā cuts through Agastya's arguments and has her way. Both achieve immortality; spiritual and tangible (through children).
- *Magical spells*. Incantations and spells are generally absent from the *maṇḍala*s of *Ṛgveda*; the exception is *maṇḍala* X. In this *maṇḍala* there are (few) examples of both benign and malign spells; to heal, to conceive, have healthy children, overcome rivals, delay death, etc.

The Hindus regard the *Ṛgveda Saṃhitā* to be the bed-rock of their religion; to them it is the fount of all wisdom and knowledge. But those Hindus who cannot read and understand Sanskrit, i.e. the majority, can only drink from this fount through the English translations of the nineteenth century (Wilson 1997, Griffith 1973) or the inaccessible German (Geldner 1951-7, Witzel and Goto 2007), French (Renou 1955-69) and Russian (Elizarenkova 1972) translations of the twentieth century. *Ṛgveda Saṃhitā*, in its entirety, does not appear to have been translated into a single modern South Asian language, a magnificent failure of Hindu scholarship.

3.2. *YAJURVEDA*

The verses of the *Ṛgveda* presuppose an elaborate ritual system, which is the subject of the *Yajurveda* (the name is taken from *yajus* – sacrificial formulae). The Saṃhitās of this Veda contain liturgical formulae required for ritual practices. The subject matter of the Saṃhitās of various *śākhās* (schools) that have come down to us is similar. The core of a Saṃhitā is the *Darśapūrṇamāsa* or full and new moon sacrifices (the model for all sacrifices of the *iṣṭi* type), the sacrifice to *soma,* the deific drink (*uttamaṃ havis* or *paramāhuti* – supreme oblation) that plays a central role in Vedic rituals and the construction of the great fire altar (*agnicayana*). Other rites, like those of animal sacrifice are scattered among the core sacrifices. Some rites like the royal consecration are added as appendices to each Saṃhitā. Some rites, rituals for special advantages (*kāmyeṣṭi*), are unique to some *śākhās*. The surviving Saṃhitās of the *Kṛṣṇa* (Black) *Yajurveda* (so called because the Saṃhitās of this Veda contains both the liturgical *mantra* and explanatory (Brāhmaṇa notes) are: *Taittirīya*, *Kaṭhaka, Kapiṣṭhala-Kaṭha*, and *Maitrāyaṇī*. In these Saṃhitās, the portions containing directions and injunctions for ceremonies are similar. The *Taittirīya* appears to have arranged and systematized their traditions at an early stage and their Saṃhitā is more complete than that of the other *śākhās*. The *Śukla* (White or Clear) *Yajurveda*, on the other hand, consists entirely of *mantra*s to be recited at sacrifices and is free of explanatory matter, i.e. there are no Brāhmaṇa portions. The Saṃhitā of this *Yajurveda* namely, *Vājasaneyi Saṃhitā,* has survived in two recensions; the *Kāṇva*

Saṃhitā and the *Mādhyaṃdina Saṃhitā*, each composed of forty chapters (*adhyāya*). The *mantra*s of many of these *adhyāya*s are Ṛgvedic verses and in this *Yajurveda* the Ṛgvedic verses are more numerous than in the *Kṛṣṇa Yajurveda*. The Mādhyaṃdina recension of this Veda is more complete and more systematically arranged than the Kāṇva recension. It has therefore attracted much more attention.

3.3. *SĀMAVEDA*

The *Sāmaveda Saṃhitā* is a collection of the *sāman* (melodies). Only Saṃhitās of two *śākhā*s of this Veda have come down to us, they are those of Kauthumas and of the Jaiminīyas. The better known *Kauthuma Saṃhitā* consists of 1,810 *ṛcā*s arranged in two *ārcika*s (collections). All except 76 *ṛcā*s are taken from the *Ṛgveda Saṃhitā*, mainly from *maṇḍala*s VIII and IX. These *ṛcā*s were intended for singing and some verses differ from those in the *Ṛgveda Saṃhitā* as the words have been altered to be set to music. The arrangement of the first collection (*Pūrvārcika*) follows roughly that of the *Ṛgveda Saṃhitā*. Thus, the first two set of *ṛcā*s are addressed to Agni and Indra respectively, and the last to *soma*. The second collection (*Uttarārcika*), on the other hand, is concerned only with the *Soma* sacrifice. These two *ārcika*s give the texts in their spoken form; the melodies with musical notes are given in song-books (*gāna*). The *gāna*s identify the modifications to the *ṛcā*s when sung to a melody, such as lengthening of syllables, repetitions, breaking up of words, etc. The *Jaiminīya Saṃhitā* (or *Talavakāra Saṃhitā*) – a less well-known *śākhā* that has survived mainly in south India – is very similar to that of the Kauthumas except its deviations from the *Ṛgveda* are less numerous.

3.4. *ATHARVAVEDA*

The *Atharvaveda Saṃhitā* is mainly a collection of spells – auspicious and hostile. There are also passages that have no magical elements and parts that include theological and cosmological speculations. Two recensions of this Veda have come down to us; the Paippalāda and the Śaunaka. The Śaunaka recension is better preserved and better known. Its text, like that of *Ṛgveda Saṃhitā*, is mostly metrical. Its

730 *sūkta*s are divided between twenty books (*kāṇḍa*s); books XIX and XX appear to be supplementary additions at a later date. The order of the hymns in the first eighteen books follows a pattern and suggests a carefully constructed editorial plan. The books and hymns deal with various topics, for example, book XIV has nuptial hymns, book XVIII has funeral hymns and books XV and XVI (composed in prose) have an explanatory (Brāhmaṇical) style. The contents of the Paippalāda recension are similar to those of the Śaunaka recension but the arrangement of the material is significantly different. This recension has a large number of stanzas not found in the Śaunaka version and has material not found in any other text. The arrangement of the Paippalāda recension is rather rudimentary and the material has more in common with *Ṛgveda*. Such differences have suggested the higher antiquity of this recension.

3.5. Brāhmaṇas

The Brāhmaṇas, i.e. 'comments upon *brahman* or the Veda' are voluminous works concerned with the sacrificial (*śrauta*) rituals. The aim of these texts is not to describe, but to explain the origin, meaning, and justification of the rituals. These texts also attempt to prove the validity and significance of the *mantra* and the chants employed and the connection between the ritual acts and the phenomenal world. The composers of these works expected their audience to be familiar with the course of the ceremony, its terminology, and the techniques involved. Thus, most of the detail is omitted and a modern reader has to be familiar with the complicated sacrificial rituals to understand these works. The detail, presupposed in the Brāhmaṇas, is found in the ritual *sūtra*s, compiled to guide the sacrificer, who is always a male. Although women/wives take part in the ceremonies, they are never the principle sacrificer. The Brāhmaṇas maintain that the rituals, if properly understood and accurately performed, would save the sacrificer from evil and misfortune in this world and beyond and would benefit him materially and spiritually. This belief in the efficacy of rituals is based on the conviction that there exists a connection between ritual acts and the natural and divine forces. The ritual also attempt to emulate the natural order (*ṛta*) that the ritualists saw around them. The sacrificial rituals had to be performed with full knowledge of the technical intricacies and a full understanding of its procedures and significance. No interruptions were permitted, as there are no interruptions in the divine order. In the Brāhmaṇas the exposition of the rituals is given with detailed and intricate arguments with frequent speculative digressions. The Brāhmaṇas represent the intellectual efforts of a priestly class to transform the older form of belief and worship into a highly complicated system of sacrificial ceremonies. Whatever the motives for this transformation, underlying it is a desire to place all things in a causal context wider than that provided by common sense. There is here a desire to identify the unity, order and regularity that underlie the apparent diversity and disorder. In these texts, we witness the foundation of analytical and rational thought in South Asia.

At present two Brāhmaṇas are known to be associated with the *Ṛgveda* (Table I.2), both intended for the *hotar* (the priest who recites hymns of the *Ṛgveda*). The *Aitareya Brāhmaṇa* is written in forty chapters (*adhyāya*s) that are grouped into eight books called *pañcikā* because each contains five *adhyāya*s. Tradition has it that this is the work of one person, Mahidāsa Aitareya. He is more likely to have been the redactor, as this text does not seem to be the work of a single person; *pañcikā* VII and VIII (dealing with the animal sacrifice, expiation and the *Rājasūya* – royal consecration) are later additions. The main theme of the *Aitareya Brāhmaṇa*, covered in *pañcikā* I-VI, is the *Soma* sacrifice and variations of this in *Agniṣṭoma, Gavām ayana, Dvādaśāha, Agnihotra*.

The second Brāhmaṇa of *Ṛgveda*, the *Kauṣītaki Brāhmaṇa* consists of thirty *adhyāya*s. It is brief and without repetition, more harmonious, uniform and systematic in character. This suggests that it is later then *Aitareya Brāhmaṇa*. The first six *adhyāya*s are devoted to the establishment of the sacred fires (*Agnyādhāna*), the *Agnihotra, Ḍarśapūrṇamāsa, Cāturmāsya*, etc. The exposition of the *Soma* sacrifice starts from *adhyāya* VII.

The *Yajurvedas* being concerned with the sacrificial formulae, their Brāhmaṇas follow the ritual closely and are intended for the Adhvaryu (who was responsible for the 'mechanics' of the sacrifice). The *Taittirīya Brāhmaṇa* follows the style of the *Taittirīya Saṃhitā* and contains both *mantra*s and prose. It consists of three books called *aṣṭaka* because each book has eight chapters (*prapāṭhaka*). The contents of this Brāhmaṇa can be summarized thus: *aṣṭaka* I deals with *Agnyādhāna*, *Gavām ayana*, *Vājpeya* and the royal consecration;

aṣṭaka II deals with *Agnihotra*, *Daśahotra*, *Sautrāmaṇī*, special animal sacrifices and some subsidiary sacrifices; *aṣṭaka* III deals with sacrifice to the *nakṣatra*s (stars and asterisms), *Darśapūrṇamāsa*, human sacrifice (with symbolic human victims) and some special ceremonies. This Brāhmaṇa only considers topics not dealt with in the Saṃhitā; thus, only special fire-altars are described as the normal one is considered in the Saṃhitā (see Chapter III, Section 2.8). Only fragments (of the original hundred chapters) of the Brāhmaṇa of Kāṭhakas have survived. The Maitrāyaṇīs have not left us an independent Brāhmaṇa. The prose sections of the Saṃhitā of this *śākhā* serves the purpose of its Brāhmaṇa. The Brāhmaṇa of the *Śukla Yajurveda* is the *Śatapatha Brāhmaṇa* (the Brāhmaṇa of Hundred Paths). Two recensions have come down to us – the Mādhyaṃdina and the Kāṇva. The Mādhyaṃdina recension consists of 100 *adhyāya* (from which the Brāhmaṇa gets its name) grouped into fourteen books (*kāṇḍa*). This is the most elaborate and best known Brāhmaṇa. The first nine books follow closely the first eighteen chapters of the 'parent' Saṃhitā, the *Vājasaneyi Saṃhitā*. They deal with the fundamental elements of the *Yajurveda* namely *Darśapūrṇamāsa*, the establishment of the sacred fire, *Agnihotra*, *cāturmāṣya*, the *Soma* sacrifice and the construction of the fire altar. In books I to V, Yājñavalkya is frequently quoted and his opinions on ritual matters is final. In books VI to X, Yājñavalkya is never mentioned and the figure of authority on ritual matters is Śāṇḍilya. These books are primarily concerned with the construction of the great fire-altar or *Agnicayana*. In the last four books, the figure of authority is again Yājñavalkya. In his dialogues with King Janaka, we witness the gradual transition from preoccupation with ritual to philosophical speculation.

The Brāhmaṇas of the *Sāmaveda,* intended for the *udgātṛ* (the priest responsible for chants), have come down to us in three versions; *Pañcaviṃśa* (or *Tāṇḍyamahā*), *Jaiminīya* (or *Talavakāra*) and *Ṣaḍviṃśa*. The material of *Pañcaviṃśa* (consisting of twenty-five chapters – hence its name) is similar to that of *Jaiminīya* but the two Brāhmaṇas differ in style and presentation. The *Pañcaviṃśa* presents the technicalities of ritual in a rather dry, matter-of-fact manner and only states what is absolutely necessary. It is often so succinct that some portions are incomprehensible. The *Jaiminīya*, on the other hand, is 'wordy' with a number of mythical tales and liturgical anecdotes. The first five chapters of *Pañcaviṃśa* contain, instead of the Brāhmaṇa text, a collection of *yajus* formulae for the chanter. The rest of the Brāhmaṇa is concerned with various *śrauta* rituals in the context of the *Sāmaveda*. The *Ṣaḍviṃśa* (the twenty-sixth), as its name suggests, is a kind of supplement to the *Pañcaviṃśa*. It consists of five lessons (*prapāṭhaka*) concerned with Sāmavedic material on various *śrauta* rituals.

The only Brāhmaṇa of the *Atharvaveda* that has come down to us is the *Gopatha Brāhmaṇa*. Unlike other Brāhmaṇas, this Brāhmaṇa does not follow its own Saṃhitā. It consists of two parts, with five and six chapters respectively. The second part has portions that are similar or identical to some sections of the other Brāhmaṇas. Like a 'regular' Brāhmaṇa, this half has discussions on various *śrauta* rituals like *Darśapūrṇamāsa*, *āgrāyaṇa* (oblation of the first fruit), etc. The first part discusses the duties of a brahmin disciple (*brahmacārin*), the rules concerning consecration (*dīkṣā*), the *sahasradakṣiṇa* sacrifice (through which one obtains enduring results), the three daily *soma* pressings, etc. It is impossible to tell if this Brāhmaṇa is the text of the Paippalāda or the Śaunaka *śākhā*.

3.6. Śrauta Sūtras

The Śrauta Sūtras contain rules for the setting-up of sacrificial fires required for the *Agnihotra*, *Darśapūrṇamāsa* sacrifices, *cāturmāsya* sacrifices, *Paśubandha* sacrifices and the *Soma* sacrifice and its various forms. These *sūtras* are the most important source for understanding Vedic sacrifices. However, they describe the sacrificial rituals with extreme brevity and little or no commentary. The Śrauta Sūtras of the *Ṛgveda Saṃhitā* of the Śākalya *śākhā* and *Aitareya Brāhmaṇa* is *Āśvalāyana Śrauta Sūtra*. It describes the ritual duties of the *Hotṛ* priest. The oldest Śrauta Sūtra is the *Baudhāyana Śrauta Sūtra*. It is part of the *Taittirīya* corpus and resembles a Brāhmaṇa in many respects. The most copious description of the Vedic sacrifices is given in the *Āpastamba Śrauta Sūtra*, also of the *Taittirīya* corpus. Various quotations from other ritual texts suggest that this is a later work. The Śrauta Sūtras of the *Śukla Yajurveda* is *Kātyāyana Śrauta Sūtra*. Although the subject of the *Atharvaveda* is not the *śrauta* rituals, it nevertheless has a Śrauta Sūtra namely the *Vaitāna Sūtra*.

A *śrauta* ceremony was performed at a well-defined time and the layout of the sacrificial ground and the fire-altars had to follow well-defined rules. The determination of the appropriate time for a ceremony is discussed in *Vedāṅga Jyotiṣa* (see Chapter V). The rules for orientation of the fire-altars and their geometry are given in Śulbasūtras (N.I.5). These are the oldest South Asian texts on calendrical science and geometry respectively and their significance to the history of science and mathematics cannot be underestimated.

3.7. Gṛhya Sūtras

The Śrauta Sūtras present the rules of the 'major' or 'corporate' sacrifices, whereas the subject of the Gṛhya Sūtras is the customs and ceremonies pertinent to the everyday life of an Ārya or rituals belonging to the house (*gṛhya*). These texts describe exhaustively observances associated with conception, birth, conducting a pupil to a teacher (*upanayanam*), parting of a pupil from a teacher, the customs of courting, engagement, and marriage and even those relevant to death (funeral customs) and beyond (offerings to *pitṛ*s). The customs and ceremonies connected with house construction, cattle breeding and agriculture also are described. The magic rites that are supposed to ward off diseases, omens signifying misfortune, and charms for exorcism are also the subject of Gṛhya Sūtras. The Gṛhya Sūtras mention the 'five great sacrifices' (described in the Brāhmaṇas), so called because their performance is part of the duties of every householder. These consist of placing a stick of wood on the holy fire of the hearth, a small offering of food, and a libation of water to the gods, demons and *pitṛ*s, entertaining a guest (considered an 'offering to men') and the daily reading of a chapter of Veda (considered an 'offering to the Brāhmaṇa'). The Gṛhya Sūtras provide an insight into the domestic life of the Āryas and are an invaluable source of information on the everyday culture of the Vedic people.

4. THE GEOGRAPHY OF THE VEDIC TEXTS

The Vedic texts are the products of a number of composers from widely separated locations. Moreover, these texts were composed over an extended period. Some texts are the work of a particular *śākhā* that may well have been localized to a particular locale and a particular time. However, some texts have been put together by comparatively late redactors and the constituent parts of such texts may have originated in periods and at locations well removed from that of the redactors. The geographical location of a text or a particular part of a text can be identified from the rivers, mountains, and other locations mentioned in the texts. There are dangers here as the names of features can change or be transferred from other locations, but the major geographical features appear to have retained their names, for example, Sindhu, Gaṅgā, etc. The geographical location(s) of texts can also be identified from the flora, fauna, and the direction of the flow of rivers and the weather described in the text. It is necessary to exercise caution in interpreting this information as rivers can change their direction of flow and their course can even disappear. The climate of the Vedic times may not necessarily have been similar to the current climate and this would influence the interpretation of the described flora and fauna. A number of attempts have been made in last hundred or so years, to identify the geography of Vedic texts; for recent studies see Witzel (1987 and 1997).

The geographical location of *Ṛgveda Saṃhitā* is believed to be the land of Sapta Sindhu or land of seven rivers (the Indus and its five tributaries plus Sarasvatī) and can be identified from the rivers and mountains mentioned (and not mentioned) in the Saṃhitā. The western edge of this region is defined by the rivers Kubhā, Kurum and Gomatī (Kabul, Kurram and Gomal rivers, western tributaries of the Indus) in east Afghanistan and west Pakistan. In the east, the principle river was the Sarasvatī (N.I.2) but the eastern edge is defined by the region close to the rivers Yamunā (mentioned three times) and Gaṅgā (mentioned only once, in *maṇḍala* X). The northern border of this region is defined by the Himālayas, known in the Saṃhitā as Himavat. The southern border is not so well defined and may have stretched as far south as the Indian Ocean (*samudra*), but the Vindhya mountains of central India are not mentioned in this Saṃhitā. Thus, the land of the *Ṛgveda Saṃhitā* was eastern and northern parts of Afghanistan, Pakistan, and Panjab, Haryana, parts of Rajasthan and Uttar Pradesh in India. This is also the geography of Indus Valley Civilization (Figure I.1).

Witzel (1987) has investigated the geography of the *śākhā*s of the post-Ṛgvedic Saṃhitās and the Brāhmaṇas. This is the vast region from Pakistan to Bengal, between the foothills of the Himālaya and the Vindhya. Some of the texts have been composed in two locations, for example, judging from grammar and content, *Aitareya Brāhmaṇa*, one of the oldest Brāhmaṇas, is composed in two parts. The older part (*pañcikā* I-V) can be located in the land bound by the Himālaya in the north and the desert on the banks of Sarasvatī in the south, and from eastern Punjab to the Yamunā-Gaṅgā Doab. The locale of the later parts of this Brāhmaṇa (*pañcikā* VI-VIII) is the whole of northern and central India, including some eastern regions. Similarly, the late Brāhmaṇa, *Śatapatha Brāhmaṇa* (Mādhyaṃdina recension) can be shown, as mentioned above, to have three constituent parts. The first part, *kāṇḍa* I-V, can be located in the eastern Gangetic plain, the Bihar region. There are also indications of movement and settlements south of the Yamunā River. The second part, *kāṇḍa* VI-X, can be located in western Uttar Pradesh. These *kāṇḍa*s of the Brāhmaṇa appear to have been imported at a later date by the eastern ritualists. The third part, *kāṇḍa* X-XIV, can also be located in the eastern parts of India. The *Jaiminīya Brāhmaṇa* appears to have originated in the southern parts of northern India but still north of the Vindhyas, although some penetration of this mountain range may have occurred. The Sūtra text-style is very much a development of the eastern regions of northern India. The *Baudhāyana Śrauta Sūtras*, the oldest text of its kind, suggests a wide geographical horizon that includes all of northern India and even parts of the Deccan.

These texts, the Saṃhitās, the Brāhmaṇas, and the Sūtras, indicate a gradual expansion of Vedic thought and rituals from the regions of eastern Afghanistan, northern Pakistan and north-western India to lands to the east and south. Although the land south of the Vindhyas is mentioned, there does not appear to have been major incursions in southern India. Similarly, eastern India is not included in the Vedic fold. It is impossible to tell if this spread of the Vedic ritual practice indicates a movement of large groups of people or small adventurous bands who inducted and assimilated the indigenous people into the Vedic ritual practices.

5. THE VEDIC AGE

Dating the Vedic Age is a daunting task. Lacking archaeological artefacts, the tried and tested methods of dating cannot be brought to bear on this period. The Vedic canon was known to and was quoted by the Sanskrit grammarian Pāṇini (estimated to be between 600 and 400 BCE). This date is partially corroborated by the Buddhist texts in Pāli, which have knowledge of the Vedic corpus. This suggests that the Vedic corpus pre-dates Buddha, whose birth is estimated to be around 400 BCE. This limiting date is one of the two dates (the second date is discussed below) that can be assigned unambiguous to the Vedic Age. All other attempts to obtain absolute chronology of the Vedic texts and in particular that of the *Ṛgveda Saṃhitā* are based on either speculative identification of archaeological structures with texts or on 'intelligent guesses' guided by preconceptions.

The *Ṛgveda Saṃhitā* is the oldest Vedic texts. Hindus believe that the *Ṛgveda* is a revealed text (*śruti*) and before the late eighteenth century, no South Asian scholar had attempted to date it. The first attempt to date the Vedic Age was made by William Jones. Based on the narratives of the Old and New Testaments, the date of Creation had been computed by Archbishop James Ussher (1581-1656) to be 4004 BCE. The age of earth determined scientifically by geologists and evolutionary biologists, was still 100 years away. William Jones (and other Christians) had absolute faith in the sanctity of the two Testaments and he was concerned that some eighteenth-century European Christian sceptics were challenging the veracity of these texts. William Jones turned to the Sanskrit texts for an independent corroboration of Genesis. He concluded that 'Indian civilization' was about 3,000 to 4,000 years old, that is, safely within the date of creation, computed by Archbishop Ussher. In 1805, Henry Thomas Colebrooke (1765-1837) used the stellar constellations mentioned in the Vedic texts at the start of spring and therefore at vernal equinox, and the constellations at the beginning of spring in his own time to estimate that the Vedas could not have been composed before 1400 BCE (Colebrooke 1805). This was consistent with the date of the Flood (which William Jones had concluded occurred at about 2350 BCE). Allowing about 1,000 years for the tribes to disperse and migrate after the Flood before undertaking the composition of the *Ṛgveda*!

The nineteenth-century Sanskrit scholar Max Müller (1823-1900) attempted to estimate the date of composition of *Ṛgveda Saṃhitā* just from the textual sources. He arrived at 1200-1000 BCE for the date of composition of this text (Müller 1859, 1892). The lack of rigour in his analysis is obvious from an examination of his work. Müller identified a fictional character, in a ghost story (*Kathāsaritsāgra* by Somadeva) written in the twelfth century, as Kātyāyana, a possibly historic figure in the fourth century BCE. He then assumed that this Kātyāyana was the author of a Śrauta Sūtra. Müller thus believed that he had established the epoch of composition of the Sūtras. Assuming that the Brāhmaṇas preceded the Sūtras, which preceded the Saṃhitās, Müller arrives at the date of composition of *Ṛgveda Saṃhitā*. An examination of Müller's numbers suggests that he was aiming at this date for the composition of this text. Challenges to this dating were immediate and robust. It should be said in fairness to Müller that he was fully aware of the arbitrary nature of his assumptions in dating *Ṛgveda* and he willingly retracted the rather precise chronology he had suggested. Although a number of other dates were suggested for the composition of *Ṛgveda* (Gonda 1975), it became a habit, particularly among Western Indologists, to claim, 'Max Müller had *proved* 1200-1000 BCE to be the date of *Ṛgveda*'. Over the last 100 years, this date has proved to be remarkably resistant to re-evaluation. In mid-twentieth century, the Oxford Sanskrit scholar Burrow (1955) placed the composition and compilation of *Ṛgveda* to 'round about the period 1200-1000 BCE'. Similarly, the Dutch Sanskrit scholar Gonda (1975) concluded, '. . . as far as the *Ṛgveda* is concerned [Max Müller's] computation is not unreasonable'. A re-evaluation of the chronology of *Ṛgveda* only started in the last decade of the twentieth century.

The *Ṛgveda Saṃhitā* that has come down to us was redacted probably more than 1,000 years after the start of the composition of the early *ṛcā*s (Witzel 1997) and the history of its composition can be reconstructed only because of the meticulous care with which its original language, content, and structure have been preserved. The most recent estimate of the relative and absolute chronology of the *Ṛgveda* has pushed the composition of the earliest *maṇḍala*s to about 1700 BCE (Table I.3) and stretched the period of composition of the 'family books' and *maṇḍala*s I to about 500 years. As noted above, *maṇḍala*s IX and X were later additions. These dates are a 'chronological hypothesis' and should not be considered to have been 'established'. The *maṇḍala*s of the *Ṛgveda* can also be arranged in a relative chronological order by identifying the genealogies of

Table I.3: Relative and absolute chronology of Ṛgvedic *maṇḍala*s (Witzel 1999)

Period	*Maṇḍala*
Early 1700-1500 BCE	IV, V, VI, (II?)
Middle 1500-1350 BCE	III, VII, VIII(1-66), I(51-191)
Late 1350-1200 BCE	VIII(67-103), I(1-50), VIII(49-59)

the composers (*ṛṣi*s or *kavi*s), the identification of the students of composers (who may have composed within the family of their teacher) and traits in the hymns. A relative chronology based on this type of analysis is shown in Table I.4. (Talageri 2000). This chronology differs from that obtained from analysis of the language of the *maṇḍala*s (Table I.3). However, for the current state of analysis of *Ṛgveda*, a consensus should not be expected.

The *Ṛgveda* mentions the priests of *Yajurveda* and *Sāmaveda* suggesting that the Vedic sacrificial ceremonies either existed before the composition of the Ṛgvedic *mantra*s or were developed in parallel with the *mantra*s. Either way, the sacrificial formulae would have been formulated and the chants composed as the sacrificial rituals evolved. These early *Yajurveda* and *Sāmaveda* are now lost as the language of the current Saṃhitās of these two Vedas suggests a post-Ṛgvedic development. The structure of the language permits the arrangement of these Saṃhitās in an approximate (relative) chronological order; shown schematically in Table I.5 (Witzel 1997). An absolute chronology of these texts is difficult to determine, as datable chronological markers have not been identified in these texts. The exception is the *Atharvaveda*, which mentions iron — *śyāma ayas* (black metal/iron, *AV*.XI.3.7). Smelted iron has been archaeologically attested at a number of sites, including those in northern parts of South Asia, from about the thirteenth century BCE onwards (Bryant

Table I.4: Relative chronology of the Ṛgvedic *maṇḍalas* (Talageri 2000)

Period	*Maṇḍala*
Early	III, VI, VII
Middle	II, IV, I
Late (a)	V, VIII, I
Late (b)	IX
Final	X

2001). This puts the *Atharvaveda* in the post-1300 BCE period. This is the second of the two unambiguous chronological markers of the Vedic texts. The language of *Ṛgveda* though different from that of *Atharvaveda* is not very different and the interval between these two texts cannot have been many centuries. The reference to iron suggests the (current) consensus of 1700-1200 BCE bracket for the composition of *Ṛgveda* (Table I.3). This time-bracket should be treated with caution; it is impossible to estimate the rate of change of a language, especially when only one or two texts are available. The bracket of 1700-1200 BCE for the composition of *Ṛgveda* should be seen as an 'intelligent guess' rather then a proof.

The language of the Brāhmaṇas suggests that these texts developed after the Saṃhitās, and the distinctive Sūtra style suggests that the Śrauta and Gṛhya Sūtras followed the Brāhmaṇas. This relative chronology is based on the language of the texts that have come down to us, that is, on the language of the final redaction of the texts. However, the composite character of several texts and the probability that part of them have been recast or have been long in making (Gonda 1975) make precise chronology largely uncertain. The Yājñavalkya teachings of *Śatapatha Brāhmaṇa* border on the period of the early Upaniṣads, which precede Buddha (whose birth is estimated to be around 400 BCE). It is possible that Pāṇini (estimated to be between 600 and 400 BCE) knew of *Aitareya* and *Kauṣītaki Brāhmaṇa* as he mentions two Brāhmaṇas with forty and thirty chapters (Gonda 1975). However, the language of the two Brāhmaṇas

Table I.5: Relative chronology of the post-Ṛgvedic Saṃhitās (Witzel 1997)

Ṛgveda Saṃhitā					
Early *Yajurveda Saṃhitā*	*MS*				
		KS,Kp-KS			
			TS		
				VS(M)	
					VS(K)
Early *Sāmaveda Saṃhitā*			*JS, KauthS*		
Early *Atharvaveda Saṃhitā*			*PS*		
				ŚS	
	——— Time ———→				

is significantly different from the language described by Pāṇini, and there must have been a long interval between the redaction of these Brāhmaṇas and Pāṇini. Considering these dates as the upper limit, and the reference to iron in *Atharvaveda*, as suggesting the lower limit, it can be concluded that the surviving post-Ṛgvedic texts may have been composed/redacted from about 1300 BCE to 500 BCE. Thus, the period from the compilation of the *Ṛgveda Saṃhitā* to about 500 BCE was a period of intense intellectual activity, a number of *śākhās* were founded in northern South Asia, and texts of some of these have survived to the present day.

The Āryan homeland and the models of Āryan migration are not pertinent to the subject of this book which is the calendar of the Vedic Period. However, the geographic and temporal extent of Indus Valley Civilization and the revised chronology of *Ṛgveda Saṃhitā* leads to the inescapable conclusion that there was close contact between the (pre-)Āryas, Āryas and the Indus Valley people over a large area and over hundreds of years. This close contact between the Āryas and the Harappans has to be born in mind in an exploration of the Vedic texts and the Vedic culture. Even if it were assumed that the Ārya arrived suddenly in South Asia around 1500 BCE, they would have encountered a flourishing Bronze Age culture of Late Harappa period (Table I.1). Although this culture had passed the peak of its urban development, the Āryas nevertheless would not have encountered just the ruins of a civilization. Evidence of a close contact between the Āryas and other population(s) of South Asia is provided by the loan words in *Ṛgveda Saṃhitā* (Kuiper 1991; Witzel 1999b), the use of fired bricks in Vedic rituals (Chapter III, Section 2.8), etc. It is possible that two (or more) linguistic cultures with different belief systems co-existed within the same material culture identified by the Indus Valley Culture. The Vedic corpus thus enables us to sample one of the earliest strata of South Asian history, it may not be the oldest stratum but it is probably close to it.

NOTES

1. Staal (2004) has argued that the Vedic *śratua* rituals have an obsessive emphasis on 'the east'. In a lengthy article, he presents an impressive amount of information on nuances of *śratua* rituals, but his basic thesis is incomprehensible. He asserts that the *śratua* ritual enclosure roughly

represents a historical map and the 'moving east' in the enclosure corresponds to movement of the Āryan tribes to the eastern lands.

The Sun was perhaps the first 'object' to be adored and deified by the human beings. This should not be surprising, for early man depended for his survival on the daily and seasonal changes of weather and he would have turned to worship the great force that (appeared) to regulate these changes, namely the sun. It would also not have been surprising for early man to face east (where the sun rises) to worship his (first) god. The organized religions that followed adopted this worship of the sun, e.g. *Ra* in Egypt, *Shamash* in Mesopotamia, *Apollo/Helios* in Greece. This was also true of Aztecs of Mexico, the Incas of Peru, and many indigenous Americans. The worshippers and the officiating priests of these religions would also have faced east to begin their rituals. Aspects of sun-worship can also be detected in the rituals of the Jewish Second Temple (516 or 350 BCE to CE 70). The Jerusalem Temple was oriented east-to-west, and at the feast of the Tabernacles, the priests walked from the sanctuary in the west to the open square before the east gate, to address the laity (I Esdras 9.38). Similarly, aspects of sun-worship can be discerned in the architecture of Christian churches. The nave of a Christian church is oriented east-west with the opening to the east to let in the first rays (or image) of the sun on the (Saint's) day of the patron saint of the church.

Religious ceremonies are a theatre, the aim is to impress and ultimately control the gullible. These ceremonies are not blueprints for projected military expeditions and conquests. Just as the east-west orientation of the nave of the Christian church does not indicate a Christian desire to conquer the East, the *prāṅmukhaḥ* (facing east) of the Vedic priest did not indicate the desire of the Vedic warriors to roll their war-chariots from the BMAC to the Punjab and further east. Both the Christians and the Āryas were repeating the actions of their ancestors – turning towards their first god.

2. In all books of *ṚV* (except the fourth), Sarasvatī is a long and mighty river that receives a number of tributaries and flows from the mountains to the sea (*girībhya ā samudrāt* – *ṚV.*VII.95.2) and nourishes the five Āryan people. It is a bringer of wealth and knowledge and all epithets of this river are in the superlative. The middle to late *RV* books (Books III and VII; Table I.3) mention the confluence of rivers Vipāś and Śatudrī (modern Beas and Sutlej) and Sarasvatī would seem to have lost a significant portion of its water (Witzel 2001). In *Pañcaviṃśa Brāhmaṇa* (*PB*.XXV.10.1), a text of the Middle Vedic Period, river Sarasvatī is 'lost in sand'. *Atharvaveda Saṃhitā* describes an abundant crop of barley (*yava*) on the banks of Sarasvatī (*AV.*VI.30.1; Whitney 1962). This either refers to the upper reaches of Sarasvatī, which was fertile during the middle to late *ṚV* period (Witzel 2001) and later, or this is a verse from an earlier version of *Atharvaveda* that has been reproduced in a later redaction. The Vedic texts thus seem to know the history of river Sarasvatī from a might river that reached the sea to a river that does not have enough water to reach the sea and has dried out. It is now believed that this

river is the modern river Ghaggar-Hakra, east of the Indus River system. The present river Ghaggar-Hakra acts as rain water drainage for the Shivaliks hills and dries out in the Thar Desert; it does not have enough water to reach the sea. Fieldwork suggests that in earlier times, the Ghaggar-Hakra was a far bigger river (Possehl 2002), and its course can indeed be traced to the ocean in the Rann of Kutch. This is because many small and two large rivers (Sutluj and Yamunā) emptied into Ghaggar-Hakra. Environmental and geological changes at some epoch diverted the Sutluj and Yamunā away from river Ghaggar-Hakra. The chronology of the archaeological complexes in the dry river bed suggests that Ghaggar-Hakra could only have been a mighty river before about 1900-1700 BCE. The proponents of the invading or migrating Āryas attempt to explain the two divergent descriptions of river Sarasvatī in the Vedic texts by placing the mighty Sarasvatī somewhere to the west (e.g. Helmand province of Afghanistan, Kochhar 2000), the presumed earlier home of the Āryas. The migrating or invading Āryas are supposed to have carried this name with them when they went east. No evidence has been presented for this hypothesis, and its only aim appears to be to preserve the 'accepted' (i.e. 1500-1200 BCE) chronology of *Ṛgveda Saṃhitā*.

3. *armaká* ṚV.I.133.3; (MW93a) n. rubbish, ruins (but other meaning is possible, see Thompson 2004).
   vailasthānaka ṚV.I.133.1, 3; (MW1025c) mfn. Situated in a hole or lurking place or a pit. (But see Thompson 2004).
   (MW: Monier-Williams, 1899).
   These words occur in only one *sūkta* (ṚV.I.133.) of ṚV and have been analysed by Burrow (1963). He interpreted *vailasthānaka* as a proper noun, the name of a city (A.I.4.). His analysis is based on his belief that the urban civilization of Indus Valley had ended (long) before the arrival of the invading Āryas. The analysis reaches two mutually contradictory conclusions – that the Āryas encountered only ruins and they fought and laid waste the cities of the inhabitants.
4. Vedic Sanskrit is archaic or old Sanskrit as opposed to Classical Sanskrit, which was defined by the rules laid down by the grammarian Pāṇini (*c.* 600-400 BCE). Classical Sanskrit was a scholarly language that had to be studied and mastered whereas Vedic Sanskrit was a vernacular. It is the oldest attested language of the Indo-Iranian branch of the Indo-European language family. It is closely related to the language of Avestan, the oldest preserved text in the Iranian language. Some syllables of Vedic Sanskrit are either missing or have changed in Classical Sanskrit.
5. Śulba Sūtras are part of one of the appendices of the Vedas. The appendices aid a sacrificer or a Vedic priest to understand the Vedas and guide him to correctly perform a ritual. For full efficacy of a sacrifice, the altar and the ground where the ceremony was performed had to conform to precise shape and measurements and geometric accuracy was of the utmost importance. The Śulba Sūtras give rules for constructing the fire altar(s) (Chapter III, Section

2.8). Four recensions of these Sūtras have survived as independent treatises; these are the *Baudhāyana*, *Mānava*, *Āpastamba*, and *Kātyāyana Śulba Sūtras*. In addition, *Laugākṣi*, *Varāha*, *Vādhūla* and *Hiraṇyakeśin Śulba Sūtras* have survived as chapters of the respective Śrauta Sūtras. Śulba Sūtras contain a large amount of geometrical knowledge, but rules are given without a proof. Some of the rules, such as the method of constructing a square of area equal to a given rectangle, are exact. Others, such as constructing a square of area equal to that of a given circle, are approximate. The 'rule' known today as the 'Pythagoras's Theorem' is stated in these texts. Śulba Sūtras were first translated into English by Thibaut (1875). By comparing the methods employed by the Greeks and those described in Śulba Sūtras, he came to the conclusion that there must have been an independent development of geometry in South Asia. The academic establishment of the West was outraged since it was held as established that the 'Indians' had borrowed geometry and mathematics from the Greeks. Thibaut recanted by proposing a date for Śulba Sūtras that would not offend Western academics. The date he proposed for Śulba Sūtras was fourth-third century BCE. To his credit, he maintained the possibility that mathematics could have developed independently in South Asia. Seidenberg (1978), a historian of science, has called Thibaut's revision of the date of Śulba Sūtras 'A terrible statement!' Passages on altar construction given in *TS* and *ŚB* suggest that the geometry and mathematics of the Śulba Sūtras must predate 800 BCE that is, predate the development of geometry in Greece by a large margin. For over 100 years, the Śulba Sūtras were denied their rightful place in the history of mathematics. Scholars no longer believe that Vedic geometry and mathematics were borrowed from the Greeks.

II

Seasons

The annual revolution of the earth around the sun and the tilt of the earth's axis relative to the plane of this revolution results in changes in the atmospheric conditions (climate) that divide a year into natural periods (N.II.1). These divisions are the seasons and these are the foundational elements of a calendar. Early man would have noticed the birth, flowering of life and death in the cycle of seasons.

> **Box 1**
>
> *stately in form was the man of seventy winters . . . that maiden of seventeen summers*
>
> – 'Evangeline'; Longfellow (1807-82)

Not surprisingly many (early) cultures (and nineteenth-century poets, Box 1) defined the age of a person by seasons. The survival of a species and perhaps more importantly the survival of a tribal and a family group depends on food gathering and food production, both at the mercy of the weather. Food accumulation, whether through hunting or farming, is a communal undertaking and aims at maximizing food stock and minimizing the effort involved in creating this food stock. There would therefore have been 'social pressures' to predict the best time(s) for gathering food(s), go hunting or tend the crops or to put it differently, a society would have demanded and required the creation of a seasonal calendar.

The origins of the division of a year into seasons in the Āryan society are lost in the mists of history. Vogel (1971) has speculated that the earliest division of a year was into just two seasons, the cold and the warm. With the development of agriculture, a third season was added, this period of growth and ripening was called *śarad* (from roots *śrā, śrī, śṛ,* to boil, to become cooked, become ripe). A further division of the year, to identify the transition from winter to summer was made, and called *vasanta* from root *vas* 'to become light' and refers to the change in the weather. This scheme implicitly assumes that the division of the year into seasons followed the development of agriculture. However, seasons are as important to hunter-gatherers as they are to farmers (N.II.2) and the division of the year into seasons must have occurred long before man broke the sod with a hoe or a plough. The Āryas were aware that the sun divides the year into seasons (A.II.1 and A.III.5). Also, the words *vasanta* and *śarad*, identifying seasons, are encountered in the *Ṛgveda*, and the Āryas must have identified and classified the year into seasons long before the *ṛcās* and *sūkta*s of this Saṃhitā began to be composed. The Ṛgvedic society had a well-developed priestly class that was stratified into principal (consisting of four priests) and secondary (consisting of twelve priests) levels. These priests were called *ṛtvij*. The root of this word is *yaj*, which means 'honour or adore' but can also mean a 'sacrificer' and *ṛtvij* (*ṛtu* + *yaj*) can mean a 'season sacrificer'.

The Sanskrit word for 'season' is *ṛtu* and is mentioned frequently in the *Ṛgveda*, and sometimes in plural (*ṛtubhiḥ*). In this Saṃhitā, *ṛtu* can also mean time, time suitable for a ritual or a proper time for a ritual. The identification and use in *Ṛgveda Saṃhitā* of names of seasons (*vasanta*, *grīṣma*, *śarad* and *hemanta*) found in later texts suggest that, by Ṛgvedic times, the Āryas had identified seasons as the annual changes in the weather and to them *ṛtu* denoted not just a specific time but also the passage of time. This is confirmed by verses in *Ṛgveda Saṃhitā* (e.g. A.II.2, 3, 4), which illustrate that the Āryas had used seasons as (meteorological) markers to indicate the elapse of a year, and this year is an interval between successive autumns or springs or any other season, that is, it is a tropical/seasonal year. In later texts, particularly in the Brāhmaṇas the Āryas had transformed the earlier forms of beliefs and worship into a complicated system of sacrificial ceremonies. These texts highlight the crucial role seasons played in the ritual, social, and intellectual life of the Āryas. This is emphasized in *Śatapatha Brāhmaṇa*, where the Kurupañcālas, the performers of the *Rājasūya* ceremony, say (*ŚB.* V.5.2.5, Eggeling 1894) 'It is the seasons that, being yoked, draw us, and we follow the seasons, thus yoked'. Seasons are both the glue that holds the year together (*ŚB.*VIII.7.1.3; *ŚB.*VIII.7.1.1, Eggeling 1885b) and the bricks that make up the year (*ŚB*.VIII.2.1.3). Seasons come and go (*ŚB*.VIII.5.2.10) but not in a random fashion, there in a cyclical order to them (*ŚB*.XII.8.2.35, Eggeling 1900). This order is also continuous and a sacrificer is enjoined not to have breaks in the performance of rituals as this would not reflect the natural cycle or order (e.g. *ŚB*.I.3.5.16, Eggeling 1882).

Zimmer (1879), Dīkṣita (1896), Thibaut (1899), Oldenberg (1988), Renou (1948-9), Vogel (1971), and Raghavan (1972) have given brief descriptions of seasons in the Vedic texts.

1. SEASONS IN THE *ṚGVEDA SAṂHITĀ*

The word *ṛtu-* (season) occurs fifty-five times in *Ṛgveda Saṃhitā* (Lubotsky 1997) and is mentioned in almost every *maṇḍala*. Two *sūkta*s, *ṚV*.I.15 and *ṚV*.II.37, are dedicated to *ṛtu*. The first is recited when making a libation to seasons during the *Agniṣṭoma* ceremony. The second (along with *ṚV*.II.36) is recited on the sixth day of the twelve day (*dvādaśāha*) ceremony. In some hymns reference is made to three dawns; Hillebrandṭ (1980) has suggested that this may refer to three seasons. He has also suggested that the earliest Vedic names of the seasons may have been *gauḥ* (cow or cattle), *āpaḥ* (waters), and *svar* (sun or light), which he regarded to be synonymous with *vasanta* (spring), *varṣāḥ* (rains) and *gharma* (heat, summer) respectively. In the *Ṛgveda Saṃhitā gā*, *apáḥ* and *súvar* appear together in *ṚV*.V.14.4 (A.II.5) and in this order, and this hymn is addressed to *agni* or the Sun which suggests that seasons may, indeed, have been intended.

The frequency of the words *vasanta* (spring), *grīṣma* (summer), *hemanta* (winter), *śarad* (autumn) and *prāvṛṣ* or *varṣāḥ* (rains) in the *Ṛgveda Saṃhitā* and *Atharvaveda Saṃhitā* is given in Table II.1. To compare these frequencies, it is best to ignore *maṇḍala*s I and X. The number of hymns in these two *maṇḍala*s is considerably larger than in the other *maṇḍala*s, the subject matter is heterogeneous, and hymns and verses are (mostly) late additions to the Saṃhitā. The 'family *maṇḍala*s' (II to VII), on the other hand, are homogeneous and have (statistically) equal number of hymns (the mean number of hymns in these *maṇḍala*s is 71±20). In Table II.1, the numbers involved are small, and conclusions should be drawn with caution. In all family *maṇḍala*s, a year is denoted by *Śarad* (i.e. a year that was measured from autumn to the following autumn). Winter seems to

Table II.1: Frequency of season names in *Ṛgveda Saṃhitā* and *Atharvaveda Saṃhitā*

ṚV. maṇḍala	*Vasanta*	*Grīṣma*	*Prāvṛṣa/ varṣāḥ*	*Śarad*	*Hemanta*
I				8	2
II				3	
III				2	
IV				4	
V			2	1	1
VI				4	5
VII			3	6	
VIII					
IX					
X	2	1		8	2
AV	1		1	22	5

have made a significant impression only on the composers of *maṇḍala* VI and the locale of this early *maṇḍala* may have been significantly different from that of the other five family *maṇḍalas*. In only one verse, the words *vasanta*, *grīṣma*, and *śarad* occur together, and they do not imply a year; this verse will be discussed later. The words for rains are *prāvṛṣa* and *varṣāḥ*; the word *prāvṛṣa* occurs three times in a single hymn (*Ṛ*V.VII.103.9, e.g. A.II.6) and *varṣāḥ* occurs only twice (Lubotsky 1997). This word also occurs a few more times in compounds that refer to rainfall. Unlike the names of other seasons, in *Ṛgveda Saṃhitā* the words *prāvṛṣa* and *varṣāḥ* are never used to denote a year and their use does not suggest the passage of time.

Analysis of verses of the *Ṛgveda Saṃhitā* suggests that the Ṛgvedic Āryas divided a year into three, five or six seasons. The verses of the Saṃhitā from which the number of seasons in a year can be inferred are given in Table II.2. Three seasons in a year are inferred from references to 'three-wheeled' or 'three-axled wheel of' a chariot that brings prosperity (A.II.7, 8, 9, 10, 11, 12, 13). The reasoning here is similar to that of Yāska (around 600 BCE) in his interpretation of a three-navelled wheel as the year with three seasons; summer, rains, and winter (*Ṛ*V.I.164.2; *Nirukta* IV.27; Sarup 1920-7). Similar arguments have been used by others (e.g. Vogel 1971), to infer three seasons, from some of the verses given in Table II.2. The words *vasanta*, *grīṣma*, and *śarad* (spring, summer and

Table II.2: Verses of *Ṛgveda Saṃhitā* analysed to identify seasons in a year. The relative chronology is given in brackets.

Seasons	*ṚV* verses
Three	I.34.2 (M); I.34.5(M); I.118.2(E); I.157.3(E); I.164.2(E); I.164.48(E); IV.36.1(E)
Five	I.20.7(L); I.164.12(E); I.164.13(E); VII.87.4(M); X.90. 15(L)
Six	I.23.15(M); I.164.12(E); I.164.15(E); VII.87.5(M); VIII. 68.14(L)

The chronology of the hymns is according to Witzel (1999):
E – Early Ṛgvedic period, 1700-1500 BCE.
M – Middle Ṛgvedic period, 1500-1350 BCE.
L – Late Ṛgvedic period, 1350-1200 BCE.

autumn) appear together and in this order only in the Puruṣa Sūkta (*Ṛ*V.X.90.6, A.II.21). The verse refers to items of a sacrifice; *vasanta* is identified with *ghī* (clarified butter), *grīṣma* with fuel and *śarad* with oblation. This 'human sacrifice' was developed into an elaborate ceremony in the *Śatapatha Brāhmaṇa* but the order of the offered oblations and their association with seasons is the same as that in the *Ṛgveda*. It is surprising that the rainy season is not mentioned. This verse is repeated in *Atharvaveda Saṃhitā* (*AV*.XIX.6.10). In the Ṛgvedic verse (*Ṛ*V.X.90.7), the sacrifice (*Puruṣa*) is anointed on the sacred grass (*barhi*) but in *AV* the sacrifice is anointed with rain (*prāvṛṣa*). Raghavan (1972) has combined the second hemistich of *ṚV*.X.90.6 and the first hemistich of *ṚV*.X.90.7 to argue that rain (*prāvṛṣa*) is 'implicit' in these verses. It would appear more likely that the composer(s) of the *Atharvaveda Saṃhitā* (a text later then *ṚV*) has incorporated a fourth season (rain or the rainy season) in the verse of the *Puruṣamedha* that the *ṛṣi*s of *Ṛgveda Saṃhitā* did not feel compelled to include.

Evidence of tripartite division of a year during the Vedic period is also implied in the *cāturmāsya* sacrifices (see Bhide 1979 for details). These sacrifices have not been identified in the *Ṛgveda Saṃhitā*, but are described in the post-Ṛgvedic texts. These three sacrifices, performed at four-monthly intervals (*ŚB*.II.6.4.2, 3, 4; Eggeling 1882) presuppose the division of a year into three periods or three seasons. In the post-Ṛgvedic texts, the 'dates' of the performance of the *cāturmāsya* sacrifices are indicated by the coincidence of the full moon and a specific *nakṣatra*, this is discussed in Chapter IV. The post-Ṛgvedic texts rarely mention three seasons and (as discussed below) five and six seasons are the norm in these texts (Figure II.2). The choice of just three seasons (out of five or six) for special sacrifices suggests that three seasons in a year had a particular significance, and it is possible (and highly likely) that the *cāturmāsya* sacrifices were performed during the Ṛgvedic period when a year was possibly divided into three seasons.

Some verses of *Ṛgveda Saṃhitā* suggest five seasons in a year (Table II.2; A.II.14, 15, 16, 17, 18). In later texts, there are frequent references to twenty-one items of a sacrifice (or aspects of Prajāpati). These are justified by 'twelve months, five seasons, three worlds and the sun or Prajāpati' (e.g. *TS*.V.4.12; *AB*.i.30(v.4); *TB*.3.8.10; *KB*.xi.6; *PB*.VI.2.2; *ŚB*.I.3.5.11). Similar references to twenty-one items of a sacrifice, 'thrice seven names' and 'thrice seven sticks of firewood' can be identified in the *Ṛgveda Saṃhitā* (A.II.17, 18). A degree of caution has to be exercised when interpreting Ṛgvedic verses with the aid of post-Ṛgvedic texts. It is, however, possible that the 'twenty-one' of the Ṛgvedic verses has the same interpretation as that given in the Saṃhitās and the Brāhmaṇas and these Ṛgvedic verses suggest five seasons. Gonda (1976) has discussed 'three times seven' in the Vedic texts and has shown that other interpretations for this number are also possible.

Verses implying six seasons in a year can also be identified in the *Ṛgveda Saṃhitā* (Table II.2; A.II.15, 19, 21). In the later Saṃhitās and the Brāhmaṇas, the twelve months of the year are divided between six seasons with two months per season (Chapter III). Some verses of *Ṛgveda* (A.II.21) suggest that this practice may have originated in the Ṛgvedic times. In verse A.II.21, the 'six form pairs' (almost certainly) refers to six seasons of two months each. The 'seventh single-born' is the intercalary month (which has no associated season) introduced to synchronize the quasi-synodic ritual year and seasons (Chapter VI). Analysis of other Ṛgvedic verses (Renou 1948-9) also suggests that the Āryas had divided the year into six seasons of two months each.

The analysis presented above suggests that during the Ṛgvedic period the year may have been divided into three, five and six seasons. It is impossible to determine if there was a gradual progression from three seasons to six seasons over the entire region occupied by the Āryas during the Ṛgvedic times or different groups of people (in different locations) divided the year into a different number of seasons at about the same time (see N.II.2). A corroborated relative chronology of the Ṛgvedic verses is required to determine if the verses suggesting five and six seasons were composed after the verses suggesting three seasons. Arnold's (1905) chronology of Ṛgvedic hymns gives ambiguous results; the verses suggesting three seasons cover the 'archaic' to 'popular' range or early to late period. The verses suggesting the five and six seasons also cover this range. The recent chronology (Witzel 1999a) suggests that a significant number of verses that suggest three seasons may have been composed in the early Ṛgvedic period; those suggesting five and six seasons cover the early, middle and late periods. The complexity of identifying the epoch of different seasons from the hymns and verses of *Ṛgveda Saṃhitā* is highlighted by *sūkta* *Ṛ*V.I.164. In this one hymn, there is evidence for three, five and six seasons in a year and confusingly verse *Ṛ*V.I.164.12 refers to *both* five and six seasons in a year. Houben (2000a, 2000b) has discussed the *Gharma-Pravargya* ritual that forms the core of this hymn. He has suggested that this ritual has evolved over the course of time. Although a direct link between seasons and the *Gharma-Pravargya* ritual has not been established, it is possible that new verses were added to this hymn as the *Gharma-Pravargya* ritual was elaborated. During this period, the climate of South Asia may have changed or groups of Āryas may have moved into different regions, and the number of seasons in a year may have been increased from three to five to six and corresponding verses added to the hymn without deleting the old verses.

The climate of South Asia is dominated by two different weather systems that sometimes overlap; winter cyclonic system of the western highlands and the summer monsoon system of the peninsula. There is no reason to believe that the climate the Āryas experienced in South Asia was similar to the climate in this region today. However, the current weather patterns of this region provide a good base-line for comparison with the climate that could be inferred from *Ṛgveda Saṃhitā*. The monthly average (from 1961 to 1990) precipitation (rainfall and snowfall) at four locations in the region where the *sūkta*s of *Ṛgveda Saṃhitā* may have been composed is shown in Figure II.1.

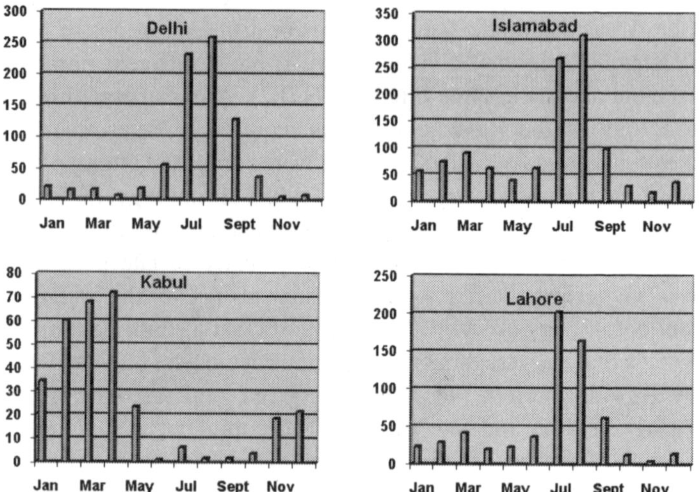

Figure II.1: Precipitation (rainfall and snowfall) in millimetres at four locations in South Asia. These locations are in and around the land of Sapta Sindhu, the land occupied by Ṛgvedic Āryas. The data are monthly averages from 1961 to 1990. *Source:* World Meteorological Office.

In Delhi (the eastern limit of the Ṛgvedic region), the temperature peaks in June followed by high rainfall from July to September, this is the west and south-west monsoon period. Judging from the annual rainfall, the climate of Lahore is similar to that of Delhi, except there is slightly higher rainfall from January to March. This pattern is repeated in Islamabad (off-centre of the Ṛgvedic region) with enhanced rainfall from January to April. The climate of Kabul (on the western edge of the Ṛgvedic region) is decidedly different from that at the other three locations. The bulk of the precipitation (rain and snow) is from January to April and rain does not follow the high temperature from June to August as at other three locations. If it is assumed that the climate of northern India/Pakistan during Ṛgvedic period was similar to the present climate of this region then it is possible to divide a year into three seasons, namely spring, rains and autumn and this has indeed been suggested by Zimmer (1879), Thibaut (1899) and Vogel (1971). This would suggest that rains/rainfall should be a principle season during the Ṛgvedic period as it is in South Asia today. It is therefore, surprising that a rainy season is relatively infrequently mentioned in the *Ṛgveda Saṃhitā* and never used to indicate a year as the other seasons (and *śarad* in particular) have been. Although it is not entirely obvious if an agricultural cycle is woven into the rituals of *Ṛgveda Saṃhitā*, rain as bringer of prosperity appears to play an important part in the culture of this people. A large number of hymns in this Saṃhitā are in praise of Indra, the provider of water, both in the rivers and the rain. The most heroic act of Indra is the victory over Vṛtra and the liberation of waters (Oldenberg 1988), although other interpretations of this myth have also been suggested (Kuiper 1983). A number of verses in the ninth *maṇḍala* compare the flow of *soma* with raindrops or rainfall. *Soma* has a central role in the ceremonials of the Āryas; it grants fertility and confers upon the sacrificer the riches of heaven and earth. This suggests that the Āryas recognized the strong connection between rain, fertility, and wealth. Hymns *Ṛ*V.V.83 and *Ṛ*V.VII.101 (respectively an early and middle period hymn) are addressed to Parjanya (the rain-cloud). Hymn *ṚV*.V.83 is a powerful evocation of the monsoon or a storm. The verse *ṚV*.V.83.4, refers to the fertilization of the land by showers and plants springing up. In *ṚV*.V.83.8, 'the rivers flow unimpeded to the east'. This suggests a location on the eastern edge of the 'Vedic region', exactly the region where well-defined monsoon rainfall would be expected. It is tempting to assume that, in this hymn, the Āryas have eulogized the spectacular phenomenon of the onset of the monsoon rains in South Asia. However, in this hymn or *ṚV*.VII.101, there is no indication that the composer is referring to an annual phenomenon nor is there an allusion to a season. Only one hymn (A.II.22) suggests a link between rainfall and a season. Similarly, only *sūkta ṚV*.VII.103, suggests a cyclical nature of rainfall (Vajracharya 1997). Although the verses and the hymns of this Saṃhitā suggest that water was crucial to these people, it is surprising that when three seasons are mentioned (*ṚV*. X.90.6), the 'rainy season' is not one of these three.

When the year was divided into five and six seasons in this Saṃhitā, it is very likely that the rainy season was one of these seasons. This is certainly true of the later Saṃhitās and the Brāhmaṇas in which the year is also divided into five or six seasons. It is possible that the rain-related verses and hymns discussed above were composed when a rainy season was well defined, and the Āryas had divided the year into five (and later) six seasons.

1.1. Ṛgvedic climate and palaeoclimatic data

Climate has influenced and changed civilizations since the end of the last Ice Age, about 11,000 years ago. There are well documented cases of changes in climate affecting societies of West Asia and elsewhere (Fagan 2004). The discussion, over last 150 years, of seasons in the Vedic texts is based on the assumption that the climate in South Asia during the Vedic period was similar to the present climate. However, it is unlikely that the climate of South Asia has been static for last 3,000-4,000 years, changes must have occurred, and the finger prints of these changes may have been left in the Vedic texts. The discussion on three seasons presented above suggests that parts of *Ṛgveda Saṃhitā* may have been composed during a period of prolonged drought or scarce rainfall in north-western South Asia.

The development of palaeoclimatology techniques enable an independent study of the climate of South Asia during the Vedic period. Programmes to decode the palaeoclimate of South Asia have been undertaken since mid-twentieth century and data are gradually being accumulated to present both a global picture and a picture of localized variations (e.g. Agrawal and Pande 1976). The analysis of oxygen-isotope record of stalagmites from southern Oman (Fleitmann et al. 2003) indicates that, after 6000 BCE there was a gradual reduction in monsoon precipitation and by 1000 BCE the precipitation had more than halved. These data also indicate a sharp drop in precipitation around 1500 BCE, the putative age of the *Ṛgveda Saṃhitā.* An abrupt drop in rainfall around 1500 BCE is also indicated from the analysis of accumulated marine deposits off the coast of Karnataka (Bentaleb et al. 1997). These deposits record palaeoclimatic signals representative of a large region of southern India. The selected parameters, oxygen-isotope record and marine deposits, require precise climatic and environmental conditions for their optimal development and can therefore provide reliable clues to climatic conditions. These periods of low precipitation lasted for almost 200 years. A 'long drought' or dry period around Bikaner in Rajasthan from about 2000 to 0 BCE is also inferred from pollen profiles from sediment cores taken from the bottom of dry or seasonally dry lake-beds (Bryson 1997). Analysis of oxygen and carbon isotopes in the micro-fossils in the sediment cores collected from the eastern Arabian sea (Sarkar et al. 2000) suggest arid periods around 2500, 1800 and 500 BCE. A dry period after 2000 BCE is also suggested from analysis of annually laminated marine sediment collected west of Karachi, that is, from the north-eastern Arabian Sea (von Rad et al. 1999). This sediment is deposited by river Indus and contains a record of precipitation in the land occupied by the Āryas. Extensive and detailed surveys are required to obtain palaeoclimatic data from northern India and Pakistan, before a complete picture of the climate of the Vedic period can be constructed. However, a definitive comparison of climate determined from the palaeoclimatic data and that inferred from Vedic texts will not be possible without an independently corroborated chronology of the Vedic texts.

The brief description of palaeoclimate of South Asia given above sheds little or no light on the number of seasons in a year alluded to in *Ṛgveda Saṃhitā*. It could be speculated that some of its *mantra*s were composed during one of the arid periods when rainfall would have been infrequent and precipitation limited, there would have been no rainy season. The scarcity of water may have prompted the large number of hymns and prayers addressed to Indra (and other deities e.g. Maruts) to provide water. The Āryas would have been dependent on (some) rain but mostly on river water, as suggested by the Saṃhitā. With the passage of time, the arid period would have ended and both the frequency and intensity of rainfall would have increased. Equally likely, the people could have moved further east into regions watered more by rainfall rather than by rivers. Due to the end of the arid period, or migration of people, the rainfall would have become regular and this may have prompted the Āryas to divide the year into more than three seasons that included a rainy season. The composition of the later Saṃhitās and the Brāhmaṇas may have taken place around this time, and this would account for the similarity in the number of seasons in a year and their justification in these diverse texts.

It has also been proposed (Falk 1997) that the Ṛgvedic *mantra*s can be explained in terms of the geographical location of the Āryas. The *mantra*s may have been composed in three locations:

- the lands between Gurgan and Arachosia, that have scarce rainfall and seasonal rivers with uncertain supply of river water.

- the land of Sapta Sindhu and Punjab with its perennial supply of river water.
- the land east of Punjab with its uncertain monsoon rainfall.

The implicit assumption here is that 3,000 to 4,000 years ago the climate of these lands was not very different from what it is now. If this (unrealistic) assumption is accepted, then this proposition is consistent with the frequency of seasons described above. The division of a year into three seasons may have been made in the first location and there would have been no rainy season. The division of a year into a larger number of seasons would have been made in the second and the third location and would have included a rainy season. It is possible that *Ṛgveda Saṃhitā* was compiled in the later two locations and included the *mantras* composed in the first location. The early versions of other Saṃhitās and the Brāhmaṇas may have also been composed at about the time of compilation of *Ṛgveda Saṃhitā*.

2. SEASONS IN THE POST-ṚGVEDIC SAṂHITĀS AND BRĀHMAṆAS

In the post-Ṛgvedic Saṃhitās and Brāhmaṇas three, five, six and seven seasons are mentioned. For illustration, the total number of times three, five and six seasons is enumerated in a sample of Saṃhitās and Brāhmaṇas and *Śatapatha Brāhmaṇa* is shown in Figure II.2. The reason for showing the seasons in *Śatapatha Brāhmaṇa* separately will be discussed below. Three seasons are mentioned rather infrequently in these texts, and when these are mentioned, it is primarily with respect to *cāturmāsya* sacrifices. These sacrifices were an essential part of the annual ceremonies of the Āryas. Vogel (1971) has suggested that these sacrifices were originally associated with the three Ṛgvedic seasons, namely, winter, warming-growth (spring), and rain. He contends that, at a later date (i.e. the period of *ŚB*), they were performed at the commencement of spring, the rains and autumn. However, if these sacrifices were performed at the beginning of these three seasons, then the interval between the last two sacrifices would not be four months, as the name of the sacrifices implies. This interval would have been two months, because in the post-Ṛgvedic Saṃhitās and Brāhmaṇas the rainy

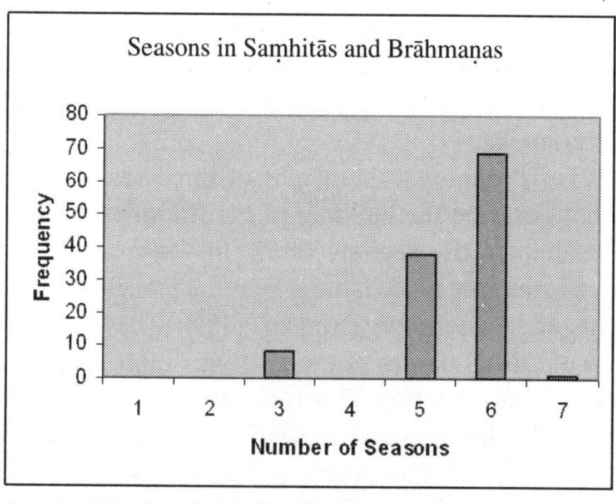

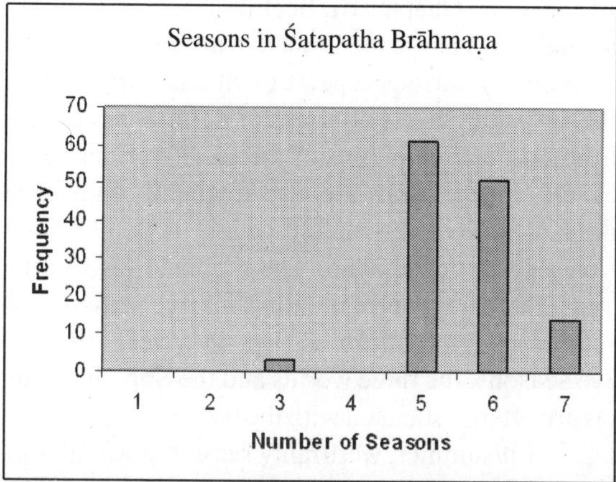

Figure II.2: Frequency of enumeration of seasons in a selection of Saṃhitās and Brāhmaṇas, and *Śatapatha Brāhmaṇa* (*ŚB*).

season and autumn are each allocated two months (see discussion below). The three seasons suggested by Vogel (1971) would also be in disagreement with the *nakṣatra*s 'under' which the performance of the three sacrifices was prescribed (Chapter IV, Section 5). Judging from the frequency of seasons (Figure II.2), the division of the year into three seasons was abandoned in the post-Ṛgvedic age, and the *cāturmāsya* sacrifices were just three sacrifices, four months apart and not necessarily at the beginning of a prescribed season (the timing of these sacrifices determined from astronomical data will be discussed in Chapter IV, Section 5). The few references to three seasons seem to have been introduced (or retained) in order to continue the tradition of these ceremonies.

In *Taittiriya Brāhmaṇa*, the names of three seasons are given in passages that describe the building of the fire-altar, *Sāvitra Cayana* (*TB*.3.10.1, Chapter III, Section 2.8.2). In the first circle from the centre, three bricks are placed, these bear the names of three seasons; *agni*, *sūrya*, and *candramā* (Dumont 1951). These names are different from the usual names of the seasons (Table II.3), and are not encountered in any other text. This is the only text in which a fire altar is described with three associated seasons; in all other texts, the fire altar is associated with six seasons and these are 'mapped' as layers of bricks of the altar (Chapter III, Section 2.8.).

Five seasons in a year are mentioned in all Vedic texts, and they are introduced to justify various aspects of rituals, for example, pairs of bricks in the fire-altar, layers of bricks of a fire-altar, aspects of Agni, leaders of the host and chieftains of the sacrificer, metres and verses of liturgies, etc. Five seasons are also frequently introduced via the seventeenfold or the twenty-onefold action or items of a sacrificial ceremony or aspects of Prajāpati. The rationale for seventeenfold is that, in a year 'there are twelve months and five seasons' and that for twenty-onefold, as given above, is that, in a year 'there are twelve months, five seasons, the three worlds and the Sun'. In these texts, the five seasons are often associated with the five Vedic cardinal directions (east/spring, south/summer, west/rainy season, north/autumn and the upper region/winter and dewy). Unlike *Ṛgveda Saṃhitā,* the post-Ṛgvedic texts give lists of names of seasons, and two sets of names of five seasons are given in Table II.3. The names in column 2 are only found in *TS*.II.6.1.1 (Keith 1914), *ŚB*.I.5.4.1 and *ŚB*.II.1.3.1 (Eggeling 1882). *KB*.iii.4 (Keith 1998) gives the correspondence between these names and the usual names (given in all Saṃhitās and Brāhmaṇas) given in column 3. The order in which the names of the seasons are given is the same in all texts (the exception is *AV* and this will be discussed below), and spring is always the first season. Two months are allocated to each season but the names of these months are not given, also this leaves two months in a year unaccounted for. This is also reflected in a degree of ambiguity in the name of the fifth

Table II.3: The names of the seasons, and months. The association of months and seasons is for six seasons only.

Five seasons				Six seasons		
Seasons	*TS*.II.6.1.1 *ŚB*.I.5.4.1 *ŚB*.II.1.3.1	Vedic Seasons	Devatās	Seasons	Vedic Seasons	Months
Spring	Samidh	Vasanta	Vasus	Spring	Vasanta	Madhu Mādhava
Summer	Tanūnapād	Grīṣma	Rudras	Summer	Grīṣma	Śukra Śuci
Rains	Iḍ	Varṣā	Ādityas	Rains	Varṣā	Nabhas Nabhasya
Autumn	Barhiḥ	Śarad	Ṛbhus	Autumn	Śarad	Iṣa Ūrja
Winter	Svāhā	Hemanta	Maruts	Winter	Hemanta	Sahas Sahasya
				Cool/Dewy	Śiśira	Tapas Tapasya

season; in some passages, the fifth season is called 'winter and dewy or cool (*hemanta-śiśira*)' but in others it is called just 'winter (*hemanta*)'. Vogel (1971) has suggested that in this scheme the fifth season would have been assigned four months. The dual name of this season and (perhaps the later) division of this season into two seasons, *hemanta* and *śiśira*, of two months each, would support this suggestion but there is no corroborating textual evidence for this. A winter of four months would be inconsistent with the current weather system of South Asia. Only a detailed survey of the palaeoclimate of Madhyadeśa would throw light on the length and severity of winters in the post-Ṛgvedic period. It would be more logical to allocate four months to the rainy season (*varṣā*), if the climate of Madhyadeśa were similar to the present climate, but there is no textual evidence for this either. Krick (1982) has proposed a privileged role for *śiśira* as the season connecting the old and the new year. She has also suggested that these two months were 'switching months to reconcile the solar and the lunar year'. It is not clear how Krick (1982) arrived at this conclusion. The synchronization of the tropical (seasonal) year and the synodic/lunar year is discussed in Chapter III.

Like the five seasons, six seasons in a year are also given in all Vedic texts. The seasons are included in the rituals through the number of obeisance, syllables of a metre, the number of sacrificial victims, the number of gifts, the number of case-forms of Agni, the number of offerings, the number of priests, the number of circumambulations, etc., and (surprisingly) colours of smoke (*VS*.XXIV.11) and types of birds (*VS*.XXIV.20). A correspondence between the ritual practices and the agricultural practices of the time is also made in these texts. The six formulae recited in baking an offering equivalent the six seasons in which a seed is planted and ripens to food. The names of the six seasons given in these texts are given in Table II.3. The *devatās* (the associated deity) of the seasons (*TB*.2.6.19) are also given in this table, surprisingly there is no *devatā* for the sixth season. In this scheme, each season lasts for two months. The name of a month (Table II.3, Column 7) of each season describes the season (*TS*. IV.4.11; *AB*.iv.26 (xix.4); *ŚB*.IV.3.1.14-20) and is probably the earliest name of the month. The spring months are Madhu (honey – the usual symbol for fertility and rain in the Vedas) and Mādhava, because in the spring plants sprout and the trees are brought to ripeness. The summer months are Śukra (clear) and Śuci (bright) because in these months 'it burns fiercest'. The months of the rainy season are Nabhas (mist, cloud) and Nabhasya because in these months 'it rains from yonder sky'. The autumn months are Iṣa and Ūrja because in these months food (*ūrj*) and juice (in plants) ripens. The winter months are Saha and Sahasya because winter, by force (*sahas*), brings creatures into his power. The months of the dewy season are Tapas and Tapasya because in these months it freezes most severely (this suggests latitudes north of 30° north). The names of these months are unique to the post-Ṛgvedic texts and cannot be identified in the *Ṛgveda*, however, the names of the first six months are mentioned in the *Ṛgveda* as *devatās* of *sūkta* *ṚV*.II.36. In the post-Ṛgvedic texts there is also a correspondence between the seasons and the apparent northern and southern transits (*ayāna*) of the sun (the apparent motion of the sun is discussed in Chapter III). Spring, summer, and rains represent the gods (*deva*) and autumn, winter, and the dewy season the deceased fathers (*pitṛ*). This is a reference to *devayāna* and *pitṛyāna* (discussed in Chapter III) and suggests a division of the year from equinox to equinox (Chapter III) because, in northern South Asia, spring would be around the vernal equinox. Only in one statement, the identity of the word for the rainy season (*varṣāḥ*) and year (*varṣa*) is introduced (*ŚB*.VIII.3.2.6, Eggeling 1885b). The word *varṣa* for the year and its six seasons are recognized in India even today.

The consistency in the order in which the five seasons are listed in all post-Ṛgvedic texts is not followed in *Atharvaveda Saṃhitā*. In some verses (*AV.*VIII.2.22) the names of the five seasons are the same as those given in Table II.3, but the order in which these names are given is different, the first season being autumn, the last being rain. This verse implies a calendar in which the start of the year is in autumn and the year ends with rains. It is possible that some followers of this Saṃhitā will have used *varṣa* to indicate a year. This inconsistency and lack of regularity in the list of seasons also extends to the six seasons given in this Saṃhitā. In some verses, the names of the seasons are the same as those given in Table II.3, but the order is summer, winter, cool season, spring, autumn, rains. This cannot be a logical sequence of seasons in a year in South Asia because, in South Asia, the rains do not follow autumn. Vedic ritualists were aware of this and in the *Śatapatha Brāhmaṇa* (*ŚB*.II.2.3.7) it is stressed that rains precede autumn. This is 'mapped' on to the layout of the bricks of the fire altar (Chapter III, Section 2.8). Somewhat unusual names of the seasons are also encountered in this Saṃhitā, like *naidāga* (scorching/summer), *kurvanta* (rains), *samyanta* (cold), *pinvanta* (winter), *udyanta* (dewy) and *abhibhūva* (spring). These names are not found in any other text considered here.

The prominence given to five and six seasons in a year in all the post-Ṛgvedic Saṃhitās and Brāhmaṇas suggests that these texts may be co-eval. This may explain why, apart from the five and six seasons, the same set of ideas (e.g. the allocation of two months to a season, the names of the seasons, the same astronomical events to time the same ceremonies) occur in all these texts. These ideas may have been formulated in the late Ṛgvedic period, and the post-Ṛgvedic Saṃhitās and Brāhmaṇas may have adopted these ideas and elaborated them. Zimmer (1879) has suggested that this similarity in the number of seasons and the ritual concepts are the product of a uniform culture from north-west to south-east Madhyadeśa. The differences in the linguistic structure of these texts (Gonda 1975, Witzel 1987) suggest that the redaction of the texts occurred at different times and in different locations. However, the similarity of the number of seasons in a year and the ritual concepts suggests that there was a single source of these texts and this must have been the *Ṛgveda Saṃhitā*.

Irrespective of how and when these Saṃhitās and Brāhmaṇas were composed, the retention of mutually redundant five and six seasons is surprising. There are no logical reasons for including both five *and* six seasons in the same text. However, as described above, there are ritual reasons for either division of a year. The rituals justified by five seasons could have been abandoned or modified after six seasons had become the norm. This suggests that there are old and new layers in these texts. The old statements with five seasons were not deleted after statements with six seasons were introduced. Unfortunately, there is no clear division between these layers and it is not possible to identify the old and new parts. It might be interesting to identify the nuances in the language of the statements with different number of seasons.

It seems very likely that the primary division of the year was into five seasons. In the late Ṛgvedic period, the year was divided into five seasons, and the names of these (Table II.3) may have been carried over in to the post-Ṛgvedic period. The number 'five' had a central

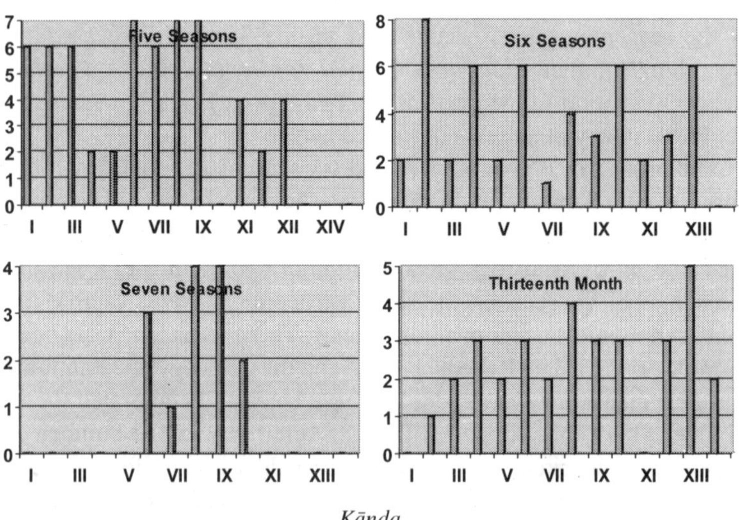

Figure II.3: The frequency of five, six, and seven seasons and the thirteenth month in *Śatapatha Brāhmaṇa*.

role in the Vedic thought. For example, in the *Agnicayana* (Chapter III) the principal layers of the fire altar are the first five layers. The names of the five layers are the same as those of years in the five-year Vedic intercalation period, the *yuga* (Chapter III). The *Agnicayana* begins with the immolation of five victims. The *Aśvamedha* is a five-day ritual. The anomalous name of the fifth season has been pointed out above. The prominent status of 'five' in Vedic ceremonies may have started in the Ṛgvedic times. In the *Ṛgveda Saṃhitā* there are a number of references to five items in a ritual: five Adhvaryus (*ṚV*.III.7.7), five steps of the sacrifice (*ṚV*.X.13.3), five oblations (*ṚV*.X.124.1). Kuiper (1983) has attributed the prominence of 'five' to the five cardinal directions (*pañca pradīśaḥ* *ṚV*.IX.86.29) identified by the Āryas. No rationale can be identified in these texts for 'six' except in the *Rājasūya* ceremonies described in *Aitareya Brāhmaṇa*. In a year of six seasons, two months are allocated to the rainy season. This is inconsistent with the current climate of South Asia. At present, the monsoon or the rainy season lasts for at least three months (July to September) and for four months in some locations. The allocation of two months to the rainy season is artificial unless the climate of South Asia during the Vedic times was fundamentally different from the current climate. Thibaut (1899) had also noticed this anomaly and came to a conclusion similar to that given here. The division of a year into six seasons may have been made to achieve symmetry of seasons between the (apparent) northern and southern passage of the sun; by this division, there are three seasons in *devayāna* and three in *pitṛyāna*. This symmetry would not be possible in a year of five seasons. In a year of six seasons, there is also symmetry in the number of months per season (two months for every season). Oldenberg (1988) suggested that the division of a year into six seasons could only have been made in order to have two months per season. Vogel (1971) has also suggested that six seasons were introduced to 'systematize the division of the year'. The allocation of two months to each season links (the motion of) the moon to the year; 'to the moon this year is linked by means of the seasons' (*ŚB*.VI.7.1.19; Eggeling 1894), and indirectly to Prajāpati. Thus by dividing a year into six seasons the Āryan theologians achieved a link between entities that are central to their thought, namely, the sun, the moon, the seasons, the year and, of course, Prajāpati who represents Time.

Seven seasons in a year are only mentioned in *Atharvaveda Saṃhitā* and *Śatapatha Brāhmaṇa*. The verses in *Atharvaveda Saṃhitā* (*AV.*VIII.9.16,17,18.) suggest that a 'discussion' may have taken place before the inclusion of a seventh season in a calendar. Verse *AV.*VIII.9.16 identifies the importance of six (items, actions, etc.) in rituals. Verse *AV.*VIII.9.17 notes that there are six cold months and six hot months (this refers to the apparent southern and northern passage of the sun), and then poses the question 'tell ye us the seasons, which one is in excess' (Whitney 1962). This leads to *AV.*VIII.9.18 and seven seasons. This suggests a society that was identifying new concepts and was attempting to incorporate these into its belief system but not without challenging them. A year of seven seasons is mentioned in the *Śatapatha Brāhmaṇa* but only in books VI to X, the Śāṇḍilya books (Figure II.3). In about half the passages of this Brāhmaṇa there is a correspondence between the seven seasons and the seven layers of the fire altar. This correspondence appears somewhat superficial as no further significance is attached to these layers. In the remaining passages, the seven seasons are introduced to justify quantities of ritual entities like the *pada* of a metre, and the number of verses or as the redactor puts it in *ŚB*. IX. 5.2.8, 'whatever else there is of seven kinds' (Eggeling 1885b). The names of the seasons are not given in any passage that mentions seven seasons and the name of the seventh season is not given in any Brāhmaṇa or Saṃhitā. It has been suggested that the seventh season was introduced to accommodate the thirteenth intercalary month and (or) because of the 'Vedic predilection with number seven' (Macdonell and Keith 1912). Both these reasons seem unlikely. The frequency of five, six, and seven seasons and the thirteenth month in the fourteen *kāṇḍa*s of *Śatapatha Brāhmaṇa* are shown in Figure II.3. The five and six seasons and the thirteenth month are mentioned in almost all *kāṇḍa*s, but the seventh season, as noted above, is mentioned only in books VI to X or the Śāṇḍilya books of this Brāhmaṇa. The thirteenth month is also mentioned in all other Saṃhitās and Brāhmaṇas, but these texts do not mention the seventh season (except once in *Atharvaveda Saṃhitā*, as noted above). It is worth noting that this season has no name, although there is a name for the thirteenth month (Chapter III). It is possible that the Śāṇḍilya School introduced the seventh season to have a season associated with the intercalary thirteenth month, but this certainly was not a common practice among all Vedic schools.

3. SEASONAL NEW YEAR

The high frequency of the word *śarad* (autumn) in the *Ṛgveda Saṃhitā* and the use of this word to quantify the age of a person, or a physical period suggests that it denoted a year (tropical/seasonal). It therefore, seems highly likely that at an early age in the Vedic Period, the year commenced in *śarad* (autumn). Unfortunately, no Vedic text provides any information to identify the epoch when the year may have commenced in *śarad* (autumn). In the post-Ṛgvedic texts, the list of seasons (Table II.3) is always headed by *vasanta* (spring). This suggests that, in the later Vedic period, the New Year's Day was in *vasanta* (spring). Dīkṣita (1896) has also concluded that the Vedic seasonal cycle started with *vasanta*. This is echoed in '*Vasanta* is the mouth of the seasons' (*TB*.1.1.2.6). However, a level of uncertainty is also echoed in 'the season-vessel has mouths on both sides, for who knows where is the mouth of the seasons?' (*TS*.VI.5.3, Keith 1914). Tilak (1893) has conjectured that initially the Vedic year commenced on the winter solstice and was later moved to the vernal equinox. A caveat should be added here, the New Year Day might have been moved to spring, there is no evidence that it was moved to the vernal equinox. The Vedic New Year Day is discussed in detail in Chapter III and revisited in Chapter IV.

NOTES

1. If the earth's axis of rotation were at a right angle to its orbit around the sun (the ecliptic), each point on earth's surface would receive the same amount of sunlight and heat throughout the year, and day and night would be of equal length. However, the axis of rotation of the earth is tilted by 23.5° from the perpendicular to the ecliptic. Therefore, the angle between earth's axis and the earth-sun line changes throughout the year. Twice a year, at spring/vernal equinox (about 21 March) and autumnal equinox (about 23 September – the exact dates may vary a bit) the axis and the earth-sun line are perpendicular. Twice a year, at summer solstice (about 21 June) and winter solstice (about 21 December) the angle between the axis and the earth-sun line is maximum, that is 23.5°.

On the earth's surface, the boundary between the illuminated (day) part and the shadowed (night) region is always perpendicular to the earth-sun line. Because of the tilt of earth's axis, in summer the northern hemisphere receives more light and heat than the southern hemisphere. In winter, it is the opposite; the southern hemisphere receives more light and heat than the northern hemisphere. The seasons are thus a consequence of this tilt of earth's axis. Nowadays, 21 June is considered to be the beginning (and not the middle) of summer in the northern hemisphere. On this day, the northern hemisphere receives the maximum amount of sunlight but the maximum temperature is reached about a month and a half later, because the oceans take longer (than the ground) to heat-up. Similarly, on 21 December, the northern hemisphere receives the lowest amount of light and heat and this is considered the beginning of winter but the coldest day is (on average but not always) about a month later.

2. It has been generally assumed that the division of a year into seasons and a calendar followed the development of agriculture (Vogel 1971). However, the knowledge of seasons was as important to pre-agriculture hunter-gatherer societies, as it was to the agricultural societies. The availability of fruit and root and the migration of the game are seasonal, and early man would have realized that this resource was dependent on the local climate. These societies would have developed a calendar based on local climatic conditions, long before the development of agriculture. Examples of this can be found in the present indigenous Australian cultures. These people were hunter-gatherer and never developed into an agricultural society but they have a well defined seasonal calendar. The Jawoyn Calendar (Katherine Region, Northern Australia) has six seasons. About 180 km to the west, the Wardaman Calendar has four seasons. Further south, the Yanyuwa Calendar (on the Gulf of Carpenteria) has five seasons and around the middle of the continent, the Walabunnba Calendar (approx. 300 km north of Alice Springs) has, not surprisingly, only two seasons, hot and cold (rain is relatively rare). It is important to note that societies respond to local conditions and people separated by even short distances can have different seasonal calendars to reflect the difference in the climatic conditions.

   Source: Australian Government, Bureau of Meteorology
   www.bom.gov.au/iwk/index.shtml

III

Days, Months and Years

All civilizations have revered the luminaries in the sky, the sun, moon, planets and the stars. The realization of the regular movement of these bodies and the certainty of sunrise and sunset and the waxing and waning of the moon would have provided the ancients with anchor points to regulate their lives. Early attempts to keep a record of the movement of the heavenly bodies can be identified on a bone fragment, from Upper Palaeolithic (from about 40,000 to 10,000 years before present), found in a rock shelter site in Blanchard (Dordogne), France (N.III.1). This seems to be a representation of the phases of the moon and may be a lunar calendar. A 13,000-year-old eagle bone, also from the Dordogne Valley, has seven marks incised below a series of regular cuts. This may also be a record of the phases of the moon – new, quarter, full, quarter and back to new. These bone fragments are very similar to the 'calendar sticks' of a number of ancient (and some modern) societies (N.III.2). These artefacts suggest that, by about 10,000 years before the present, man had acquired the fundamentals of a calendar. These early hunter gatherers would have recognized the intimate link between the behaviour of the heavenly bodies and the seasons and this link would have become more obvious as man progressed from hunter gatherer to agricultural communities. This link between the heavenly bodies and the food resource may have convinced early man that the heavenly bodies were gods who had control over their lives. The religious monuments and mythologies of heavenly bodies found all over the world confirm that all societies had appreciated the need to assuage these heavenly gods.

This (probably) brought about the uneasy alliance between religion and astronomy. The increasing sophistication of the belief-systems of different cultures would have driven a corresponding advance in the measurement of time as it would have been necessary to determine the time when a sacrifice had to be offered or 'fix' the time when a cycle of ceremonies had to commence. Nobody would have wanted to incur the wrath of god(s) by missing or delaying an appropriate ritual or a ceremony. This desire to remain in favour with the gods probably led to the making of the calendar (Box 1).

Box 1

God made the days and nights but man made the Calendar.

– Richards, 1998

The shortest naturally occurring unit of time is the day – the period of one revolution of the earth about its axis, or from the human perspective, the time taken by the sun to return to the same point in the sky. This interval is punctuated by alternating day and night (except in the Polar Regions). This rhythm of time is deeply embedded in the human psyche and physiology. Early man was aware of the powerful influence the sun had on his daily life and the regularity and the inevitability of the recurring day and night. Another unit of time occurring in nature is the lunation or the month – the cycle of the moon as it moves in its orbit round the earth. The regular waxing and waning of the moon had grabbed the attention of the humans a long time ago (as noted above). Two obvious and visible chronological markers, the sun and the moon, respectively define the day and the month. The third naturally occurring unit of time – year – can be measured from the cycle of the seasons, and this is probably how it was first identified (Chapter II). The year is the time taken by the earth to complete one orbit around the sun, or as the ancients would have seen it, for the same season to recur. This does not give an accurate length (in days) of the year as the start and end of a season are determined by complex and varying local climatic conditions. The accurate length of the year was only established after humans recognized the solstice and equinoctial points (discussed below). The calendar is a scheme to group the days into months and the days and months into years and as will be discussed later this is not as straightforward as it might at first appear.

Today we know that the earth revolves around the sun and rotates on its own axis. The ancients believed, following common sense, that the earth was stationary and flat. Surrounding the observer was a circle, the horizon, which defined a plane with the observer at its centre. The sky or the firmament was a hemisphere of arbitrary extent centred about the observer (Figure III.1). The horizon was defined by the four cardinal points (east, south, west, and north), and the hemisphere was defined by the zenith or the overhead position. In modern astronomy, this is called the horizon reference system. In this reference system, the position of an object on the hemisphere is defined by its azimuth and altitude. The azimuth is the (angular) position along the horizon measured towards east facing north (i.e. in angular measure, the azimuth of the north is 0°, east is 90°, south is 180° and west is 270°). The altitude is the (angular) distance measured vertically upward from the horizon. The position on the

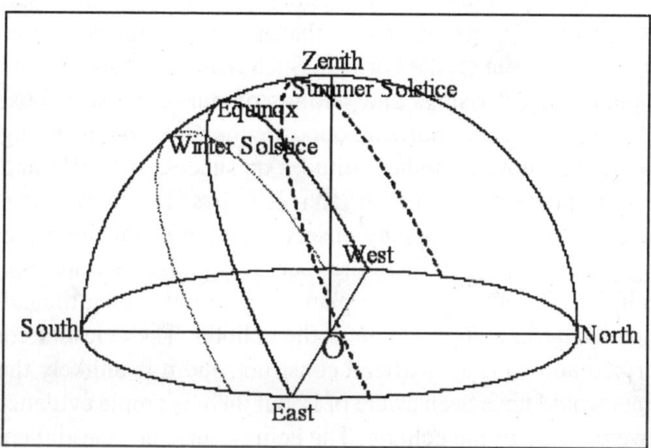

Figure III.1: The 'common sense' view of the earth, and the sky. An observer at O will 'see' the surrounding circle of the horizon and the enclosing hemisphere of the firmament. An observer in northern latitudes, facing south, will see the sun rise on the left (east) and set on the right (west). During the year, the sun will appear to move south, reaching the southernmost position at the winter solstice, when it will appear to rise south-of-east of the observer. The sun will then appear to move north and pass through the spring equinox, the sun will then rise directly east of the observer. The sun will reach the northernmost position at the summer solstice when it will appear to rise north-of-east of the observer. It will then begin its southward movement.

hemisphere of the firmament is sometimes also defined by 'zenith distance'; this is the angular distance measured from the zenith (altitude = 90° – zenith distance). The azimuth and the altitude are equivalent to longitude and latitude on earth; the equivalent of the equator is the horizon. The azimuth and altitude of an object in the sky are not fixed but change with the position of the observer and the time of observation. A more stable reference system is obtained by projecting earth's equator on the sky and measuring the position of celestial objects along and from this plane; this is known as the equatorial coordinate system. The point to bear in mind is that, in this reference system, the positions (coordinates) of stars are fixed but those of the sun, moon and the planets are not as these objects move relative to the stars. The celestial coordinate systems used by astronomers are described in N.IV.7.

During the day, the stars are invisible, and at night, the sun is hidden from view. However, just before sunrise or just after sunset it is possible to observe a set of stars that are close to the position of the setting or rising sun on the horizon. Such risings and settings of stars are called heliacal risings and settings of stars. The stars observed above (say) the eastern horizon, opposite the direction of setting sun, will appear to move to higher altitudes on successive nights and new stars will appear above the horizon. It is possible to map the path of the sun among the stars, by observing these heliacally rising and setting stars everyday through the year. This path of the sun is called the ecliptic. The moon and the planets also move along the ecliptic, their paths are at a slight angle to the ecliptic. The celestial equator (described above) is an artificial construct, and it is unlikely that the ancients would have been aware of it, but there is ample evidence that they were aware of the ecliptic. The ecliptic and the celestial equator are inclined at an angle of about 23.5°. These great circles intersect at two points called the vernal (spring) and autumnal equinox. In the course of a year, the sun will cross the equator at the vernal equinox (from Latin *aequus* – equal and *nox* – night) in spring at northern latitude; it will continue to move along the ecliptic and away from the equator till it reaches its furthest point, at summer solstice. Thereafter the sun begins to move towards the equator and will eventually cross the equator once more at equinox in autumn. The sun continues to move along the ecliptic till it reaches the (second) furthest point from the equator, the winter solstice. The sun will then start its journey back to the vernal equinox. The word solstice means 'the sun stands still' in Latin and at the two solstices the sun does indeed appears to halt before reversing its motion along the ecliptic. The orbit of the earth around the sun is not circular but elliptical and it does not move at constant speed in its orbit, therefore, the seasons are not of equal length. The time (in integer days) taken by earth to move along the four segments of the ecliptic is given in Table III.1. This makes the seasonal or tropical year 366 days long, the actual length is 365.24 days.

The sun rises in the east and sets in the west everyday. At the same time, the earth moves in its orbit around the sun. The result of these two motions is that the rising and setting positions (on the horizon) of the sun will change slightly from day-to-day. The daily path of the sun on two equinoctial days and two solstice days are summarized in Figure III.1. On spring and autumn equinox, the sun will rise exactly at the east point (azimuth 90°) and will set exactly at the west point (azimuth 270°). After the spring equinox, as the year progresses towards summer (at northern latitude) the rising and setting position of the sun on the horizon will gradually move north and will reach the northernmost point on the summer solstice. At noon on this day, the sun reaches its greatest altitude. The sunrise and sunset points on the horizon then begin to move southwards. On the autumn equinox, the sunrise is again exactly due east (azimuth 90°), and the sunset is exactly due west (azimuth 270°). The sunrise and sunset points reach the southernmost point on the winter solstice when the sun will be at the lowest altitude. These positions of sunrise and sunset points and the local time (for longitude 77.20° E and latitude 28.58° N, i.e. Delhi) of sunrise and sunset are summarized in Figure III.2.

Table III.1: Duration (in whole days) of sun's transit of four sectors of the ecliptic

Section	Length (days)
Spring equinox to summer solstice	93
Summer solstice to autumnal equinox	94
Autumnal equinox to winter solstice	90
Winter solstice to vernal equinox	89

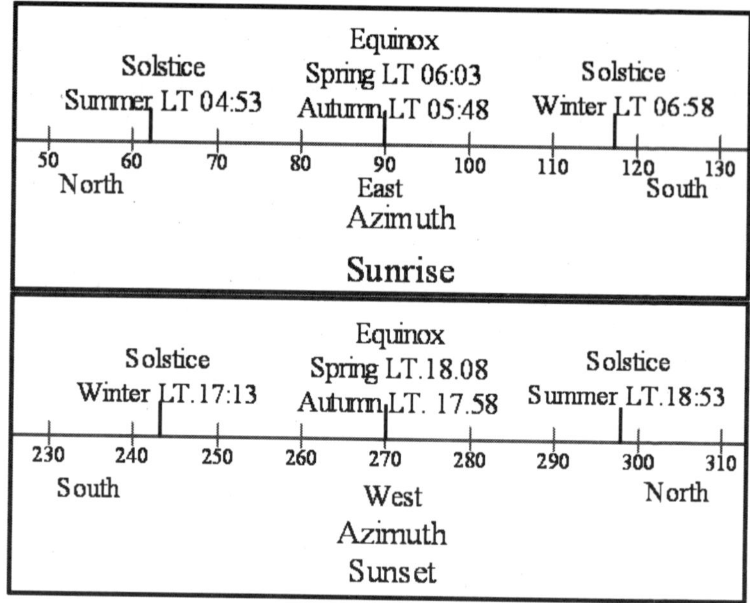

Figure III.2: The azimuth and the local time of sunrise and sunset at the spring and autumn equinox and summer and winter solstices. The data are plotted for longitude 77.20° E and latitude 28.58° N (Delhi).

The sun illuminates the moon and a casual observer can only see the lit disc of the moon. To the ancients the moon would have appeared to move on the surrounding hemisphere of the sky but with a motion that was irregular compared to that of the sun. Perhaps the obvious aspect was the lunar phases. At new moon, the moon is lost in the glare of the sun and this persists for one or two days. The new moon is first seen as a thin crescent in the west at twilight close to the position of the setting sun. On this day (Day 1), the moon is visible only for few fleeting moments as it follows the sun below the horizon. By the next night, the angle between the moon and the sun will increase (as seen from the earth), and a slightly thicker crescent is visible slightly longer in the sky. The angle of separation between the moon and the sun continues to increase and in just over seven days it will be about 90°, the moon has waxed (or grown) to the first quarter (Day 7), that is, half the lunar disc is visible. On the following days, the moon rises progressively earlier and on full moon (when the full disc of the moon is visible) night (Day 15) it appears, at sunset, opposite the point where the sun sets. The moon then wanes and on Day 21 it wanes to last quarter and again only about half the lunar disc is visible. On Day 29, the moon again passes into the glare of the sun, and the cycle repeats. In the second half of the cycle, the moon is visible only in the late evening and early morning sky. Towards the end of this cycle, the moon is visible in the daytime. The period between any two phases (say, from new moon to new moon) is the synodic month. The mean length of this month is 29.53 days (N.III.3.) but the ancients probably counted it as a period of 29 or 30 days. During the synodic month, the moon is visible mostly for 27 days and very occasionally for 28 days.

1. THE VEDIC VIEW OF THE SKY

The Vedic universe consisted of three regions, the earth (*pṛthivī*), the firmament (*antarikṣa*), and the heaven (*dyaus*). The earth was the abode of humans, the firmament that of the luminaries (sun, moon and stars), and the heaven was the abode of the gods. The Āryas saw the earth and the sky as two hemispheres (*VS*.IV.25), but their description of the earth relative to the firmament is similar to that given above; a flat earth bound by the horizon and enclosed by the vault of the firmament that supports the luminaries (A.III.1). The horizon was defined by the four cardinal points (A.III.2) and the intermediate directions (*TS*.VII.2.3). The hemisphere of the firmament was defined by the zenith or the overhead position (*KB*.vii.6; *ŚB*.V.4.1.3-7). In other words, the Āryas had defined and were using the horizon reference or coordinate system described earlier. The sun is drawn across the sky in a chariot with seven horses (A.III.3); the number seven is often associated with the sun, but it is not entirely clear what this is supposed to signify. The sun is visible to all, it is the source of light, and it defines the directions or the cardinal points (A.III.1, A.III.2) and shines throughout the firmament (A.III.4). It 'measures out' the days and nights and man rests at night and starts work at daybreak (A.III.5). The sun is referred to as the 'son of the south' (A.III.6) and suggests that to the Āryas the sun always appeared in the southern direction. The sun rises in the east, and sets in the west. To a person facing south, the sun will appear to rise on the left and set to the right. This is how the Āryas saw the apparent daily motion of the sun (e.g. *AB*.i.7(ii.1); *TS*.V.2.5; *TB*.2.1.4). The *adhvaryu* is enjoined to turn from left to right (to follow the path of the sun) when making oblations (*TS*.V.2.5; *ŚB*.VII.5.1.37). The Āryas also noticed the apparent north-south motion of the sun (sub-section 2.4) and noted that the growth of plants or the crops starts in the south (*AB*.i.7(ii.1); Keith 1998). This description of the apparent daily and annual motion and position of the sun places the Āryas north of Tropic of Cancer (latitude 23.5°) in corroboration of the geographical data in the Vedic texts (Chapter I).

The Āryas did not just describe the motion of the sun but also theorized about the nature of day and night and the sun. They noted that one half of the sun gives light and supports all existence but the other half is a mystery (*AV*.X.8.7; Whitney 1962). A more sophisticated theory was that the sun moves from east to west, making day below and night above (*AB*.iii.44(xiv.6); *TS*.VI.4.10.2, 3; *ŚB*.IV.2.1.18). In the evening, it inverts itself, making night below and day above (presumably moving from west to east). Perhaps the most intriguing speculation is in a *sūkta* by Vāmadeva Gautama. In this *sūkta* (dedicated to Agni which could signify the sun), the sun is praised for bringing light at dawn, and it refers to 'borne onwards by seven shining mares' (A.III.7). The question is then posed, 'Why does he never fall? What inner power propels him?' (A.III.8). Frustratingly (as so often in the study of *Ṛgveda Saṃhitā*), the *ṛṣi*s hazard no guesses.

The Āryas, like other ancient people, had observed the motion of the moon and recognized the central role of the moon in defining the month, a period intermediate between the day and year. They appear to have realized that the moon has no light of its own and is illuminated by the sun (*TS*.III.4.7). A symbolic explanation of the phases of the moon was that the moon was a bowl that fills with *soma*, a divine drink or supreme oblation (*uttamaṃ havis* or *paramāhuti*) that plays a central role in Vedic rituals, during the waxing phase. The bowl empties during the waning phase as gods drink the *soma*. This idea can be traced back to the *Ṛgveda Saṃhitā* (A.III.9). In the Vedic texts, the moon and *soma* are often synonymous. A 'deductive' explanation of the phases of the moon may also have been reached. At full moon, the moon is at a great distance from the sun. It then 'swims' towards the sun, 'enters into his (sun's) open mouth' at new moon; 'having swallowed him' the sun rises but the moon is not seen in the east or the west (i.e. the new moon day), he (sun) then 'throws' him (moon) out and he is seen in the western sky (*ŚB*.I.6.4.18-20; Eggeling 1882; *AB*.viii.28 (xl.5); Keith 1998). This can be interpreted as — the angular separation between the sun and the moon is greatest at full moon and gradually decreases to zero at new moon (dwelling together – *amāvāsyā*) and that a crescent moon is seen again in the western sky after the new moon.

2. THE VEDIC CALENDAR

The Vedas are liturgical and sacerdotal texts and not surprisingly, calendric material is not deliberately included in these texts. However, the texts are rich in references to the fundamental elements of a calendar namely the day, month, and year. The texts also prescribe 'dates' for performance of rituals and ceremonies by describing the seasonal and/or astronomical conditions that have to be satisfied at the performance of these rituals and ceremonies. The calendar of the Vedic period can be (re)constructed by defining a single and self-consistent link between the calendric elements and the prescribed 'dates'. The first attempt to identify the calendar of the Vedic period was made by the German Indologist Weber (1861, 1862). Thibaut (1877, 1899), Tilak (1893), and Dīkṣita (1896) continued this analysis in the last decade of the nineteenth century. In the twentieth century, this topic has been examined by Saha and Lahiri (1992), Pingree (1970-80), Jaggi (1986), Pingree (1989b), Kaṇsara (1992), Abhyankar (1993), Narahari Achar (1998b, 1998c) and Balachandra Rao (2000). The fundamental elements of a calendar, the day (and night), phases of the moon, the annual motion of the sun, and the year can be identified in all Vedic texts, from the earliest to the latest. The Āryas developed elaborate rituals and ceremonies during the performance of which they emulated the cosmos. They recognized that the days, nights, months, seasons, and the year were cyclic and continuous (e.g. *AV*.XII.2.25); there being no breaks from one day to next or from one year to next. They saw this continuity as a divine order (*ṛta*) and they stress that there should be no breaks in the rituals as these would be contrary to the divine order. Ceremonies were performed at precise and well-defined times: that is, they followed a calendar. This calendar was embedded in some of their annual ceremonies, in particular in the great sacrificial session (*sattra*) of *Gavām ayana* (cows walk/course). The daily ceremonies of this *sattra* are given in Table III.2 (*ŚB*.XII.1.2.1-22; Eggeling 1900; *PB*.XXIV.20; Caland 1931; Sen 1978; *KŚS* Ch.XIII in Ranade 1978). This *sattra* was arranged in two halves to follow the apparent annual motion of the sun. The northern and southern, or the ascending and descending course, being divided by the central *Viṣuvant* day. There are a number of other *ayana sattras* (e.g. see *LŚS*), but for calendric purposes *Gavām ayana* is the most appropriate. The correspondence between various aspects of this annual ceremony and the calendar will be discussed in the following sections.

The Vedic texts refer to all aspects of a calendar like *yuga* (long period), *saṃvatsara* (year), *ayans* (half-year), *ṛtu* (season), *māsa* (month), *adhimāsa* (intercalary month), *ardhamāsa/pakṣas* (half-months/fortnight), *aha* (day), *rātri* (night) and special nights. These aspects are discussed in the following sections (for *ṛtu* see Chapter II).

2.1. Day – *Aha*

The basic unit of a calendar, the day (*aha*), is a fully developed concept in *Ṛgveda Saṃhitā* (A.III.10) and must have been identified in the pre-Ṛgvedic times. The two naturally occurring divisions of the day, sunrise to sunset and sunset to sunrise (night – *rātri*) were recognized. Other names for the day and night may also have been used, e.g. *śabda* and *sagarā* (*ŚB*.I.7.2.26). The sun as the causative agency of the day and night was identified; 'the sun in the sky (who is) established for (both) day and night'. Similarly, the 'day and night are sisters' (A.III.11). A sunrise to sunrise (i.e. a 24-hour period in modern terms) also appears to have been recognized judging from 'sunset is the middle of the work of Sūrya' (*TB*.2.8.7; Dumont 1969). This also suggests that the start of the day was at sunrise. This period from sunrise to sunrise was called *ahorātra* (N.III.4). A verse in the late *maṇḍala* X of *Ṛgveda Saṃhitā* suggests that a day may have been designated or identified by the lunar asterisms (A.III.12, N.III.5) or the *nakṣatra* that is in conjunction with the moon on that day. The

Table III.2: The rituals, their duration and the cumulative number of days of the *gavām ayana sattra* (*ŚB*.XII.1.2.1-22; Eggeling 1890); *PB*.XXIV.20; Caland 1931; Sen 1978; *KŚS* Ch.XIII; (Ranade 1978).
The daily rituals are described in the glossary.

Ritual	Duration (days)		Total (days)
Prāyaṇīya Atirātra (opening day)	1		1
Caturviṃśa (an *ukthya*)	1		2
5 months, each consisting of 4 *Abhiplavaṣaḍaha* (of 6 days each) 1 *Pṛṣṭhyaṣaḍaha* (of 6 days)	 4×6=24 6	5×(24+6) =150	152
3 *Abhiplavaṣaḍaha* (of 6 days each)	18		170
1 *Pṛṣṭhyaṣaḍaha* (of 6 days)	6		176
1 *Abhijit* (*Agniṣṭoma* day)	1		177
3 *Svarasāman*	3		180
Viṣuvat (central day)	1		181
3 *Svarasāman*	3		184
1 *Viśvajit*	1		185
1 *Pṛṣṭhyaṣaḍaha* (of 6 days)	6		191
3 *Abhiplavaṣaḍaha* (of 6 days each)	18		209
4 months, each consisting of 1 *Pṛṣṭhyaṣaḍaha* (of 6 days) 4 *Abhiplavaṣaḍaha* (of 6 days each)	 6 4×6=24	4×(24+6) =120	329
3 *Abhiplavaṣaḍaha* (of 6 days each)	18		347
1 *Goṣṭoma* (*Agniṣṭoma* day)	1		348
1 *Ayuṣṭoma* (an *ukthya*)	1		349
1 *Daśarātra* (10 days)	10		359
1 *Mahāvrata* (*Agniṣṭoma* day)	1		360
1 *Udayanīya Atirātra*	1		361

number of *nakṣatras* (27 or 28) suggests that their primary role may well have been to identify the days because in a lunation, the moon is visible this number of days, as discussed above (*nakṣatras* are discussed in Chapter IV). The practice of designating a day by a lunar asterism seems to have continued for some time and may have continued through the Brāhmaṇa period although there are no textual clues for this.

The names of days and nights of the bright (waxing) and the dark (waning) fortnights (*ardhamāsa*) of a month are given in the passages of *Taittirīya Brāhmaṇa* that describe the construction of *Sāvitra Cayana* (a fire altar in the form of the sun, Section 2.8.2). These names are given in Table III.3 and are found only in this Brāhmaṇa. The names of the days do not follow any particular theme and appear to be purely symbolic except the 'day-names' are in the neuter gender and the 'night-names' are feminine. In contrast, the names of the nights of the waxing fortnight describe a physical phenomenon that is, the appearance of the moon as it 'grows' from new moon to full moon. These names give an indication of the close attention the Āryas paid to the accurate observation and classification of astronomical events. The theme of the names of the nights of the waning fortnight is about pressing, drinking, desire, and satisfaction, clearly referring to the preparation and consumption of *soma* juice. These names follow the 'theory' described above of the phases of the moon. It is rather surprising that while the fifteenth night of the waxing phase is called *pūrṇamāsī* (full moon), as it should be, the fifteenth night of the waning phase is not called by its appropriate name, *amāvāsyā*.

Apart from the two obvious divisions of a day into 'day' and 'night', the Āryas, in analogy with the division of a month into four lunar phases, divided the day and night into four *praharas* (or *yāma*). Further, taking their cue from the division of a month (discussed in Section 2.3) into thirty days, the Āryas divided a day (i.e. the sunrise to the sunrise period) into 30 parts. This division of the day into 30 units may have been made during Ṛgvedic times. Verse *ṚV*.VI.59.6 (A.III.13) is rather difficult to interpret, and 'the footless (i.e. dawn) traverses 30 steps' could mean either that the dawn takes 30 steps before the next dawn or that the dawn takes 30 steps in a month. However, 'shines over 30 domains (of the day)' of *ṚV*.X.189.3 (A.III.14) does indeed seem to indicate thirty parts of a day. The *Ṛgveda Saṃhitā* does not name this thirtieth part of a day, but in the post-Ṛgvedic Saṃhitās and the Brāhmaṇas this part is called a *muhūrta*. This word occurs twice in *Ṛgveda Saṃhitā* (*ṚV*.III.53.8 and *ṚV*.III.33.5) and in both verses it means a 'short time'. This word is also used in a similar sense in some post-Ṛgvedic texts (e.g. *ŚB*.I.8.3.17 and *ŚB*.II.3.2.5). However, *ŚB*.X.4.2.18 (Eggeling 1885b) describes how Prajāpati (who personifies the year) saw the 15 parts of the day and the 15 parts of the night that is, 30 *muhūrtas*

Table III.3: Names of the days and nights during the waxing and waning fortnights of a month (*TB*.3.10.1; Dumont 1951)

	Days		Nights	
	waxing	waning	waxing	waning
1	*saṁjñāna*/ consciousness	*prastuta*/ praised one	*darśā*/ just visible	*sutā*/ one pressed out
2	*vijñāna*/ understanding	*viṣṭuta*/ highly praised one	*dṛṣṭā*/ perceived	*sunvatī*/ one pressing out
3	*prajñāna*/ knowledge	*saṁstuta*/ extolled one	*darśatā*/ conspicuous	*prasutā*/ continually pressed
4	*jānat*/ knowing one	*kalyāṇa*/ auspicious one	*viśvarūpā*/ many coloured	*sūyamānā*/ one being pressed
5	*abhijānat*/ recognised one	*viśvarūpa*/ many coloured	*sudarśanā*/ lovely	*abhiṣūyamāṇā*/ prepared by being pressed
6	*saṅkalpamāna*/ wishing one	*śukra*/ pure one	*āpyāyamānā*/ beginning to swell	*pītī*/ drinking
7	*prakalpamāna*/ becomes fit	*amṛta*/ immortal one	*pyāyamānā*/ swelling	*prapā*/ giving drink
8	*upakalpamāna*/ preparing itself	*tejasvin*/ splendid one	*pyāyā*/ swollen	*sampā*/ drinking together
9	*upakḷpta*/ prepared one	*tejas*/ splendour	*sūnṛtā*/ friendly	*tṛpti*/ satisfaction
10	*kḷpta*/ one who is ready	*samiddha*/ inflamed one	*irā*/ enjoyment	*tarpayantī*/ one that satisfies
11	*śreyas*/ splendid one	*aruṇa*/ red one	*āpūryamāṇā*/ beginning full	*kāntā*/ desired one
12	*vasīyas*/ wealthy one	*bhānumat*/ bright one	*pūryamāṇā*/ becoming full	*kāmyā*/ desirable one
13	*āvat*/ coming one	*marīcimat*/ radiant one	*pūryantī*/ filling	*kāmajātā*/ daughter of desire
14	*sambhūta*/ developed one	*abhitapat*/ one that heats	*pūrṇā*/ full	*āyuṣmatī*/ possessed of vital power
15	*bhūta*/ one who has come into being	*tapasvat*/ burning one	*pūrṇamāsī*/ has full moon	*kāmadughā*/ one who yields object of desire

in a day and night (*ahorātra*). This is reinforced by the statement that a (ritual) year of 360 days (see Section 2.5) is made of 10,800 *muhūrtas* (*ŚB*.X.4.2.20). Thus, a *muhūrta* is a 'Vedic hour' and it is 48 minutes long. The Āryas also divided the Vedic hour into smaller units (*ŚB*.XII.3.2.5; Eggeling 1900) and these are given in Table III.4. In *Taittirīya Brāhmaṇa*, the Vedic 'minute' (a fifteenth part of a *muhūrta*) is called a *kṣudra-muhūrta*. A different division of the *muhūrta* into smaller units is given in *Vedāṅga Jyotiṣa*, a text of the late Vedic period discussed in Chapter V.

The *ṛṣis* of the *Śatapatha Brāhmaṇa* find a correspondence between the *muhūrtas* in a year and the number of syllables in the three Vedas i.e. *Ṛk*, *Yajus*, and *Sāman*. The reasoning behind this is as follows (*ŚB*.X.4.2.23-25; Eggeling 1885b): the Vedic ritual year of (360 × 30 =) 10,800 *muhūrtas* is Prajāpati. However, Prajāpati is five-fold (this is the Vedic predilection with number 'five') or is a *paṅkti*. The *paṅkti* metre consists of five *pāda* (feet) of eight syllables each that is, the *paṅkti* metre is composed of 40 syllables. *ŚB* contends that the *Ṛgveda* is composed of 40 × 10,800 = 432,000 syllables. The total number of syllables in the other two Vedas is also equal to this number. If this is the actual total number of syllables in these three Vedas, then in the final redaction, the *maṇḍalas*, *sūktas*, and *ṛcās* of these Vedas must have been extensively edited to deliberately establish a correspondence between the number of syllables in these three Vedas and the number of *muhūrtas* in a Vedic ritual year. *Śatapatha Brāhmaṇa* is considered to be a late Brāhmaṇa (mid-first millennium BCE); the above equation, however, suggests that the passage *ŚB*.X.4.2.23-25 must be from an earlier period and a version of *ŚB* (now lost) when *Atharvaveda* was not considered to be one of the Vedas.

Table III.4: Correspondence between the Vedic units of the day (*ŚB*.XII.3.2.5) and modern units

1 *ahorātra* (day) = 30 *muhūrtas*	24 hours
1 *muhūrta* = 15 *kṣipras*	48 minutes
1 *kṣipra* = 15 *etarhis*	3.2 minutes
1 *etarhi* = 15 *idānis*	12.8 seconds
1 *idāni*	0.855 seconds

The division of a day into thirty units (*muhūrta*s) was unique to the Āryas, and no other culture appears to have divided a day in this way. It is not known if the *muhūrta*s were of equal duration or how these (and the smaller units of time) were measured. The Vedic texts do not mention any instruments. A clue may be in *Aitareya Brāhamaṇa*; in the *Soma* sacrifice discussed in this Brāhmaṇa the ritualists are asked to recite 15 *stotra*s during the day and 15 during the night (*AB*.iv.6 (xvi.6); Keith 1998). It is not clear if this ritual was performed everyday or was reserved for a particular day. What cannot be disputed is that time-keeping (division of a day into smaller units) was part of the Āryan culture from the earliest texts. Could the Āryas have used their day-time and night-time chants as a clock in which a *stotra* lasted for one *muhūrta* or (approximately) 48 minutes?

2.2. WEEK – *ṢAḌAHA*

The seven-day week is not a naturally occurring period. This is a Babylonian construct, it is about a quarter of a lunation and it may have been introduced for religious purposes. The seven-day week was introduced into Christian Europe in CE 323 by Emperor Constantine (CE 280?-337) and appeared in South Asia in about CE 484 (it is mentioned in an inscription produced during the reign of Emperor Buddhagupta, CE 476-95?). A 'seven-day week' cannot be identified in the Vedic texts; an important event was usually identified by the phase of the moon, the *nakṣatra* coincident with the moon, and sometimes the season. However, a period of six days has a unique role in the Vedic rituals. This can be seen in the *sattra* of *gavām ayana* (Table III.2). After the first two 'opening' days of the *sattra*, each month of the following five months of the *sattra* has four *Abhiplava* ceremonies (*PB*.XXII.3; *AB*.iv.15 (xviii.1)) each lasting for six days (*ṣaḍaha* – *ṣaṣ*(*t*)+*aha* – six days). This is followed by one *Pṛṣṭhya* ceremony, also lasting for six days (*AB*.iv.29(xx.1)). This is followed by three *Abhiplava* ceremonies and one *Pṛṣṭhya* ceremony before the ceremonies for *Viṣuvat*, the central day. After the conclusion of ceremonies for *Viṣuvat*, the *Abhiplava* and *Pṛṣṭhya* ceremonies are repeated in reverse order and again each lasts for six days. These six-day ceremonies of *gavām ayana* are described in all Saṃhitās and Brāhmaṇas. Similarly, in the ceremony of the

Mahābhiṣeka (great consecration) a king is anointed for six days (*AB*.viii.19 (xxxix.5)). Perhaps the most direct evidence that a six-day period can be interpreted as a Vedic 'week' is found in *TS*.VII.5.6 and *TS*.VII.5.7 (*ṣaḍahir māsānt sampādyāhar*) – made up the month with six-day periods (Keith 1914). Similarly, in *KB*.xix.3, the sun '. . . goes north for six months; him they follow in six-day periods . . . ' and *KB*.xxvi.1, 'How many sets of six days are in the year? There are sixty six-day sets (or 360 days in a ritual year); thus the course of the year by six day sets is unbroken' (Keith 1998). That is, a Vedic month (of 30 days) was divided into five units of six days each, and the Vedic ritual year (of 360 days) was divided into 60 units of six days each. Whereas the Babylonians and then the Europeans had arbitrary number of weeks (of seven days) in a month, the Āryas had opted for aesthetically appealing equal number of 'weeks' in each (ritual) month.

2.3. Month – *Māsa*

The lunation — the period defined by one cycle of phases of the moon — is the basis of the month but the moment when each cycle starts is a matter of convention. The Sanskrit word for the month is *māsa* and follows from the root *mā* 'to measure', months were probably used to 'measure' the year (e.g. *AB*.iii.41(xiv.3)). The word for the moon is also *māsa*. This correspondence between the 'moon' and the 'month' is made in *Ṛgveda Saṃhitā*; in *ṚV*.X.138.6 (A.III.15) *māsa* means the moon and in *ṚV*.X.85.5 (A.III.16) it means the month. Thus, the moon is the 'maker of the months' (*māsakṛt*). This notion is carried over to later Vedic texts. The month is a well established concept in the earliest strata of *Ṛgveda Saṃhitā*. In *ṚV*.V.78.8, 9 (A.III.17, 18, and *ṚV*.X.184.3; *AV*.V.25.10; *AB*.vii.13 (xxxiii.1)) the gestation period of a child is given as *daśa māsa* (10 months). The month is also incorporated in Vedic rituals and references to 10-month rites (A.III.19) occur in the earliest verses of the *Ṛgveda*.

In order to anticipate the day of the start of a month, it is necessary to know the number of days in a lunation. The mean length of a lunar or synodic month is 29.53 days (N.III.3), but the Āryas rounded this up to 30 days, they thus gained half a day every month and the Vedic 'ritual' year is 360 days (N.III.9) long and not 354.37 days, the correct length of the synodic year. A month was measured from new moon to new moon, this was an *amānta* month and the waxing phase was called *pūrvapakṣa* (first fortnight). The new moon was considered a 'normal' start of the month (*TS*.III.5.1; Keith 1914). The period from full moon to full moon was also identified for some purposes; this was the *pūrṇimānta* (end of the full moon) month. The waxing and waning halves of a month were marked by the rite (*iṣṭi*) of *Darśapūrṇamāsa*. Although the Āryas had rounded-up the synodic month to 30 days to obtain a ritual year of 360 days, they were aware that this would cause problems with the performance of *Darśapūrṇamāsa* rites, that is, a rite would be performed a day after the month had ended. This can be explained thus: the mean length of a synodic month is 29.5 days, and in two synodic months there are 59, not 60, days; thus in the second month the end of month rite would be performed a day after the month had ended. To allow for this, the *ṛṣis* recommend (*TS*.VII.5.6) that 'having made up the month with six-day periods he should leave out a day' because 'if they did not leave out a day, then the year would fall asunder' (Keith 1914). This empirical correction must have been determined after a period of observations and record keeping.

The names of the months are given in the post-Ṛgvedic Saṃhitās and the Brāhmaṇas (e.g. *TS*. I.4.14; *TS*.IV.11.1; *VS*.VII.30; *ŚB*.IV.3.1.14-20) and are reproduced in Table III.5 (column 3). Although these names are not mentioned in the *Ṛgveda Saṃhitā*, the names of the first six months are mentioned in the list of *devatā*/gods of *sūkta ṚV*.II.36. These names are 'descriptive' in the sense that they describe the seasons during these months. An entirely different set of names of months is given in *Taittirīya Brāhmaṇa*, in the description of the construction of the *Sāvitra Cayana* (a fire altar in the form of the sun, *TB*.3.10.1, Section 2.8.2). These names are given in column 4 of Table III.5; they are a mix of *nakṣatra*-names (see later) and names that vaguely describe the 'environment' during the month, but not as descriptively as the names (column 3) given in other Saṃhitās and Brāhmaṇas. The names of months in column 3 are found in *TS* (and other texts as given above), but the names in column 4 are only found in *TB*. This difference within the same *śākhā* is unexpected and surprising. This *śākhā* does seem particularly prone to differences because, as will be discussed in Chapter IV, the names of *nakṣatra*s

Table III.5: The Vedic seasons, the 'seasonal' names and the *nakṣatra*-names of the months.

Seasons	Vedic Seasons	Months	*TB* months	*Nakṣatra*-names of the months
Spring	Vasanta	Madhu	Aruṇa	Caitra
		Mādhava	Aruṇrajas	Vaiśākha
Summer	Grīṣma	Śukra	Puṇḍarīka	Jyeṣṭha
		Śuci	Viśvajit	Āṣāḍha
Rains	Varṣā	Nabhas	Ābhijit	Śrāvaṇa
		Nabhasya	Ārdra	Bhādrapada
Autumn	Śarad	Iṣa	Pinvamāna	Āśvina
		Ūrja	Annavant	Kārttika
Winter	Hemanta	Saha	Rasavant	Mārgaśīrṣa
		Sahasya	Irāvant	Pauṣa
Cool	Śiśira	Tapas	Sarvauṣadha	Māgha
		Tapasya	Saṃbhara	Phālguṇa
		aṃhasaspati/ saṁsarpa	*mahasvant*	

in the Saṃhitā and the Brāhmaṇa of this *śākhā* also differ. The Āryas abandoned the subjective or phenomenon-based names of months at some stage and identified the months by the 'presiding' *nakṣatra*. That is, a month was named after the *nakṣatra* with which a full moon was conjoined. The *nakṣatra*-names of the months are given in Table III.5 (Column 5) and are discussed in Chapter IV, Section 6.

All Saṃhitās and Brāhmaṇas mention a thirteenth month, and a thirteenth month can be traced back to *Ṛgveda Saṃhitā* (A.II.22). It has been assumed that this month was added to a year to synchronize the ritual year (of 360 days) and the tropical year (of 365.25 days). However, the Vedic schemes for intercalation were not simple, as will be discussed in Section 2.7. The thirteenth month is considered to be the tail or the hump of a year (*TB*.3.8.3, *tasya trayodaśo māso viṣṭapaṃ*, Dumont 1948). It is variously called *aṃhasaspati* (lord of distress; *TS*.I.4.14, *TS*.VI.5.3; *VS*.VII.30, XXII.31), *saṁsarpa* (creeping or sliding along; *TS*.I.4.14, VI.5.3; *VS*.XXII.30; *MS*.III.12), *malimluca* (a thief, a demon or a brahmin who omits the five devotional acts; *VS*.XXII.30) and *mahasvān* (the powerful one, *TB*.3.10.1). In the *Atharvaveda Saṃhitā* the thirteenth month is called *sanisrasa* (fragile, frail; *AV*.V.6.4). The context of these names is not entirely clear. This month does not have a *nakṣatra*-name and the names in the texts, particularly the first three names above, indicate that this month was not considered to be auspicious and was either feared or despised. In some texts (e.g. *KB*.v.8), this month is associated with *śunāsīriya*, the fourth seasonal sacrifice, the starting time of which is uncertain (see Chapter IV, Section 5). The references to the thirteenth month in the earliest strata of the Vedic texts suggest that the concept of intercalation was developed early in the Vedic period.

2.4. Half-Year – *Ayana*

The word *ayana* means a path, course, or walking a road and usually refers to the annual apparent motion of the sun. The ceremonies of *Gavām ayana* (Section 2.5) are symmetric around the two solstices suggesting that the Āryan ritualists had divided the year into two equal parts; *uttarāyana*, the (apparent) northerly course of the sun (from winter solstice to the summer solstice) and *dakṣiṇāyana*, the (apparent) southerly course (from summer solstice to the winter solstice). The *uttarāyana* is also known as the *devāyana*, path of the gods, and *dakṣiṇāyana* is known as *pitrāyana*, the path of the (deceased) fathers (*ŚB*.II.1.3.3). The word *devāyana* occurs frequently in *Ṛgveda Saṃhitā*, but *pitrāyaṇa* occurs only once (A.III.20). In these verses there is no reference to the apparent motion of the sun. In his analysis of 'the Myth of the Hidden Sun' Witzel (2005) has argued that the 're-emergence' of the sun after the winter solstice is embedded in the *Ṛgveda Saṃhitā* (e.g. *ṚV*.III.3), that is, the Ṛgvedic Āryas had identified the solstice-foci of the year and had myths and probably also rites, as yet unidentified, to mark these days.

A detailed description of the apparent motion of the sun in a year is given in *KB*.xix.3 (Keith 1998). This passage clearly identifies *Viṣuvat* and *Mahāvrata*, the two foci of *Gavām ayana* ceremonies, with the two solstices; on *Viṣuvat* day the sun 'turns' south, this is the summer solstice. On *Mahāvrata* day the sun 'turns' north, this is the winter solstice. The sun spends six months in each hemisphere, and *KB*.xix.3 notes that at the end of each six month period 'he (sun) stands still', that is, it is the solstice. This correspondence between the (apparent) annual motion of the sun and the rituals of *Gavām ayana* suggests that this *sattra* was a procedure to keep the 'count of the days' of the year. The sun's apparent traverse of six months in the northern and southern hemispheres is also given in a number of passages of other Saṃhitās and Brāhmaṇas (*TS*.VI.5.3; *PB*.IV.6.17; *ŚB*.VII.4.11). Thibaut (1899) has argued, particularly on the basis of his interpretation of *KB*.xix.3 that the Āryas had not divided the year into two halves of six months. His reasoning is not entirely clear, and his conclusion is false. The Āryas had a very clear understanding of the apparent motion of the sun and they had observed that the sun appeared to 'stand still' on the two solstice days. The officiating priests of *Gavām ayana* are enjoined to emulate precisely the apparent motion of the sun, and to rest on these two days (N.III.6).

In the post-Ṛgvedic period the three seasons, *vasanta (*spring), *grīṣma* (summer) and *varṣā* (rains) were considered the seasons of the gods; *śarad* (autumn), *hemanta* (winter) and *śiśira* (cool) were considered the seasons of the *pitṛ* (*ŚB*.II.1.3.1). Similarly, spring, summer, and rains represent gods who vanquished the demon Vṛtra while autumn, winter, and the dewy season represent those (*pitṛ*) whom the gods recalled to life (*ŚB*.II.6.1.1). These passages are interpreted to mean that *vasanta*, *grīṣma*, and *varṣā* occur during *devāyana* and *śarad*, *hemanta*, and *śiśira* occur during *pitrāyana*. If this interpretation is accepted, then the identification of *uttarāyana* (the apparent motion of the sun from winter solstice to the summer solstice) with *devāyana* and that of *dakṣiṇāyana* (the apparent motion of the sun from summer solstice to the winter solstice) with *pitrāyana* becomes meaningless. This is because at South Asian latitudes, spring cannot occur at or around the winter solstice. Tilak (1893) has discussed this inconsistency, but Kane (1968) does not see a contradiction, arguing that because *vasanta* and *grīṣma* are important or principal parts of *uttarāyana/devāyana*, than 'by association with these two and for the sake of symmetry *varṣā* is also held to be a *ṛtu* of the gods'. This is wholly unconvincing.

The degree to which the Āryas were guided in their rituals by this apparent north-south motion of the sun can be judged from their emulation of this motion during their rituals. During a *Soma* sacrifice, the Adhvaryu was enjoined to exit by the southern door and the Pratiprasthatṛ by the northern door, of the sacrificial enclosure (*TS*.VI.5.3; Keith, 1914). This was because the sun goes south for six months and north for six months. In ceremonies related to the

Agnicayana (building of the fire-altar, this is described in Section 2.8), the sacrificer holds up the fire (representing the sun) towards south-east, then the north-east, and then again towards the south-east (*ŚB*.VI.7.3.9; Eggeling 1894). As the Brāhmaṇa explains, the sacrificer here is imitating the position (along the horizon) of sunrise at the winter solstice and summer solstice and the apparent motion of the sun between the two solstice points.

2.5. Year – *Śarad* and *Saṃvatsara*

Like all ancient people, the Āryas identified the annual revolution of the earth round the sun by changes in the seasons. It was also recognized that this change of seasons was a cyclical process and that the seasons recurred at intervals of a year. In addition, like all ancient people, the Āryas had recognized that year was the fundamental rhythm of rejuvenation of nature, bringing light, warmth, and the bounty of food and pasture. The Ṛgvedic Āryas often refer to an ideal life span of 'hundred years' and this is expressed as the number of autumns 'seen' by a person (A.II.2). Similarly, but less frequently, spring (*vasanta*) and winter (*hemanta*) also indicated the number of years (A.II.3, 4). The rainy season (*prāvṛṣā*) was never used in this sense in *Ṛgveda Saṃhitā*, but statements like 'the season of rains', (*prāvṛṣā*) has come after a year (A.II.6), suggests that by some, or sometimes, a year may have been counted from the start of the rainy season.

In a number of hymns of *Ṛgveda Saṃhitā* (e.g. A.III.9) *śarad*, *māsa* (month) and *aha* (day) occur in the same verse, suggesting that *śarad* was not a poetic way of expressing an interval but was meant to indicate a 'physical' year. The use of *śarad* to indicate the 'physical' age of a person or the 'physical' year suggests that in the Vedic period, a year was measured from a season to the following season, that is, the age of a person was measured in tropical years (of 365.25 days). This practice of measuring a year by the number of autumns (or other seasons) that have elapsed was continued in *Atharvaveda Saṃhitā* (e.g. *AV*.II.28.4), and in the post-Ṛgvedic Saṃhitās (e.g. *VS*.XXXV.15; *VS*.XXXVI.24) and Brāhmaṇas (e.g. *TB*.3.7.11; *TB*.2.8.6; *ŚB*.VII.4.2.31). In these texts, years were never counted by the rainy season. The first (and perhaps the only) time the rainy season is used to count a year is in the late text of *Śatapatha Brāhmaṇa* (*ŚB*.VIII.3.2.6) where the word for the year (*varṣa*) is identified with rains or the rainy season (*varṣāḥ*). In parts of South Asia today, *varṣa* means a year although the year is not necessarily measured from a rainy season.

In the *Ṛgveda Saṃhitā* the year is also referred to as *saṃvatsara* (A.II.6, A.III.21) and *parivatsara* (A.III.22). The contexts in which these words are used suggest that a tropical year (i.e. a year measured from a season to a season) is meant (N.III.8). These words for the year also occur in other Vedic texts (e.g. *AV*.III.10.5; *VS*.XXVII.45; *TB*.3.10.4; *PB*.XVII.13.17; *ŚB*.II.1.1.12; *KŚS*.4.4.26; *LŚS*.IX.1.4). However, as will be discussed in Section 2.6, these are often mentioned as a group and in a slightly different context. However, there seems to be little doubt that from the earliest time, the Āryas had recognized and used the tropical year to keep count of 'social' (as opposed to ritual) events like the birth of a child or a calf. This is summarized in *ŚB*.VI.7.1.18 ('by means of the seasons the year is able to exist').

The Āryas appreciated that the days and nights were units of the year; days and nights are variously described as 'the wheel of the year', 'these two halves produce it (year)' and 'the year, that arranges days and nights' (A.III.23; *AB*.v.30 (xxv. 5); *ŚB*.III.2.2.4; *ŚB*.VI.6.4.3). With the days and months (of 30 days) as building blocks, the Āryas 'constructed' a ritual year of 360 days (or 720 days and nights; N.III.9). They made a distinction between the year to measure the age of a person and the year as the duration of a ritual. The ritual year is described in *Ṛgveda Saṃhitā* (A.II.12 and A.III.24) and in all subsequent Vedic texts (e.g. *TS*.II.5.8.3; *AB*.ii.17(vii.7); *KB*.iii.2; *KB*.xi.7; *PB*.IX.3.5,6). Annual ceremonies were conducted with reference to this ritual year, in particular the great sacrificial session (*sattra*) of *Gavām ayana* ('cow's walk/course' *ŚB*.XII.1.2.1-22). The rituals to be performed during this *sattra* are described in detail in *PB*.XXIV.20 (Caland 1931, see also Eggeling 1885; Sen 1978; *KŚS* Ch.XIII in Ranade 1978) and are given in Table III.2. This is a *sāṃvatsarik sattra*, that is, it is an annual ceremony. The *sattra* lasts for 360 days, with the central day or the *Viṣuvat* being considered an extra day (*ŚB*.XII.2.3.6). The order of performance of the rituals after the central day is a mirror image of those in the first half of the *sattra*. This reflects the apparent reversal of the course of the sun in a year. The daily rites of the *sattra* are as follows (Table III.2):

- *Prāyaṇīya Atirātra*. *Prāyaṇīya* means, 'to go forth' and *Atirātra* means 'performed overnight', with this rite the initiated person begins the *Soma* sacrifice. This is the first day of the *sattra*.
- On the second day, a *Caturviṃśa stoma*, a chant of 24 parts, was recited.
- For next five months (of 30 days each), four *Abhiplava* and one *Pṛṣṭhya* ceremonies are performed each month and each lasts for six days – *ṣaḍaha* (see Section 2.2). Both are *Soma* ceremonies but differ in their respective chants.
- On the following 24 days, three *Abhiplava* and one *Pṛṣṭhya* ceremonies are performed.
- The following day is the *Abhijit* or *Agniṣṭoma* day. It is not clear why this should be called a 'superior victory' (*Abhijit*) day. The *Agniṣṭoma* is a ceremony in 'praise of Agni' and is a modification of the *Jyotiṣṭoma* ceremony.
- The *svarasāman* or sun (*svar*) melody (*sāman*) is chanted for next three days.
- *Viṣuvat* (having or sharing both sides equally) or the central day that divides the *sattra* in two equal parts. *Viṣuvat* can also mean a summit or a vertex and the *Viṣuvat* day can also be interpreted as 'the day on which the sun reaches its highest point' (Figure III.1) or summer solstice. Confusingly, in *Vedāṅga Jyotiṣa* (Chapter V), *Viṣuvat* is interpreted as equinox (day and night of equal length).
- For 167 days after *Viṣuvat*, the ceremonies are the same as those on the 167 days before *Viṣuvat* but they are performed in reverse order. The only change is that the *Abhijit* day is replaced by *Viśvajit* day on day four (after *Viṣuvat* day). It is again not clear why this day should be called 'world victory' day.
- Day number 348 is *Goṣṭoma* a form of *Agniṣṭoma*.
- On the day following *Goṣṭoma* an *Ayuṣṭoma* ceremony is performed. This is a *Ukthya* or *Soma* ceremony similar to *Agniṣṭoma* but three additional *stotra*s (verses that are sung) and three additional Śāstras (verses that are recited) are included in the ceremony.
- This is followed by a ceremony lasting 10 nights (*daśarātra*).
- The finale is the *Mahāvrata* or the great observance. The religious ceremonies of this day are the ceremonies of the *Soma* sacrifice. The main feature is the chanting of the *Mahāvrata sāman* by the Hotṛ at midday. This chant is followed by the recitation of the

mahad uktham or great litany also by the Hotṛ. The special feature of both these ceremonies is the supposed 'bird-like' form of both the chant and the litany. Along with the religious ceremonies, a number of secular ceremonies are also performed on this day. At the time of chanting of some of the *stotra*s, a brahmin plays a *vāṇa* (a plucked stringed instrument), the Udgātṛ sits on a chair of special wood, the Hotṛ sits on a swing, and the Adhvaryu sits on a board, the rest of the assembly sits on the grass. A brahmin and a śūdra both praise and abuse the performers; there is a symbolic fight between a Ārya and a śūdra; a prostitute and *brahmacārin* (a chaste student) abuse each other; a kṣatriya rides a chariot round the sacred ground (*vedī*) and shoots three arrows at a hanging hide and a man and a woman (both strangers) have sexual intercourse in a screened-off part. Drummers at the four corners of the sacred ground beat their drums and a hide placed over a hole in the ground is beaten like a drum. This day is 180 days after *Viṣuvat* (summer solstice) day, or it is (close to) the winter solstice (see Witzel 2005 for comparisons with the end-of-the-year ceremonies in ancient Japan).

- The 'concluding ceremony' (*Udayanīya*) follows and is performed overnight.

This *sattra*, superficially at least, appears to be a way of keeping track of the days (that is, this is a 'calendar stick' (N.III.2), used by most early cultures 'to keep track' of the year) of the ritual year or it is a ritual calendar. This *sattra* cannot be identified in *Ṛgveda Saṃhitā* and the elaborate grouping of the rituals bears the hallmark of a post-Ṛgvedic development. *Gavām ayana* is a model *sattra,* the Āryas also performed other *ayana sattra*s (*LŚS*.X.13.4-16; Ranade 1998), and not all of them lasted for 360 days. For example, a *sattra* was performed in which on even-numbered months the *abhiplava* was performed on five days instead of the usual six. This *sattra* was called *Cāndramāsa Gavām ayana* and lasted 354 days, or (about) one synodic year. It should be immediately obvious that this ritual year will not be synchronized with the tropical/seasonal year. The need for intercalation would therefore have been obvious. The schemes for intercalation developed during the Vedic period are considered in Section 2.7.

2.5.1. *Vedic New Year*

The Vedic texts do not make any unambiguous statements about the start of the Vedic year. The lists of names of seasons given in the Saṃhitās and the Brāhmaṇas (Table III.3), start with *vasanta* (spring), suggesting that the start of the Vedic seasonal year was in the season of spring. Statements like 'spring kindles other seasons', 'spring is the beginning of the seasons and beginning of the year' and 'spring is the head of the seasons' (*ŚB*.I.3.4.7; *ŚB*. I.5.3.9-13; Eggeling 1882; *TB*.3.10.4; *TB*.3.11.10; Dumont 1951; *TS*.V.7.6; Keith 1914) reinforce this suggestion. However, gods tend to be obscure (*parokṣapriyā hi devāḥ*), and the *sattra* of *Gavām ayana* suggests a different season for the start of the ritual year. This *sattra* symbolizes the apparent northern and southern course of the sun with foci at *Viṣuvat* and *Mahāvrata*, that is, at summer and winter solstice respectively. The *sattra* ends with the *Mahāvrata* day, that is, around the winter solstice. The Vedic texts insist that the *sattra* should commence in *śiśira* (cool season) in the months (with *nakṣatra*-names) of Māgha or Phālguna (*AB*. iv.26 (xix.4); Keith 1998), i.e. in the seasonal months of Tapas and Tapasya. The opening rite of *Gavām ayana*, the *Prāyaṇīya* (entrance or opening) *Atirātra* is performed after the winter solstice (*KB*.xix.3) and the *Caturviṃśa* (second day of *Gavām ayana*) is the beginning, 'by it they grasp the year' (*AB*.iv.12 (xvii.6); *KB*.xix.8; Keith 1920; *ŚB*.XII.1.3.9; Eggeling 1900). This suggests the start of the ritual year shortly after the winter solstice and in *śiśira* (the cool season). The statement that '*Viṣuvat* is the middle of the year' (*AB*.iv.18 (xviii.4)) corroborates this. Witzel (2005) has argued, on the basis of the Vala myth in *Ṛgveda Saṃhitā*, that during the Ṛgvedic period too the end of the year was on the winter solstice. He cites *PB*.VI.10.4, which enjoins that the *Mahāvrata* rite (the last rite of *Gavām ayana*) should be undertaken at the end of the year, to suggest that this practice was continued in the Middle Vedic Period in agreement with the argument presented above. Kupier (1983) has suggested that the *Mahāvrata* rite symbolized the transition from the old to the new year. He has conjectured that this practice was inherited from 'pre-Indian' times and was later replaced by a practice that reflects the monsoon of South Asia. This would be in agreement with the suggestion of Tilak (1893) that initially the Vedic year commenced on the winter solstice and was later moved to spring equinox. (A caveat should be added here. The New Year Day might have been moved to spring, but there is no evidence that it was moved to spring equinox.) In the calendar of *Vedāṅga Jyotiṣa* (Chapter V) the *yuga* (the five year intercalation period) and each year of the *yuga* commences in the month of Māgha and in this calendar, the New Year Day is clearly in the first half of the cool season (*śiśira*).

It is possible that the Āryans had two calendars; an 'agricultural/ social' calendar, from spring to spring, and an 'ecclesiastic' calendar, from solstice to solstice (N.III.7). The Āryan calendar makers may have attempted to reconcile the start of these two years. The ritualist note that if the year starts with the full moon in Phālgunīs then 'the *Viṣuvat* day falls in the cloudy (rainy) season' (*TS*.VII.4.8; Keith 1914) and they recommend that 'the full moon in *Citra* is the beginning of the year' (*TS*.VII.4.8). This would put the start of the year in the first half of *vasanta* (spring) in line with the 'agricultural/social' year. This, of course, destroys the symmetry of *Gavām ayana* around the winter and summer solstice. If the *sattra* commences at New Year in the month of Caitra then the *Viṣuvat* day (180 days later) will not be at summer solstice but in the first half of *śarad* (autumn) in the month of Āśvina, unless some rites were dropped in the first half of the *sattra*. The texts do not provide any clues as to how this problem was resolved, but the *Vedāṅga Jyotiṣa* clearly opts for the start of the year after the winter solstice in *śiśira* (the cool season). The start of the Vedic year is revisited in Chapter IV (Section 4) and Chapter V.

2.6. *Yuga*

In *Vedāṅga Jyotiṣa* (Chapter V), a lunisolar calendar is described that has a five-year period of intercalation to synchronize the synodic/ lunar year with the seasons (intercalation is discussed in Section 2.7). This period is the *yuga*. This word can be traced back to the *Ṛgveda Saṃhitā* where it occurs 33 times. The word is derived from the root *yuj* 'to yoke, join' and that is how it is also used in post-Ṛgvedic texts. In *Ṛgveda Saṃhitā yuga* has been used to mean both a yoke and a temporal interval.

- In *Ṛ V*.X.60.8 and *Ṛ V*.X.101.3 (A.III.25, 26) *yuga* follows the meaning of its root and means an implement to hold draft animals together (yoke).
- In *ṚV*.X.72.1 and *ṚV*.I.92.11 (A.III.27, 28) *yuga* means a later age and a (human) generation. However, in *ṚV*.VI.15.8 and *ṚV*.X.72.2 (A.III.29, 30) *yuga* is used in the sense of 'age after age' or the 'age of gods' (which must be long!).

The second set of verses above suggests that during Ṛgvedic times *yuga* indicated time and sometimes a long time (generations and age). Could the Ṛgvedic Āryas have also used *yuga* to indicate a fixed period of time, as was done later? It is difficult to give a definite answer, as explicit evidence has not been identified. However, in *ṚV*.I.158.6 (A.III.31) Dīrghatama is said to have grown old after ten *yugas*. A person is likely to be considered old after 50 or 60 years, which would suggest a *yuga* of five or six years. In *Ṛgveda Saṃhitā* the year is often referred to as *saṃvatsara* and *parivatsara* (e.g. A.III.21, 22). These are also the names of the first two years of a group of five (sometimes six) years mentioned in the post-Ṛgvedic Saṃhitās and Brāhmaṇas. The other names of years in this group of five years are not mentioned in the *Ṛgveda Saṃhitā*. Although a five-year intercalation period (*yuga*) to synchronize the synodic year with the seasons cannot be identified in *Ṛgveda Saṃhitā*, the *ṛṣi*s were aware of the need of intercalation for this synchronization.

In the post-Ṛgvedic Saṃhitās and the Brāhmaṇas the *yuga* continues to mean an age or a generation (e.g. *ŚB*.VII.2.4.26), but the exact length of this period is not given. In these texts, a collection of years began to be mentioned as a group, but the number of years in this group varies from four to six. In later texts (e.g. *Vedāṅga Jyotiṣa*) this group of years is called a *yuga* but not in the post-Ṛgvedic Saṃhitās and the Brāhmaṇas. For brevity, this group of years will be called a *yuga* here. The names of years in these groups of years are given in the Table III.6. The names of the four years of a *yuga* in *TB*.3.10.1, (column 1) are individually scattered in all Saṃhitās and Brāhmaṇas but only in this passage these occur as a group. An association with time is implied by these names since *prajāpati* is often identified with time. Of these names, *saṃvatsara* is a year, *mahant* (the great one)

Table III.6: The names of years in a *yuga*

TB.3.10.1	*VS*.XXVII.45; XXX.15; *TB*.3.4.11; *ŚB*.VIII.1.4.8	*TS*.V.5.7	*TB*.3.10.4
prajāpati	*saṃvatsara*	*saṃvatsara*	*saṃvatsara*
saṃvatsara	*parivatsara*	*parivatsara*	*parivatsara*
mahant	*idāvatsara*	*idāvatsara*	*idāvatsara*
ka	*idvatsara*	*iduvatsara*	*iduvatsara*
	vatsara	*vatsara*	*idvatsara*
			vatsara

can be synonymous with *prajāpati* and *ka* (who?) is also synonymous with Prajāpati. These names occur in a passage that describes the *Sāvitra Cayana* (see Section 2.8.2), a fire altar in the form of the sun. According to the speculations of the Brāhmaṇas all forms of *Agnicayana* are meant to confer a structure on time, this structure being the year (this will be considered in greater detail in Section 2.8). However, it is not clear from *TB*.3.10.1 if a *yuga* of four years is suggested. This group of four years (and a possible four year *yuga*) only occurs in this Brāhmaṇa.

The names of five years (column 2 and 3, Table III.6), as a group, are encountered a number of times in the Saṃhitās and the Brāhmaṇas. These are the five years of the Vedic *yuga* and the implications of this will be discussed in greater detail later. In *PB*.XVII.13.17, there is a correspondence between the three *cāturmāsya* ceremonies and the first three years of the *yuga* and the fourth sacrifice, *śunāsirya* (of uncertain start time), is associated with *anuvatsara*. The implications of these associations are not clear. The names of six years in *TB*.3.10.4, in connection with a bird-shaped fire altar may suggest a *yuga* of six years, but this group of years is not encountered in any other Vedic text.

2.7. Intercalation

The Āryas had identified the seasonal or tropical year as a fundamental unit of time and 'physical' time, like the age of a person, was expressed in these units. The Āryas had also devised, for their rituals, a quasi-synodic year or a ritual year of twelve quasi-synodic or ritual months of thirty days each. The length of 360 days of the ritual year is similar to the length of the tropical year; the deficit of about five days could be added-on as 'extra' (see discussion of *ṚV*.I.25.8 later). The formulation of the ritual year was perhaps the first attempt to synchronize the synodic year with the seasons (N.III.9). The observed length of the synodic month, the synodic year and the tropical year and the lengths of the Vedic ritual month and the Vedic ritual year are summarized in Table III.7. It can be noted from these lengths that a calendar devised in terms of ritual months and a ritual year will not keep in step with the seasons. For example, suppose a ceremony was prescribed for the first day of spring, then the ceremony, in the year following a fiducial year, will be performed five days earlier. If the day of the ceremony was prescribed by the conjunction of a phase of the moon and a *nakṣatra* then the discrepancy in the day of the ceremony would be noticed from the first year after the fiducial year. This discrepancy will keep on increasing if the ritual year is not corrected or synchronized with the seasonal year. There will be a similar discrepancy in the day of the ceremonies if a calendar is devised in terms of the synodic months and a synodic year. In this case, the discrepancy in the day of the ceremony, in the year following the fiducial year, will be about eleven days.

The Āryas appreciated the fundamental link between the seasons and the tropical year, and this is voiced in *ṚV*.VII.103.9 (A.II.6) in which the priests keep the ordained order of the 12 months and sacrifice at the proper seasons. The Āryas have not given us the length of the seasonal/tropical year known to them, but they were aware of the mismatch between this year and their ritual year. In *TS*.VII.1.10 the *ṛṣis* note that 'The *saṃvatsara* was alone in the world.

Table III.7: Length of the tropical year, the synodic year and month and the Vedic ritual year and month

	Definition	Duration (days)	
		Actual	Vedic
Synodic month	Full-moon to full-moon or new-moon to new-moon	29.53	30
Synodic year	Twelve synodic months	354.36	360
Tropical Year	Season to season or equinox to equinox	365.24	

He desired, 'May I create the seasons.' He saw this five-night rite; . . . Then indeed he created the seasons. The four-night rite is incomplete; the six-night rite is redundant, the correct sacrifice is the five-night rite' (Keith 1914). The reference to the 'five-night rite' may lead one to suspect the Vedic predilection for the number 'five', but this verse is very clear; the *ṛṣis* want to synchronize the 'seasons' and the 'year' and this is only possible with a 'five-night rite' as the *ṛṣis* stress the inadequacy of the 'four-night rite' and 'six-night rite'. The year to be synchronized with the seasons is obviously the ritual year and the composer of this passage of the *Taittirīya Saṃhitā* is aware that the length of the tropical year is (at least) 365 days. This need to correct the ritual year to bring it in agreement with the tropical year was recognized in the Ṛgvedic times as well. In *ṚV*.I.25.8 it is noted that (A.III.32), '*Dhṛtavrata* (Varuṇa) has knowledge of the productive 12 months, he knows that which is extra'. Although the nature of 'that which is extra' is not stated, the productive twelve months, that is, a ritual year (of twelve months of thirty days each), is clear and it is reasonable to assume that the 'extra' refers to the days required to match the length (in days) of the ritual year and the tropical year. A number of other cultures also added extra days (five) to a ritual year of 360 days (N.III.9).

It was common practice among many early civilizations to use the synodic month for ritual and social purposes. It was also recognized that a synodic year of 12 months and 354.37 days did not match the tropical year of 365.24 days. The earliest (about twenty-first century BCE) attempt to bring the lunar year in line with the tropical/ seasonal year was made by the cultures of Mesopotamia. This was accomplished by an intercalated month (known as *iti dirig*). Initially the intercalation was haphazard, according to real or imagined needs,

Box 2

Mean length of the synodic month	=	29.5306 days
Mean length of a synodic year (12 × 29.5306)	=	354.3672 days
In 19 synodic years with 7 additional months	=	*6939.6910* days
Mean length of the tropical year	=	365.2422 days
In 19 tropical years, 19 × 365.2422 days	=	*6939.6018* days

and each Sumerian city inserted a month at will. Later the empires centralized the intercalation, and by 541 BCE, it was proclaimed by a royal fiat. By about 500 BCE, the calendar in the Babylonian cities began to be regulated by the lunisolar cycle of 19 years. This was the discovery that, after 19 tropical years (235 lunations or synodic months), the moon's phase recurs on the same day of the year (this cycle had also been discovered by the Chinese, around 484 BCE). This cycle is commonly known as the Metonic cycle (N.III.10) after the Greek astronomer Meton of Athens, who introduced it in Greece around 432 BCE. By 380 BCE, all calendars in Mesopotamia were regulated by this cycle. The Metonic cycle is an arithmetic rule to make the synodic/lunar calendar follow the seasons by adding an extra synodic month to seven (synodic) years in a 19-year cycle, i.e. seven years out of 19 have a thirteenth month. The rationale for this scheme is shown in Box 2. Traditionally a month was added to the years 3, 6, 8, 11, 14, 17 and 19 of the Metonic cycle. More practically, the cycle of 19 years was approximated to a whole number of 6,940 days by 125 long months of 30 days and 110 short months of 29 days.

The Metonic cycle has been modelled to test the stability of this intercalation scheme, and the results are shown in Table III.8. Column 1 is the year number, column 2 is the total number of 'tropical days' from the beginning of year 1, i.e. at the end of the tropical year 1, 365.24 days have passed, and at the end of the tropical year 2, 730.48 days have passed and so on. Similarly, column 3 is the total number of 'synodic days' from the beginning of year 1. Column 4 is the difference between the total number of 'tropical days' and the total number of 'synodic days' at the end of each year. Column 5 is the total number of synodic days, from the beginning of year 1, after an extra synodic month is added to the 3rd, 6th, 8th, 11th, 14th, 17th, and 19th year, as described above. Column 6 is the difference between the total number of 'tropical days' and the total number of 'intercalated days' at the end of each year. If the synodic/lunar calendar is not synchronized with the tropical year (or with the seasons), then the difference (in days) between the tropical year and the synodic year will increase monotonically as shown in column 4. For a stable intercalation scheme, although the difference in the number of days in the tropical year and intercalated year will change from year to year, it will not diverge with time, or it will not increase monotonically. This is

Table III.8: Results of modelling the Metonic cycle

Yn	T	S	T-S	I	T-I
1	365.24	354.37	10.88	354.37	10.88
2	730.48	708.73	21.75	708.73	21.75
3	1095.73	1063.10	32.63	1092.63	3.09
4	1460.97	1417.47	43.50	1447.00	13.97
5	1826.21	1771.84	54.38	1801.37	24.84
6	2191.45	2126.20	65.25	2185.26	6.19
7	2556.70	2480.57	76.13	2539.63	17.06
8	2921.94	2834.94	87.00	2923.53	-1.59
9	3287.18	3189.30	97.88	3277.90	9.28

Yn – Year number
T – Accumulated number of days in tropical years
S – Accumulated number of days in synodic year
T-S – Difference (in days) between tropical and synodic years
I – Accumulated number of days in intercalated years
T-I – Difference (in days) between tropical and intercalated years

shown in Figure III.3, in which the data from column 6 of Table III.8 are plotted for a period of 500 years. Although every year there is a difference between the number of days in the tropical and the (corrected or intercalated) synodic year, this difference does not diverge over hundreds of years. The great advantage of the Metonic scheme was that it offered a simple rule for intercalating months in a lunar calendar to keep it in step with the cycle of seasons. The scheme was so successful that it formed the basis of the calendar adopted in the Seleucid Empire (Mesopotamia) and was used in the Jewish calendar and the calendar of Christian Church.

It is not known if the Āryas had also developed an arithmetic scheme for synchronizing their ritual year and the tropical/seasonal year. Gondhalekar (2008b) has discussed in detail the possible intercalation schemes that the Āryas may have developed and used. The Saṃhitās and the Brāhmaṇas are full of references to a thirteenth month. The thirteenth month can also be traced back to *Ṛgveda Saṃhitā*; in *ṚV*.I.164.15 (also *AV*.X.8.5; A.II.22) six pairs of two months (for each season) and an extra thirteenth month are implied (Section 2.4). This long hymn by Dīrghatamā Aucathya is classified as a 'riddle hymn' and attempts to understand it have kept Vedic scholars

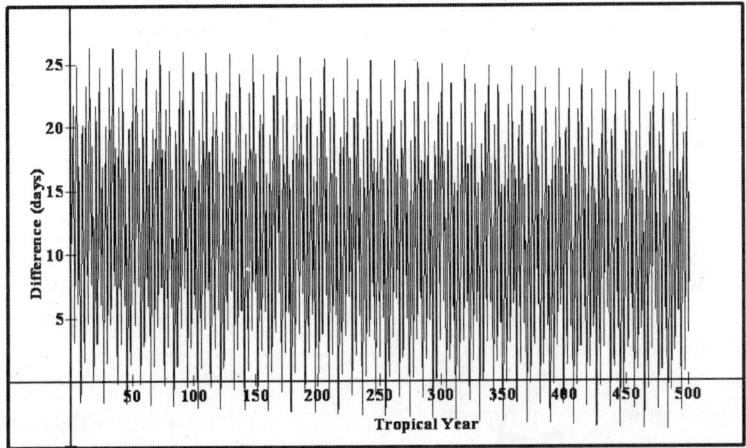

Figure III.3: The intercalation scheme of Metonic Cycle. The difference every year, between the accumulated number of days in the tropical years, and the corresponding intercalated synodic years is shown for a period of 500 years. The data shown are the data of column 6 of Table IV.8.

occupied for over 100 years. In this hymn, many questions have been explicitly stated and others are implied through obscure symbolism. Recently Houben (2000a, 2000b) has carried-out a detailed analysis of this hymn and has shown that *Gharma-Pravargya* ritual forms the core of the hymn; this is one of the two rituals (the other being the *Soma* ritual) that have been identified in *Ṛgveda Saṃhitā*. A number of other themes can also be identified in this hymn and one of these is the 'calendar'. All components of the Vedic ritual calendar; seasons (five and six), months (twelve), days (360) and days and nights (720), can be identified, and it is not surprising that it should also include a reference to the thirteenth month. Unfortunately, the Āryas have not left us the algorithm they might have used to incorporate a thirteenth month in their calendar. It has been assumed that the Āryas would have followed the arithmetic approach typified by Meton and other ancient calendar makers. In the simplest arithmetic scheme for intercalation, a month of 30 days could be intercalated after '*n*' ritual years, to get the ritual years to follow the seasons or the tropical year. This scheme is formulated in Box 3; the tropical and the ritual years would be perfectly synchronized if a 30-day month were intercalated after 5.73 years (about five years and nine months). However, the

Box 3

$$365.24 \times n = 360.00 \times n + 30$$
$$n = 30 \div 5.24$$
$$n = 5.73$$

Vedic texts make no references to such intercalation. As noted above (Section 2.6), a group of five years is frequently mentioned in the post-Ṛgvedic Saṃhitās and the Brāhmaṇas. In *Vedāṅga Jyotiṣa*, a group of five years is called a *yuga* and is the period of intercalation (discussed in Chapter V). Early investigators of the Vedic calendar conjectured that the Āryas intercalated a thirteenth month in the fifth year of a *yuga* (Zimmer 1879). The consequences of this form of intercalating are formulated in Box 4. By intercalating a thirteenth month of thirty days in the fifth year of a *yuga*, the Āryas would have obtained a seasonal year of 366 days. However, after five years, the tropical/ seasonal year would have lagged behind the ritual year by 3.8 days and this would have accumulated to thirty days after about 39 years and six months. The mismatch between the seasons and the 'date' (as obtained from the ritual year) would have been noticed long before 39 years. The difference in days between the accumulated number of days in the tropical years and the accumulated number of days in the corresponding intercalated ritual years over a period of 500 years is shown in Figure III.4. These data are equivalent to the data given in column 6 of Table III.8. The difference between the intercalation scheme of the Metonic cycle and the 'five-year-*yuga*' cycle (Figure III.4, Curve #1) is dramatic, and it should be obvious that the Āryas could not have followed this rather crude scheme of intercalation. Thibaut (1899) has speculated that, at irregular intervals, corrections were made to the intercalated calendar to bring it in line with the observed beginning of a season.

Box 4

Tropical: $5 \times 365.24 = 1826.20$ days
Ritual: $5 \times 360 + 30 = 1830$ days
Difference (T-R) = -3.8 days

The Āryas could have used a 'six-year *yuga*' for intercalation, adding a thirteenth month of 30 days every sixth year. The only reason for considering this scheme is a possible reference to a six-year *yuga* in the *Taittirīya Brāhmaṇa* (Section 2.6). This scheme is formulated in Box 5. After six years, the tropical year would have gained 1.44 days on the ritual year, and this would have accumulated to 30 days after

125 years. This scheme of intercalation results in an average year of 365 days. This scheme is far more stable than the 'five-year *yuga*' scheme nevertheless it is not particularly satisfactory over long periods. Somewhat different schemes of intercalation are suggested in *Śatapatha Brāhmaṇa*. In *ŚB*.IX.1.1.43; and *ŚB*.IX.3.3.18, a thirteenth month of 35 days is suggested and in *ŚB*.X.4.3.8, *ŚB*.X.4.3.19 and *ŚB*.X.5.4.5 a thirteenth month of 36 days is suggested (Eggeling 1885b). Weber (1861) had identified the intercalary months of 35 and 36 days in the Vedic texts, but he erroneously assumed an intercalation period of six years. The reasoning behind the two possible schemes in *ŚB* is formulated in Box 6. The result of intercalating a thirteenth month of 35 days in the seventh year is very similar to that of intercalating a thirteenth month of 30 days in the sixth year. That is, the difference of 1.68 days, in the number of days in the tropical year and the intercalated ritual year, will accumulate to 30 days after 125 years. In this scheme, the average length of the year would be 365 days. However, in the scheme of intercalating a thirteenth month of 36 days in the seventh year, the difference, in the number of days between the tropical year and the intercalated ritual year is 0.68 days. This will accumulate to 30 days after 308 years and 10 months and the average length of the year would be 365.14 days.

Box 5

Tropical: 6 × 365.24	=	2191.44 days
Ritual: 6 × 360 + 30	=	2190.00 days
Difference (T-R)	=	1.44 days

Box 6

Tropical: 7 × 365.24	=	2556.68 days
Ritual: 6 × 360 + 360 + 35	=	2555.0 days
Ritual: 6 × 360 + 360 + 36	=	2556.0 days

It is possible to make a simple correction to the 'five-year *yuga*' scheme of intercalation to obtain a very satisfactory synchronization of the seasons and the Vedic ritual year. This correction was suggested by Faddegon (1926) and involves subtracting 30 days every 40 years, that is, every fortieth year the intercalated thirteenth month of 30 days is not added. The results of this scheme of intercalation over 500 years are shown in Figure III.4 (Curve 2; the data equivalent to column 6 of Table III.8 are plotted), and it is obvious that this is an exceptionally stable scheme. Yet there is no evidence in the Vedic

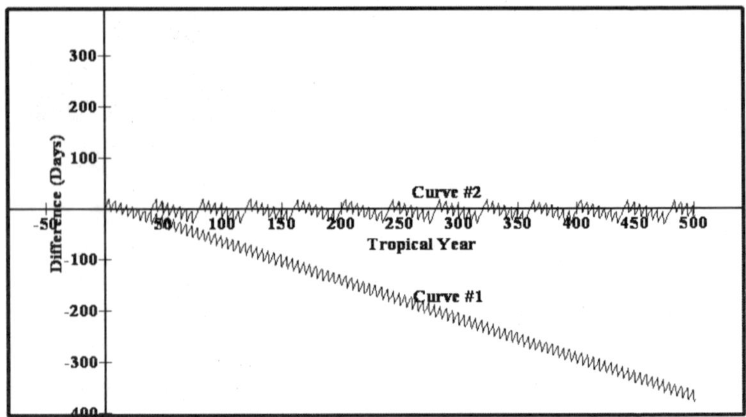

Figure III.4: The annual difference in the accumulated number of days in the tropical years and the corresponding intercalated ritual years. The 'simple' intercalation in which a month of 30 days is added every fifth ritual year is shown in Curve 1. For Curve 2, the same correction is made every fifth year except every fortieth year no correction is made, i.e. the 30-day month is not added.

texts that the Āryas followed this scheme. This cannot have been due to their lack of ability to perform simple arithmetic operations because in the construction of their fire altars the Āryas demonstrate a high degree of mathematical skill (Sarasvati Amma 1979).

The Saṃhitās and the Brāhmaṇas present a picture of people who were meticulous in their religious observances. These people took exceptional care to preserve their ritual texts and even formulated rules for the correct pronunciation of every word and every syllable in these texts. In these texts, the rituals and ceremonies, both simple and elaborate, are described in considerable detail and there are copious notes on the rationale for these rituals. The sacrificer is constantly urged (and warned) that the full efficacy of a ritual is only accrued by following procedures to the letter and these rituals and ceremonies were performed for hundreds of years. The ceremonies like *darśapūrṇamāsa* (new and full moon) and *cāturmāsya* (four monthly or the three seasons) would have required an ability to forecast and identify appropriate days for the performance of ceremonies. It is inconceivable that these would not have been performed at the appointed and appropriate times. A closer look at the Vedic texts indicates that the Āryas were aware of problems encountered in getting the times right for their rituals. For example, *AB*.vii.11 (xxxii.10) and *KB*.iii.1, describe ceremonies to be performed at new moon. In *ŚB*.IX.1.5.1, the sacrificer is warned that when he thinks it 'is the day of the new moon (when the moon should not be seen) the moon is seen', i.e. the moon rises a day earlier than the calculated new-moon day (Eggeling 1885b). Similarly, *ŚB*.IX.1.4.1, recommends expiation rituals if the moon is seen on a new-moon day 'on account of not having ascertained properly' the correct day of the new moon. Passages like this suggest that the Āryas were aware that there could be discrepancies between the calculated and observed times of astronomical events. To avoid expiations (*prāyaścitta*) if time-critical rituals were not performed at the prescribed times, the priests would have relied on careful astronomical observations rather than on mechanical application of an (inaccurate) rule.

The only Saṃhitā that explicitly describes an intercalation scheme is *MS*.I.10.8 (A.III.33). This text suggests two classes of ritualists; the *ṛtu-yājins* and the *cāturmāsya-yājins* (Faddegon 1926). The *ṛtu-yājins* celebrated the *cāturmāsya* (seasonal) sacrifices on the first full-moon day that followed the beginning of spring, summer, and autumn. These *yājins* were not concerned with the number of days between the seasons or the length of the lunation, but relied entirely on observations to perform the required rituals. The *cāturmāsya-yājins*, on the other hand, performed the seasonal sacrifices after a fixed number of lunations. They had to follow (arithmetic) rules to keep their synodic calendar in step with the seasons. The scheme suggested is as follows: the *yājins* celebrated the *cāturmāsya* ceremonies for three years (or 36 synodic months) at fixed intervals (four synodic months). They then intercalated a synodic month. The *cāturmāsya* ceremonies were then performed for further two synodic years (or 24 synodic months) followed by intercalation of a second synodic month. A possible scheme of alternate lunation of 30 and 29 days respectively is shown in Box 7. Thus, the *ṛtu-yājins* followed the seasonal year whereas, in the

Box 7

Synodic: 18 × 30 + 18 × 29 + 30 =	1,092 days
Synodic: 12 × 30 + 12 × 29 + 30 =	738 days
Total	= 1,830.0 days
Tropical: 365.25 × 5	= 1,826.25 days

calendar of the *cāturmāsya-yājin*s the synodic/lunar year followed the seasons in a period of 62 lunations (i.e. they had an arithmetic calendar). In *MS*.I.10.8, this intercalation scheme is 'interpreted' in the context of Vedic ritual year. That is, 62 synodic months or 1,830(.86) days are equivalent to five ritual years of 360 days each (total 1,800 days) plus a thirteenth month of 30 days (making 1,830 days). This is the Vedic 'thirteenth month' that figures in all Saṃhitās and Brāhmaṇas. This thirteenth month is not added *ad hoc* to the synodic calendar to harmonize it with the seasons, but it is a 'ritual interpretation' of a synchronizing scheme based on 62 synodic months. As is shown later and in Chapter V, this period of 1,830 days is at the core of the Vedic calendar. The Saṃhitās and the Brāhmaṇas do not explain how this scheme and this period of 1830 days was arrived at, but possible explanation may be in the interpretation of *TS*.VII.4.8 as shown later. However, as shown in Box 7 the ritual year of the *cāturmāsya-yājin*s would have gained on the tropical year and an additional correction to harmonize the synodic year with the seasons would have been required. No such correction is mentioned in this text. In *MS*.I.10.8, the period of 62 lunations over which the *cāturmāsya-yājin*s correct their ritual calendar is not called a *yuga*. However, this is the only Vedic text that unambiguously assigns a role to a group (or a period) of five years, in a calendar.

The *sattra* of *Gavām ayana* follows the apparent annual motion of the sun and has two foci, the *Viṣuvat* and *Mahāvrata*. These correspond respectively to the summer and the winter solstice as explained in *KB*.xix.3. It is, however, unlikely that the new cycle of the *sattra* was started two days after the *Mahāvrata* as the start of the Vedic ceremonies was usually on either the day after the new moon or on a full moon day. The day of the start of the *sattra* is not given in *KB*.xix.3, but is given in *TS*.VII.4.8 (A.III.34) and with minor variations in the wording and additional information in *PB*.V.9. The importance of this passage has been recognized for a long time and this passage has been discussed extensively by *mīmāṃsaka*s and by Sāyaṇa (died about CE 1387). In the nineteenth century, the passage again came under scrutiny, this time of Weber (1861), Tilak (1893) and Dīkṣita (1896). More recently, Chakravarty (1975) has discussed this passage. The *mīmāṃsaka*s attempted to interpret the passage and understand its implications. The nineteenth- and twentieth-century scholars used this analysis to discuss the start of a new ritual year in the Vedic calendar. Weber (1861) has given a rather convoluted exposition of various passages, in different Vedic texts, relevant to the start of the Vedic ritual year, but in these passages no single marker can be identified for the start of the ritual year. According to *KB*.xix.3, 'on the new moon of Māgha he (sun) rests, being about to turn north' (Keith 1998) and the *Mahāvrata* would be performed on this (or around this day) and the ceremony concluding *Gavām ayana* (*Udayanīya Atirātra*) would be performed on the following day (Table III.2). Thus, the new ritual year must have started (soon) after the winter solstice. The conclusion of the discussion of the *mīmāṃsaka*s (Jha 1934) is that the start of the *sattra* is on the full moon following the *ekāṣṭakā* immediately after the winter solstice. This would be the full moon of the month of Māgha. This does not make any sense, as *ekāṣṭakā* is the eighth day after the full moon, i.e. the last quarter. However, Śabara (Jha 1934) is of the opinion that 'every eighth day of every month is *ekāṣṭakā*' (*sūtra* 33). The (Vedic) month starts on the new moon day (the waxing phase is *pūrvapakṣa* – first half of a month) and according to Śabara the eighth day would be the first quarter. It is impossible for winter solstice to coincide with a new moon every year and the requirements of *KB*.xix.3 are unrealistic. Guided by the *mīmāṃsaka*s (the first half of) *TS*.VII.4.8 could be interpreted thus. The *sattra* of *Gavām ayana* ends on or around the winter solstice (which could be the new moon of Māgha, but not every year), (re) initiation is on the (following) first quarter, and the new *sattra* starts on the following full moon, which is the full moon of Māgha. The *ekāṣṭakā*-period is described as 'distressed/pained (*ārtaṃ vā*)', and has been explained as the period when one *sattra* has ended, but a new one has not started (Sāyaṇa in Tilak 1893), i.e. an 'interregnum'. This period has also been described as being 'reverse (*vyastaṃ vā*)'. Śabara (Jha 1934) explains this as the period when the *ayana* has changed, or the sun has 'reversed' its course. Both explanations are consistent with the interpretation proposed above. The crucial point is that the start of the *sattra* is locked to the winter solstice, that is, to the start of the tropical year (the tropical year, by definition, is from solstice to solstice or equinox to equinox). In this scheme, the 'date' of the start of a ritual year does not depend on the number of lunations in the previous ritual year. The start of a ritual year is determined by the phase of the moon at the winter solstice because this establishes the interval between the winter solstice (i.e. the start of the tropical year) and the start of the ritual year or the intercalated period. The synchronization of the ritual year with seasons, if, over a number of years, the ritual year starts at full moon after *ekāṣṭakā* following the winter solstice can be investigated computationally.

- Determine the date of the winter solstice. The tropical year ends on this day and the new tropical year starts on the following day. The concluding ceremony (*Udayanīya Atirātra*) of the *sattra* is on the following day and the 'interregnum' starts on the day following the concluding ceremony.
- Determine the date of the following first quarter or *ekāṣṭakā*.
- Determine the date of the following full moon. The new ritual year starts on this day and it is the first day of the new *sattra*.

The results of this analysis are shown in Figure III.5. The difference between the accumulated number of days in the tropical years and the accumulated number of days in the ritual year (including the

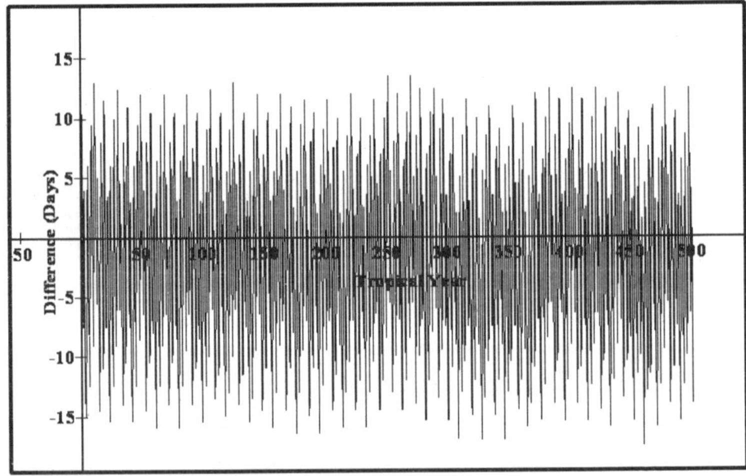

Figure III.5: The intercalation scheme deduced from *TS*.VII.4.8 and *PB*.V.9. The difference every year, in the accumulated number of days in the tropical years, and the corresponding intercalated ritual years is plotted. The data plotted are equivalent to those in column 6 of Table IV.8.

intercalated days) over a period of 500 years is plotted. The data plotted are equivalent to the data in column 6 of Table III.8. These results were calculated from 1500 to 1000 BCE, but the epoch of calculation is not important, any other epoch would have similar results. The data suggest that this scheme of intercalation is as stable as the Metonic cycle. A closer look at these results reveals a number of interesting aspects. The length (in days) of the ritual year (including the intercalated days) in a 50-year interval is shown in Figure III.6. The length of the individual years is not the same; this varies from 354.37 days (12 synodic months) to 383.89 days (13 synodic months). However, there is a clear pattern; in an interval of five years, the total number of days is 1,830 and this is true for five (sometimes six) consecutive periods of five years, the sixth (sometimes seventh) period of five years is 1,800 days long. The 1,830 days are obtained by intercalation of two synodic months in a period of five synodic years through the *ekāṣṭakā* condition and not through arbitrary

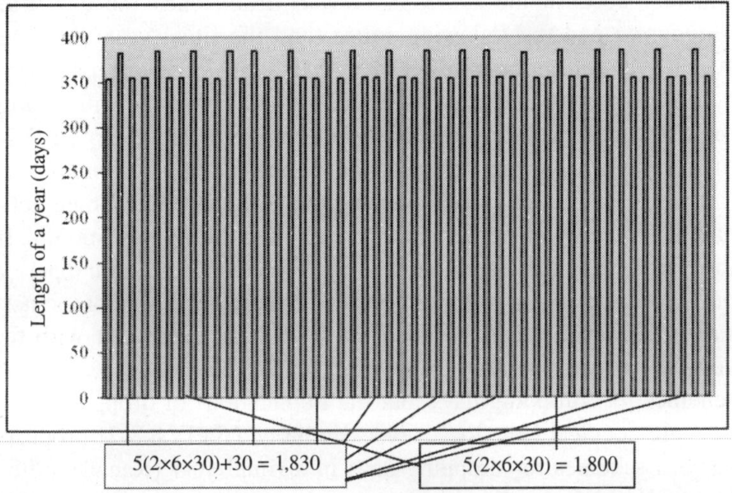

Figure III.6: The length (in days) of the ritual years in a 50-year interval. The total number of days in five-year periods is shown in the two boxes. Five consecutive periods of 62 lunations (about 1,830 days) are followed by a period of 61 lunations (about 1,800 days). That is, an intercalation month is dropped, to correct for the accumulated excess number of days in 25 ritual years (see text for detail). This pattern is repeated and results in a stable calendar.

intercalation, as in the scheme described by *MS*.I.10.8. That is, the period of 1830 days over which the ritual year follows the seasons is obtained through the *ekāṣṭakā* condition. The period of 1,830 days can be resolved into six seasons of two months, each of 30 days and a thirteenth month of 30 days. This is entirely consistent with the Vedic texts. Even in this scheme, the tropical year would lag behind the ritual year by 3.8 days after five years (see Box 4, 7). However, this is corrected in the sixth (sometimes seventh) period of five years when the thirteenth month gets omitted.(again due to the *ekāṣṭakā* condition) and the total number of days in this five-year period is 1800. Note that it is not necessary for the deficit to accumulate to a complete lunation; just enough days have to accumulate for the *ekāṣṭakā* condition to be 'switch-in'. In this scheme the start of the ritual year, which is determined by the phase of the moon at the winter solstice, is locked to the motion of the sun by *TS*.VII.4.8 (and *PB*.v.9), i.e. this is an astronomical calendar.

This scheme introduces period(s) of 'interregnum' in the Vedic calendar because the *sattra* of *Gavām ayana* ends the day after *Mahāvrata* and the following *sattra* (and the ritual year) does not start till the full-moon of Māgha (or Phālguna) which will not always be on the second day after a winter solstice. This may explain why the thirteenth 'month' is variously referred to as *aṃhasaspati* (lord of distress), *saṃsarpa* (creeping or sliding along), and *malimluca* (a thief, a demon or a brahmin who omits the five devotional acts) and why no religious observances are allowed during this 'month'. In the Vedic texts, there are no explicit statements regarding the period between the end of a *sattra* in one year and the beginning of the next *sattra*. However, the Āryas may not have been unfamiliar with the concept of interregnum. Both *TS*.VII.5.7.1 and *PB*.v.10, propose schemes for 'dropping days' but the implications of dropping these days are not clear from these texts. Similarly, *LŚS*.IV.8.8-21 proposes various schemes for dropping days in a ritual year (Ranade 1998). In *LŚS*.IV.8.1-7 ritual years longer or shorter than the canonical 360 days are mentioned, not unlike the length of the years obtained in the scheme proposed here. These passages from the Vedic texts do not have a direct bearing on the present scheme, but they demonstrate that the Vedic ritualists were familiar with a ritual year of different lengths and the need to adjust the length of their ritual year. Within a *sattra*, different groups of ritualists may have followed their own particular ways of timing the start (or the end) of rituals and ceremonies, as was done by the *ṛtu-yājin*s and *cāturmāsya-yājin*s of *MS*.I.10.8.

2.8. Piling of the Fire-altar – *Agnicayana*

The Vedic rituals are 'fire based' in the sense that *agni* (fire) is both a god and an agency to convey the sacrifice to the gods. It is, therefore, not surprising that the place where the fire is kindled should receive serious consideration in Vedic texts. However, with the Ārya, nothing is straightforward and *Agnicayana* (*agni* – fire, *cayana* – piling) is no exception. The ritual space (*vedī*), the ritual of piling and the altar are all multidimensional with social, cosmogenic, theological and calendric aspects. Only the calendric aspects will be considered here. The ritual of *Agnicayana* (or brick-built altars) cannot be identified explicitly in the *Ṛgveda Saṃhitā* but there are references to fire altars (*citi*s) in this text. For example, in *ṚV*.IX.97.54 'non-believers' are called *acitaḥ* – those who do not have a fire altar. The three sacrificial fires – *gārhapatya*, *āhavanīya* and *dakṣiṇāgni* – are frequently mentioned in *Ṛgveda Saṃhitā* and although these fires have independent functions, they also have a role in the ceremony of *Agnicayana*. The full exposition of *Agnicayana* is in the post-Ṛgvedic texts (particularly in the Yajurvedic texts) and these allude to various types of *Agnicayana* or methods of building the fire altars. Vedic fire altars are traditionally divided into two classes: the *mahā-cayaya* or the major piling, and the *kṣudra-cayana* or the small piling. The *mahā-cayana* is discussed in the *Taittirīya Saṃhitā* (*TS*.IV and V) and the Śāṇḍilya books (*ŚB*.VI-X) of the *Śatapatha Brāhmaṇa*. This fire altar is built of fired bricks and consists of five (principal) layers but can have additional layers. It is three-dimensional. This is considered the *prakṛti* (the normal form of) *Agnicayana*. There are five types of *kṣudra-cayana* and these are discussed in the *Taittirīya Brāhmaṇa* (*TB*.3.10-12) and in various Śrautra-Sūtras. These are normally two-dimensional altars and are considered secondary forms of *Agnicayana*. The *kṣudra-cayana* are:

1. *Sāvitra Cayana* – fire altar in the form of Savitṛ or the sun
2. *Nāciketa Cayana* – fire altar according to the sage Naciketā

3. *Cāturhotra Cayana* – fire altar built with the formulae symbolizing the four *hotṛs*
4. *Vaiśvasraja Cayana* – fire altar as it was built originally by the creator of the universe
5. *Aruṇaketuka Cayana* – fire altar as it was built by sage Aruṇaketu

The first three *kṣudra-cayana* have calendric features built into their form and in the performance of the rituals. Only these three *cayanas* will be described here.

According to *Śatapatha Brāhmaṇa*, *agnicayana* is essentially the restoration of Prajāpati, the creator god, who created the world through self-sacrifice (This concept can be trace back to the Puruṣa Sūkta, *ṚV*.X.90 of the *Ṛgveda Saṃhitā*). In the Vedic world-view, Prajāpati is equivalent to Time, which is identified by the year (Gonda 1984; *KB*.vi.15: *sa eṣa prajāpatir eva saṃvatsaraḥ*). Thus in *Agnicayana*, Time (year) itself, in the form of its various units, is restored. The structure of the altars suggests that the Āryas mapped onto the altars all features of their calendar, including intercalation.

2.8.1. *Mahā Cayana*

The *śruta* ceremonies involve three fire altars, namely;

- *Gārhapatya* – (householder's fire) a circular altar representing the earth, situated west of the ritual space.
- *Āhavanīya* – (offering fire or the fire in which the offerings are made) a square altar representing the sky, sited to the east of the ritual space.
- *Dakṣiṇāgni* – (southern fire) a semi-circular altar representing the firmament, sited to the south of the ritual space.

For 'high ceremonies' like the *Soma* sacrifice, the *Āhavanīya* is transferred to *uttaravedī* (high altar) in *mahāvedī* (great ritual space). The *uttaravedī* is then called *Āhavanīya*. It has been suggested (Eggeling 1885b) that the ritual of building this altar was not part of the original Vedic sacrificial system but was probably developed independently and incorporated into the Vedic rituals at a later date. Staal (2004) has argued that the Vedic fire 'cult' evolved as Indo-

Europeans/Āryans 'marched' from the Iranian plateau and BMAC/ Pirak (regions of Central Asia) to the Indus Valley and further east, assimilating rituals of the people they met (N.I.6). It has also been argued (Converse 1974) that this ritual was borrowed by the Āryas from the indigenous people, the people of the Late Haṛappā Period. The main points of this argument are:

- This ritual uses a large number of fired bricks (of different sizes and shapes) but fired bricks are not mentioned in *Ṛgveda Saṃhitā*. In the post-Ṛgvedic Saṃhitās and the Brāhmaṇas fired bricks are mentioned only in connection with *Agnicayana*. The Haṛappans on the other hand were masters of (kiln-fired and sun-baked) brick making.
- The size and shape of the bricks used in altar construction are remarkably similar to the Haṛappan bricks.
- The Sanskrit word for a brick is *iṣṭakā* and that for an oblation is *iṣṭ*. Thus to the Āryas a brick represented an oblation and was something special. This was not so for the Haṛappans for whom the brick was an 'everyday object'.

These points have a common theme; the Āryas (particularly the Ṛgvedic Āryas) were unfamiliar with brick making technology and brick culture in general whereas the Indus Valley people were familiar with both. The ex-Vedic origin of *Agnicayana* has been challenged (Staal 1978; Kashikar 1981) and this issue cannot be considered to have been settled. It does highlight the need to look more closely at the exchange of ideas between the Āryas and the people of the Late Haṛappā Period, rather than between the Āryas and cultures of Western Asia (a more usual approach of Western scholars) with whom the contact was at best tenuous. In 1975, Staal (1983) documented in detail an *Agnicayana* (performed by Nambudiri brahmins of Kerala, India) and (probably) every aspect of the modern-day *Agnicayana* is now on record. This was also (perhaps) the first *Agnicayana* witnessed by those not involved in the ceremony.

The liturgical formulae during construction of the altar are given in Chapters XI to XVIII of *Vājasaneyī Saṃhitā*, but the detailed exposition of *Agnicayana* is given in the *Śatapatha Brāhmaṇa*, which traces the 'knowledge' of *Agnicayana* from Śāṇḍilya upwards to Tura

Kāvaṣeya, who received it from Prajāpati who in turn received it from (the impersonal) *Brahman* (Eggeling 1885b). The huge *Āhavanīya* was an altar of different shapes, e.g. eagle (*suparṇa*), hawk (*śyena*), etc. (in *ŚB* there is an elaborate explanation of why this should be so, but this is not of relevance here). The altar and the *vedī* had to conform to strict geometrical principles explained in *Śulbasūtras* (N.I.5). For this ceremony, the infinite Time had to be reduced to finite proportions, that is, to the complete revolution of the seasons or a year. Hence, the complete ritual takes a year from the time the *yajamāna* (sacrificer) lights a fire in a special pot called *ūkhā* (which he carries around with him) to the culmination of the final 12 day ritual. The site for *Agnicayana* was chosen carefully; most importantly, it had to have an unobstructed view eastwards because the earthly fire was identified with its celestial counterpart, the sun. The ground was ploughed with a team of six (because there are six seasons) oxen, and it was ploughed 'from left to right . . . they turn with the turning of the sun', 12 furrows are ploughed because 'the year has twelve months' (*TS*.V.2.5; Keith 1914; *ŚB*.VII.2.2.14; Eggeling 1894). The number of vessels and chants involved in this ceremony are frequently justified by references to various aspects of the Vedic ritual calendar like the number of seasons, number of days in a month and year, and the number of months (including the thirteenth month) in a year.

The *Āhavanīya* altar represents the year and it is built of 396 *yajusmati* bricks, that is, 360 bricks to represent the Vedic ritual year and an additional 36 to represent the thirteenth (intercalary) month (Section 2.7). There are also 10,800 *lokampṛṇā* bricks (equal to the number of *muhūrta*s in the Vedic ritual year, *ŚB*.X.4.3.8, Section 2.1). The bricks are laid in a strict configuration in five layers and the names of these layers are the same as those of the five years of the *yuga* (column 2 of Table III.6). The year is held together by seasons (*ŚB*.VIII.7.1.3) and each layer includes bricks for a season. Two bricks are laid in the first layer to represent Madhu and Mādhava, the two months of *vasanta* (spring). The two bricks in the second layer represent Śukra and Śuci, the two months of *grīṣma* (summer). The third layer has four bricks, for Nabha and Nabhasya (the two *varṣā*/rainy months) and for Iṣa and Ūrja (the two *śarad*/autumn months). These bricks are of half thickness and the thickness of the two bricks together matches the thickness of the single brick for other seasons. A leaf of an *avaka* plant (*ŚB*.VIII.3.2.6; Eggeling 1885b) is placed between these bricks in the third layer to represent water/rain; the Brāhmaṇa stresses the point that it rains before autumn and never after autumn (the autumn bricks are placed on top of the leaf). The two bricks of the fourth layer represent Sahas and Sahasya, the two months of *hemanta* (winter). The two bricks of the fifth layer represent Tapas and Tapasya, the two months of *śiśiraś* (cool/dewy). Just as the third layer is the middle of the altar, the rainy and autumn seasons are the middle of the year as represented by this altar. This is not in agreement with *Gavām ayana*; in this *sattra* the middle of the year is at summer solstice, the beginning of the rainy season in South Asia. Whereas the *sattra* of *Gavām ayana* follows the course of the sun, the layers of *Āhavanīya* follow the course of the seasons with *vasanta* (spring) as the first season (Section 2.5.1). It is possible that this is an indication of the epoch of development (or origin) of the *Agnicayana* ceremony in the Āryan sacrificial system.

2.8.2. *Sāvitra Cayana*

A person seeking to attain 'the heavenly world' had to perform *Sāvitra Cayana* (building of the fire altar in the form of the sun). A detailed exposition of this ceremony is given in *Taittirīya Brāhmaṇa* (*TB*.3.10.1; Dumont 1951). The sacrificer needs

- 185 'bricks' of gold or pieces of gold the size of the last bone in the finger (or toe) or an equal number of small pebbles anointed with *ghī* (clarified butter).
- Four naturally perforated bricks.
- An unlimited number of space-filler bricks, these are packed-earth bricks.

The ceremony was performed for the sacrificer by the Adhvaryu who drew, on the sacrificial ground, a circle the size of a chariot wheel. Inside this circle were drawn further nine concentric circles. He then piled the altar with the 185 pieces of gold or pebbles by placing them (with suitable incantations) as follows:

- On the ninth (outermost) circle, 15 pieces that represent the 15 days of the *śukla pakṣa* (waxing/bright fortnight) of the month (the names of these days are given in Column 2, Table III.3). In between, these pieces were placed 15 bricks that represent the 15 *muhūrtas* (the Vedic hour of 48 minutes) of each day of the *śukla pakṣa*.
- On the eighth circle were placed 15 pieces that represented the 15 nights of the *śukla pakṣa*. The names of these nights follow the phases of the moon (Column 4, Table III.3.) as it just becomes visible and waxes to full moon. Between these pieces were placed 15 bricks that represented the 15 *muhūrta*s of each night of the *śukla pakṣa*.
- On the seventh circle were placed 15 pieces that represented the 15 days of the *kṛṣṇa pakṣa* (waning/dark fortnight). The names of these days (*TB*.3.10.1; Column 3, Table III.3) evoke splendour and brightness. Between these pieces were placed 15 bricks that represented the 15 *muhūrta*s of each day of *kṛṣṇa pakṣa*. The names of these *muhūrta*s (*TB*.3.10.1) evoke heat and beauty.
- On the sixth circle were placed 15 pieces that represented the 15 nights of *kṛṣṇa pakṣa*. These names (Column 5, Table III.3) have the theme of pressing, drinking, and satisfaction and clearly imply drinking of *soma* and the consequent emptying of the drinking vessel. Between these pieces were placed 15 pieces that represented the 15 *muhūrta*s of each night of *kṛṣṇa pakṣa*. These names suggest calm, satisfaction, and power obviously achieved by drinking *soma*.
- On the fifth circle were placed 12 pieces that represented the 12 *śukla pakṣa*s of the year. The names of these fortnights suggest purification, glory, and life.
- On the fourth circle were placed 12 pieces that represented the 12 *kṛṣṇa pakṣa*s of the year. The theme of these 12 names is power, victory, and riches.
- On the third circle were placed 13 pieces that represented the 13 months of the year. The first 12 names vaguely suggest the seasons of the year. The thirteenth month is *mahasvant*, the powerful one. On the same circle, the Adhvaryu placed eight different kinds of sands (with eight incantations). The names of these sands have no

well defined theme and it is not clear how these fit into the scheme of a calendar that has been followed in this altar.

- On the second circle were placed 15 pieces that represented 15 *kṣudra-muhūrta*s (small hours) of each *muhūrta*. The theme of the names of these bricks is short time and speed.
- On the first circle were placed nine pieces; of these, six represented the six *kratu*s (great sacrifices) namely *Agniṣṭoma* (euology to *agni*), *Ukthya* (a liturgical ceremony forming part of the *Jyotiṣṭoma*), *atirātra* (an optional part of *Jyotiṣṭoma* performed over-night), *dvirātra, trirātra, caturātra*. These three names mean rituals lasting two, three, and four nights respectively and were probably optional part of *Jyotiṣṭoma*. The three other pieces on this circle bear the mystical names of three seasons, namely, *agni* (fire), *sūrya* (sun), and *candramā* (moon).
- On the centre of the circle, the Adhvaryu placed four pieces with names of the year (Column 1, Table III.6). Considering that included in this altar are all the components of a calendar it is worth speculating that these four bricks represent the *yuga*. As noted above (Section 2.6), this is the only Vedic text in which a 4-year *yuga* may be implied (but see below).

After putting-down these bricks, the Adhvaryu placed the four perforated bricks (on the circumference of the outermost circle?), at four cardinal directions (east, south, west, and north). He then recited a litany in praise of the sun as the giver of light and heat. This is followed by formulae that include the five years of the *yuga* (Column 2 of Table III.6), seasons (Column 2 of Table II.5), and fortnights.

2.8.3. *Nāciketa Cayana*

The building of the *nāciketa cayana* resembles that of the *sāvitra cayaya* and most of the rites are similar. However, this altar has only one circle. To construct the altar, 21 bricks (gold or pebbles anointed with *ghī*) are placed at the centre to form a square or circle. The bricks are placed with 21 formulae that include the sky (supports the sun), sun (supports the moon), moon (support the *nakṣatra* or asterism), *nakṣatra* (depend on the moon and support *saṃvatsara*/year), the year (supports the seasons), the seasons (support the months), the months (support the fortnights) and the fortnights (support the days and nights). Unlike *sāvitra cayaya*, in this altar one brick represents an entire group of calendric features, e.g. months or seasons. Unlike the other fire altars, in the building of this altar the *nakṣatra*s are recognized as features of a calendar.

2.8.4. *Cāturhotra Cayana*

There are two forms of *Cāturhotra Cayana*, the *vyasta* (separated, simple) and *samasta* (united, combined). The *vyasta* has no calendric features in construction or performance and will not be considered here. The *sāvitra* and *nāciketa* altars are single layered alters but the *samastra* form of *Cāturhotra Cayana* has five layers:

- The first layer is the *sāvitra* altar.
- The second layer is made of space-filler bricks.
- The third layer is *nāciketa* altar.
- The fourth layer is again made of space-filler bricks.
- The fifth layer is the *vyasta* form of *cāturhotra* altar.

The calendric features of this altar are, of course, included in the *Sāvitra* and *Nāciketa* layers. The sacrificer intending to pile the *samasta* on the occasion of a *Soma* sacrifice had to first perform the *divaḥ-śyenī*s (the heaven's eagles/falcons) *iṣṭi*s (oblations). It is tempting to conclude that the form and piling of the *samasta* form of *Cāturhotra Cayana* may have been the precursor to the elaborate *uttaravedī* of *Agnicayana* ceremony.

If the piling of *mahā-cayaya* was borrowed by the Āryas from the (Late) Harappan people, as has been suggested and if the Āryas had also borrowed the symbolism that is built into this altar, than there is here a glimpse of the calendar of the Indus Valley Civilization. The calendric features built into this altar can be traced back to the *Ṛgveda Saṃhitā* and if the Āryas had indeed borrowed their calendar from the Harappans then this must have occurred at a very early stage in the history of the Āryas. The only feature of the Vedic calendar not incorporated in these altars (except in the *Nāciketa Cayana*) but found in the post-Ṛgvedic Saṃhitās and the Brāhmaṇas and *Vedāṅga*

Jyotiṣa (Chapter V) are the *nakṣatras* and these will be discussed in the next chapter.

NOTES

1. A late Palaeolithic reindeer bone found in 1911 at Abri Blanchard, Sergeac, Dordogne, France, appears to record the phases of the moon (Marshack 1972). This artefact was inscribed about 30,000 years ago.
2. A tally stick is an ancient device to record and document numbers, quantities and even messages. Tally sticks first appeared in late Palaeolithic as notches carved on animal bones. (Perhaps) the earliest tally/calendar stick is a baboon bone found in Swaziland (Lebombo bone). This bone, about 37,000 years old, is inscribed with 29 distinct notches and may have served as a lunar calendar. This calendar stick is similar to the calendar sticks still in use by the indigenous people of Namibia (Williams 2005). Calendar sticks can be of varying complexity. A 'calendar stick' from Sumatra is a square piece of wood with 30 holes around its edges and a count of the days was kept by threading a string through the holes (Richards 1998). A more elaborate calendar stick is that carved by the chief of Winnebago (indigenous North Americans). The stick of square cross-section has notches on each side representing the days in six months; each month contains 30 or 29 days. The phases of the moon are marked for each month (Richards 1998).
3. The average interval between two successive new moons is the synodic period of the moon, and a single cycle is sometimes called a lunation (month). The actual length of lunation can vary from 29.25 days to 29.82 days because of tidal interaction between the moon, the earth and the sun (and to a lesser degree, with other planets).
4. *Ahorātra* occurs only once in *ṚV*.X.190.2. This word has replaced *ājarasāya* (Oldenberg 2005).
5. It is not clear, if in this verse, the *nakṣatra*s refer to the day or the month. Months named after the *nakṣatra*s or referred to by the presiding *nakṣatra*s was a common practice in the Vedic texts (Chapter IV, Section 7.). Superficially, at least, this verse is about a wedding and the departure of a bride. This is more likely to take place on consecutive days rather then months and the *nakṣatra*s most likely identify the days.
6. A number of early cultures observed the annual apparent motion of the sun and noted the four cardinal days, namely the vernal and autumnal equinox and summer and winter solstice. For example, the megalithic passage tomb at Newgrange (Ireland) was built around 3200 BCE. The 19-m inner passage leads to a cross-shaped chamber. At sunrise on the winter solstice, a shaft of sunlight shines through the 'roof-box' over the entrance and penetrates the passage to illuminate the chamber. This dramatic event lasts for about 17 minutes and is repeated for few days either side of the winter solstice. The

Nebra sky disc was discovered/found in 2002 at Mittleberg, about 60 km west of Leipzig, Germany. The 30-cm diameter bronze disc has been (associatively) dated to 1600 BCE. Inlaid in gold are symbols for the sun or the full moon, a lunar crescent and stars including the Pleiades. Two arcs at opposite edges of the disc enclose an angle of 82°; this is the angle between the positions of sunset at summer and winter solstice at the latitude of Mittleberg (51° N). Whether the disc was for a calendric or a shamanistic purpose can be debated, but it does seem to be an accurate record of the apparent annual motion of the sun. These two examples show that detailed observations of the apparent annual motion of the sun were not restricted to the cultures of Mesopotamia. The Āryas have never been credited with the knowledge of equinox(es) and solstice(s). In spite of unambiguous textual evidence (*KB* xix.3) Thibaut (1899) maintained that 'Anything like a fairly accurate fixation of the sun's place among the stars at the winter solstice, cannot be imagined to have been accomplished by the people who had no approximately correct notion of the length of the year'. Thibaut could be forgiven; having incurred the wrath of the Western academic establishment for proposing a pre-Greek date for the *Śulbasūtras* (N.I.4) he may have decided to play it safe and deny Indians the knowledge of the apparent annual motion of the sun.

7. Two calendars, one for religious purposes and another for secular use (agriculture, taxation, etc.), are not unusual. The ancient Egyptians had a solar calendar side by side with a religious lunar calendar. The Jewish (Hasmonean) Kingdom of around the second century BCE considered Nisan (March-April) as the 'first month' for liturgical purposes and Tishri (September-October) as the 'first month' for civil/political events. The (Christian) Europeans and Americans use the Gregorian calendar for secular purposes, and a lunar ecclesiastical calendar to determine the date of Easter. Similarly, in India, the Gregorian calendar is used for civil purposes but various other calendars are used to fix religious festivals.
8. *Saṃvatsara* may not always have meant a year of 360 or 365.24 days. In the Vedic texts, from Ṛgvedic Saṃhitā to the Brāhmaṇas, the period of gestation of a woman (or a cow or a mare) is stated to be ten months, but, in some texts, this period is also called *saṃvatsara*. For example, *ŚB*.XI.1.6.2 (Eggeling 1900); 'a woman or a cow or a mare gives birth in a *saṃvatsara*' and *ŚB*.IV.5.2.4 (Eggeling 1885a); 'It is ten months when the embryo becomes fully grown', similarly, *KS*.28.6; 'In ten months embryos come to birth'. It is possible that at some time in the Vedic period *saṃvatsara* meant 10 (lunar) months. In his analysis of the 'Frog Hymn' *ṚV*.VII.103, Vajracharya (1997) has suggested that a Vedic year may have consisted of two parts, a 10-month *saṃvatsara* followed by two months of rains. This seems extremely unlikely in the context South Asian climate. At present, in South Asia it rains for three months in most places and four in others. However, the Āryas did indeed arbitrarily allocate two months for the rainy season and to all other seasons.

9. A number of early cultures had an 'ideal' year of 360 days (12 month of 30 days each). For example, the early Egyptian calendar had 12 months of 30 days each that is 360 days. At the end of this 'ideal' year, five days were added to bring it in line with the tropical year. This calendar was adopted by both the Coptic and Ethiopian cultures and the five (or six) intercalated days were called the 'thirteenth month'. The Sumero-Babylonian calendar was based entirely on the recurrence of lunar phases and the earliest calendar had a year of 360 days. A month was intercalated, initially at random but later by an imperial fiat. The Avesta (Iranian) calendar had twelve months of 30 days each. That is, an Avesta year was 360 days long, to this were added five Gāthā days, each sacred to the five great divisions of the Gāthā. To allow for the quarter day lost each year, a month was intercalated every 120 years (Gray 1908-21). The Old Persian calendar was divided into twelve months of 30 days each but the year was 365 days long. Nothing is known of the method of intercalation employed in this calendar (Gray 1908-21). After the fall of the Sasanian dynasty, the Persian calendar year was 360 days long, a month was intercalated every sixth year and two months were intercalated after 120 years. Similarly, the earliest Inca (in the Andes) calendar had a year of 360 days. The Maya and Aztec civil year was 360 days long (but divided into 18 months of 20 days each) to which five days were added each year (Aveni 2001). This period of five days was called *Uayeb* and was considered unlucky. The 'ideal' year of 360 days plus five days (to synchronize the synodic year and the tropical year), of these cultures, widely separated in space and time, is similar to that suggested by *ṚV*.I.25.8 (A.III.32). The reason why these ancient cultures had an 'ideal' year of 12 month of 30 days each or 360 days is similar to the reason why the Āryas had a ritual year of this length. These cultures had rounded-up a mean synodic month of 29.53 days to 30 days. A lunation can be 29 or 30 days; by choosing a synodic/lunar month of 30 days, the synodic/lunar year can be very similar in length to the tropical year. This was the first attempt to synchronize the synodic year and the seasons.
10. After 19 tropical years, the moon's phases repeat at approximately the same calendar date or after a period of 19 years, the moon will have the same phase at exactly the same position among the stars. This is the Metonic cycle. This cycle was known to the Chinese and from about 484 BCE was in their (*sìfēn* – quarter remainder) calendar to synchronize the synodic years with the seasons. The Babylonians (and later the Maya) had also discovered this cycle and the Greeks adopted it around 432 BCE. The synchronization of the synodic calendar with seasons by the 19 tropical year 'Metonic cycle' is considered one of the high achievements of Greek calendric science. This conclusion seems to be based on the accurate value of the length of the tropical year that can be obtained from this form of intercalation. The Metonic cycle of 19 tropical years is almost 6940 days long, or 1 tropical

year is 6940 ÷ 19 = 365.26 days. Nineteenth- and twentieth-century Western scholars derived some satisfaction from this example of the pre-eminence of ancient Greek and by inference European science. The length of the tropical year may have been an intellectual curiosity for ancient Greek astronomers. However, the importance of the length of the tropical year to the ancient Greek citizens is debatable. In Athens, the Metonic scheme was never used in the civil or ritual calendar. The Greek politicians and priests clung to their calendar-making power, ignored the astronomers and administered the calendar on an *ad hoc* basis. To the Vedic calendar-makers or ritualists the significant period was 62 synodic (lunar) months or almost 1,830 days (Chapter III, Section 2.7, Chapter V, Section 2.5) in which they were able to approximately synchronize the lunar calendar with seasons. The error of a day or so in the seasons was of no importance to them. An error in the day of the prescribed phase of the moon (usually full or new moon) was corrected by observations. The calendar makers were fully aware that the tropical year was between 364 and 366 days long (*TS*.VII.1.10; Chapter III, Section 2.7). However, the actual length of the tropical year was of no interest to them, just as it was of no interest to the Athenian politician and priests. Not surprisingly, the length of the tropical year is introduced rather incidentally in a late Vedic text (Chapter V, Section 2.6). This should not be considered unusual. The Maya were accomplished astronomers and were concerned with fixing the position of the sun on the horizon on particular dates. They could have determined the length of the tropical/seasonal year with considerable accuracy from these positions. However, they failed to record the 365.24 day tropical year and continued to use a 365-day year (Aveni 2001).

An arithmetic rule is not the only form of intercalation to synchronize the synodic and tropical years. A number of cultures developed intercalation schemes based on astronomical observations similar to the scheme described in Section 2.7. For example, the people of Bali bring the lunar year in harmony with the seasons by prolonging one of their months until the Pleiades become visible at sunset.

IV

Nakṣatras

Five different stellar reference frames were independently developed by early civilizations: the 'decans' in Egypt, the 'zodiac' in Mesopotamia, the *nakṣatra*s (sometimes called the lunar mansions) in South Asia, the *Xiù* or *Hsiu* (also lunar mansions) in China and the Mayan 'zodiac' in the Yucatan (South America). Only the Mesopotamian zodiac and the Chinese *Xiù* have survived to the present day. However, the *nakṣatra*s, i.e. the 27 (sometimes 28) stars or asterisms along the sidereal path of the moon (or 27 places occupied by the moon during one sidereal rotation) had an important role in the Vedic calendar. The surviving Vedic texts tell us noting about the interest Āryas might have had in the study of stars and the preparation of star catalogues (like those found in the Mesopotamian or the Chinese texts). The Vedic texts only reveal the interest they had in the study of the position of the sun and moon to develop a calendar essential for the proper performance of their rituals. Thus, nothing is known about the interest they might have had in planets. We get only a limited view of their interest in stellar astronomy, namely that of stars and constellations close (and sometimes not so close!) to the ecliptic. The aim appears to have been to develop a (fixed) frame of reference to identify (and later) determine the position of the sun and the moon on the ecliptic. The Āryan calendar makers called their stellar frame of reference *nakṣatra*s. The etymological origins of the word *nakṣatra* are somewhat uncertain; Yāska (*Nirukta* III.20; Sarup 1920-7) derives it from *nakṣ* 'to go' but in the Brāhamaṇas (e.g. *ŚB*.II.1.2.18; Eggeling 1882) it is derived from *kṣatra*, 'power or domain'. The *nakṣatra*s appear to be unique to South Asia as, so far, there is no evidence of similar division of the sky in any other Indo-European source including the *Avesta*, the sacred book of the Zoroastrians. *Vedāṅga Jyotiṣa* (Chapter V) implicitly assumes that *nakṣatra*s are wide sectors of the ecliptic, and the moon moves through these sectors at uniform speed. In this book, the wide sectors will be called *nakṣatra*-sectors and the stars and asterisms will be called *nakṣatra*s. This is to avoid the confusion created by the Vedic texts by using the same word (*nakṣatra*) to refer to both the wide sectors and stars. In the nineteenth century Weber (1861), Whitney (1874), Thibaut (1877) and Dīkṣita (1896) carried out pioneering studies of the Vedic *nakṣatra*s.

The origins of *nakṣatra*s have not yet been established. Considerable effort was made, in the nineteenth century, to trace the origin to the Greek, Chinese, Mesopotamian and even Arab sources. These efforts continued, with slightly diminished vigour, in the twentieth century (N.IV.1). In the recent past, attempts have been made to interpret some of the pictographic symbols of the Indus script in terms of celestial bodies (Parpola et al. 1970; Ashfaque 1977; Parpola 2005). It is assumed that various figures on the Indus seals identify divisions of the sky by asterisms along the ecliptic and that these were made for calendric purpose. This has suggested a possible link between the Vedic *nakṣatra*s and the pictorial stamp seals of the Indus Civilization. The discussion about this possible link has centred on the Indus Valley seal sign of the 'fish' (shown in the box). This sign has been associated with the Dravidian word *mīn* for fish and its homophone for star. It is proposed that this fish sign represents an asterism and the qualifying vertical strokes that appear with the sign represent the number of stars in the asterism (but different interpretations of these signs are also possible, Mahadevan 1970). Thus, a fish sign with three strokes (three stars, *mu-m-mīn* in Old Tamil) would signify Vedic *nakṣatra* of Mṛgaśīrṣa ('belt' of Orion), the one with six strokes would represent (six stars, *aru-mīn* in Old Tamil) Kṛttikās (Pleiades) and the one with seven strokes would represent (seven stars, *elu-mīn* in Old Tamil) *sapta ṛṣis* (Ursa Major). These studies make further associations of various Vedic deities and planets with the Indus seal signs. This perceived association between celestial objects and the seal symbols rests on the presumed link between the language of the Indus Civilization and the present Dravidian languages. There is increasing confidence in the possible presence of people of the Dravidian family of languages in the Indus Civilization (Mahadevan 1970) and this link does not seem so far-fetched. In view of the very likely close contact between the Indus Civilization and the Āryas (Chapter I), efforts to find the origin of the Vedic *nakṣatra*s (and possibly the Vedic calendar) in the Indus Valley seem more plausible than the rather sterile attempts (over last 150 years) to find this origin in Mesopotamia.

1. *NAKṢATRA*S IN THE VEDIC TEXTS

In the Vedic texts, *nakṣatra*s indicate either the 27 (equal) divisions of the ecliptic (usually referred to as lunar mansions) or 27 (sometimes 28) stars and star groups (asterisms) with which the moon is supposed to conjoin each night. The texts do not distinguish between the two, and the real meaning has to be inferred from context. It is possible that the *nakṣatra*s as lunar mansions were a late development. The Vedic texts consider *nakṣatra*s as 'so many suns yonder' whose power is taken away by the rising sun. In the *Ṛgveda*, the word *nakṣatra* occurs about nine times and can mean both the sun (A.IV.1, 2) and asterism (A.IV.3) but examples of *nakṣatra*s as lunar mansions are absent. In this text, a star is sometimes also referred to as *tāra* (A.IV.4), and this is also true of post-Ṛgvedic texts (e.g. *TS*.VII.5.25).

In the *Ṛgveda Saṃhitā*, a number of *nakṣatra*s referred to in the post-Ṛgvedic texts are mentioned and these designate stars. A group of stars is referred to as *ṛkṣa* (A.IV.5); this asterism cannot be identified from *Ṛgveda* but that it is at a high place (*uccā naktaṃ*, high in the firmament/night) suggests that it is a northern asterisms. In *ŚB*.II.1.2.4 (Eggeling 1882), it is noted that the (constellation of) *sapta ṛṣis* was earlier known as *ṛkṣaḥ* and this suggests that the asterism referred to in the *Ṛgveda* is also that of *sapta ṛṣi*s, which is indeed a northern asterism. It is interesting to note that *ṛkṣa* can also mean a bear (*VS*. XXIV.36; Griffith 1987) and in Europe, the constellation of *sapta ṛṣi*s is known as the Big Bear (Ursa Major). It should, however, be stressed that the constellation of *sapta ṛṣi*s was never called 'Bear' in South Asia. In the *Ṛgveda*, there are further oblique references to the constellation of *sapta ṛṣi*s, for example, there are references (A.IV.6,

7) to a cask or barrel (*kavandha* or *kabandha, kośa* and *avata*) that is turned over by various gods to pour riches on the earth. In these verses, the words for the cask or the barrel can be interpreted as clouds that pour rain on the earth. However, Witzel (1995) has suggested that *Atharvaveda Saṃhitā* (A.IV.8) identifies the cask as the constellation of *sapta ṛṣis*. In the night sky this constellation does indeed look like a spoon or a pot with a handle, and during the night (and the year) it 'turns over to empty its contents'. If this interpretation is correct, then all stars of *sapta ṛṣis* must have been circumpolar (N.IV.2) when these passages were composed; otherwise, the constellation will not appear like a pot or cask at all times in a year (see Section 7).

It is likely that a number of words that were in common use in the Ṛgvedic period were later used to identify the *nakṣatras* (lists of *nakṣatras* in the Vedic texts are given in Table IV.1). For example, in *ṚV*.IV.51.2 (A.IV.9) *citra* means bright/brilliant and qualifies Dawn. In *ṚV*.X.19.1 (A.IV.10) *revatī* means wealth and *punarvasū* means (two) restorers of wealth, Agni and Soma. Similarly, in *ṚV*.V.51.14 and *ṚV*.IX.72.9 *revatī* is a wealthy female or wealth and in *ṚV*.VIII.93.13 *rohiṇī* means 'red' and refers to a cow. These names (words) identify *nakṣatras* in the post-Ṛgvedic texts. Some names (words) that identify stars (asterisms) in Ṛgvedic verses have been carried over into post-Ṛgvedic texts, for example, the asterism Tiṣya in *Ṛgveda* (A.IV.11) can be identified with *nakṣatra* Puṣya in post-Ṛgvedic texts. Two *nakṣatras* — Aghā and Arjunī — are mentioned in *ṚV*.X.85.13 (A.III.12). This verse occurs in the 'bridal hymn' where the Sun (Savitṛ) gives his daughter Suryā to Soma (Moon) in marriage. This verse is repeated in *AV*.XIV.1.13, but the *nakṣatras* are now Maghās and Phālgunīs (A.III.12). It is not clear if in the *Ṛgveda* and *Atharvaveda* the *nakṣatras* signify asterisms or lunar mansions but in the later Vedic texts Maghās and Phālgunīs denote asterisms whereas in *Vedāṅga Jyotiṣa* (Chapter V) Maghās and Phālgunīs identify lunar mansions. The names of these *nakṣatras* have clearly evolved in the period between the *Ṛgveda* and *Atharvaveda Saṃhitā* and it is possible that the conceptual nature of *nakṣatras* had changed in this period. In these verses, the *nakṣatras* identify the days, that is, a day is identified by the *nakṣatra* in the proximity of the moon on that day. This practice was continued in the post-Ṛgvedic times. Further references to stars and asterisms in the *Ṛgveda* can be

Table IV.1: List of *nakṣatra*s in the Saṃhitās, Brāhmaṇas, Sūtras and *Vedāṅga Jyotiṣa*. Their *davatā* (presiding deity), gender, number and meaning are also given.

	AV. XIX.7.2-5	*MS.* II.13.20	*KS.* 39.13	*TS.* IV.4.10	*TB.* 1.5
1	Kṛttikās	Kṛttikās	Kṛttikās	Kṛttikās	Kṛttikās
2	Rohiṇī	Rohiṇī	Rohiṇī	Rohiṇī	Rohiṇī
3	Mṛgaśirṣa	Invaka	Invaka	Mṛgaśirṣa	Invaka
4	Ārdrā	Bāhū	Bāhū	Ārdrā	Bāhū
5	Punarvasus	Punarvasus	Punarvasus	Punarvasus	Punarvasus
6	Puṣya	Tiṣya	Tiṣya	Tiṣya	Tiṣya
7	Āśreṣās	Āśreṣās	Āśreṣās	Āśreṣās	Āśreṣās
8	Maghās	Maghās	Maghās	Vicṛtau	Maghās
9	Phālgunīs	Phālgunīs	Phālgunīs	Phālgunīs	Pūrva Phālgunīs
10	Phālgunīs	Phālgunīs	Uttara-Phālgunīs	Phālgunīs	Uttara-Phālgunīs
11	Hasta	Hasta	Hasta	Hasta	Hasta
12	Citrā	Citrā	Citrā	Citrā	Citrā
13	Svātī	Niṣṭyā	Niṣṭyā	Svātī	Niṣṭyā
14	Viśākhās	Viśākhās	Viśākhās	Viśākhās	Viśākhās
15	Anūrādhā	Anūrādhā	Anūrādhā	Anūrādhā	Anūrādhā
16	Jyeṣṭhā	Jyeṣṭhā	Jyeṣṭhā	Rohiṇī	Rohiṇī
17	Mūla	Mūla	Mūla	Vicṛtau	Mūla-Barhinī
18	Aṣāḍhās	Aṣāḍhās	Aṣāḍhās	Aṣāḍhās	Pūrva-Aṣāḍhās
19	Aṣāḍhās	Aṣāḍhās	Uttara-Aṣāḍhās	Aṣāḍhās	Uttara-Aṣāḍhās
	Abhijit	Abhijit			Abhijit
20	Śravaṇa	Śroṇā	Aśvattha	Śroṇā	Śroṇā
21	Śraviṣṭhās	Śraviṣṭhās	Śraviṣṭhās	Śraviṣṭhās	Śraviṣṭhās
22	Śatabhiṣaj	Śatabhiṣaj	Śatabhiṣaj	Śatabhiṣaj	Śatabhiṣaj
23	Proṣṭhapadas	Proṣṭhapadas	Proṣṭhapadas	Proṣṭhapadas	Proṣṭhapadas
24	Proṣṭhapadas	Proṣṭhapadas	Uttara-Proṣṭhapadas	Proṣṭhapadas	Proṣṭhapadas
25	Revatī	Revatī	Revatī	Revatī	Revatī
26	Aśvayujau	Aśvayujau	Aśvayujau	Aśvayujau	Aśvayujau
27	Apabharaṇīs	Bharaṇīs	Bharaṇīs	Apabharaṇīs	Apabharaṇīs

Table IV.1. *(contd.)*

	TB. 3.1.4-5	*SGS* I.26	*VJ* *RJ*.25-28	Gender	Num	*Davatā*	Meaning
1	Kṛttikās	Kṛttikās	Kṛttikās	F	P	Agni	Cutters or shearers
2	Rohiṇī	Rohiṇī	Rohiṇī	F	S	Prajāpati	(Prajāpati's daughter) Red one
3	Mṛgaśirṣa	Mṛgaśirṣa	Mṛgaśirṣa	N	S	Soma	Head of a deer
4	Ārdrā	Ārdrā	Ārdrā	F	S	Rudra	Moist one
5	Punarvasus	Punarvasus	Punarvasus	M	D	Aditi	Restoring goods
6	Tiṣya	Puṣya	Puṣya	M	S	Bṛhaspati	Nourishing
7	Āśreṣās	Āśreṣās	Āśreṣās	F	P	Sarpāḥ	Those who embrace
8	Maghās	Maghās	Maghās	F	P	Pitaraḥ	The great one(s)
9	Phālgunīs	Phālgunīs	Phālgunīs	F	D	Aryaman	Red one(s)
10	Phālgunīs	Phālgunīs	Phālgunīs	F	D	Bhaga	Red one(s)
11	Hasta	Hasta	Hasta	M	S	Savitar	The hand
12	Citrā	Citrā	Citrā	F	S	Indra	Bright one
13	Niṣṭyā	Svātī	Svātī	F	S	Vāyu	Sword or independence
14	Viśākhās	Viśākhās	Viśākhās	F	D	Indrāgnī	Fork shaped?
15	Anūrādhā	Anūrādhā	Anūrādhā	F	S/P	Mitra	Causing welfare
16	Rohiṇī	Jyeṣṭhā	Jyeṣṭhā	F	S	Indra	The eldest
17	Vicṛtau	Mūla	Mūla	M	S	Nirṛti	The root
18	Aṣāḍhās	Pūrva-Aṣāḍhās	Pūrva-Aṣāḍhās	F	P	Āpaḥ	Invincible(s)
19	Aṣāḍhās	Uttara-Aṣāḍhās	Uttra-Aṣāḍhās	F	P	Viśve-devāḥ	Invincible(s)
	Abhijit	Abhijit					conquering
20	Śroṇā	Śravaṇa	Śravaṇa	F	S	Viṣṇu	Ear or hearing
21	Śraviṣṭhās	Dhaniṣṭhās	Śraviṣṭhās	F	P	Vasavaḥ	Most famous
22	Śatabhiṣaj	Śatabhiṣaj	Śatabhiṣaj	M	S	Indra	Hundred healer

23 Proṣṭha-padas	Proṣṭha-padas	Pūrva-Proṣṭhapadas	F	P	Aja Ekapād	Feet of a stool
24 Proṣṭha-padas	Proṣṭha-padas	Uttara-Proṣṭhapadas	F	P	Ahirbudh-nya	Feet of a stool
25 Revatī	Revatī	Revatī	F	S	Pūṣan	Wealthy
26 Aśvayujau	Aśvins	Aśvayujau	M	D	Aśvins	Horsemen
27 Bharaṇīs	Bharaṇīs	Bharaṇīs	F	P	Yama	Bearer(s) of new life

Gender: F – Female; M – Male; Num: S – Singular; D – Dual; P – Plural

inferred from the myths recounted in this text. For example, a myth is alluded to in the relatively obscure *ṛcā*s *ṚV*.X.61.5-7 (*ṚV*.X.61.7 is partially translated in A.IV.12). There is no direct reference to stars and asterisms in these *ṛcā*s but these can be deduced from the explanation of the myth given in *Aitareya* and *Śatapatha Brāhmaṇa* (*AB*.iii.33(xiii.9), Keith 1998; *ŚB*.II.2.6-10; Eggeling 1882). Briefly, Prajāpati desired sexual relations with his daughter (either the sky or the dawn). To escape this unwanted attention, the daughter assumed the form of a (red) doe (*rohit*). Prajāpati transformed himself into a roebuck. The gods were appalled at this incestuous desire of Prajāpati and to punish him, they fashioned a god out of their most fearsome forms, called Bhūtavat (or Rudra). This god shot his 'three-knotted arrow' and took off the head of Prajāpati, which became the asterism of *mṛga* (or Mṛgaśīrṣa, i.e. the head of a deer). The daughter became the asterism Rohiṇī, the 'three-knotted arrow' is the three stars of the 'belt' of the constellation of Orion, the assassin (*mṛgavyādha*, i.e. the slayer of the deer) became the star Sirius, and Prajāpati is the constellation of Orion.

There are similar legends and myths associated with other *nakṣatra*s. The Kṛttikās are the seven wives of *sapta ṛṣis* and are called Ambā, Dulā, Nitatnī, Abhrayantī, Meghayantī, Varṣayantī, and Cupuṇīkā (*TS*.IV.4.5.1; *MS*.II.8.13; *KS*.40.4; *TB*.3.1.4). There was an unfortunate adulterous union between Agni and six of the seven Kṛttikās, that is *agni* was 'in' the Kṛttikās. The result was the birth of a son called Skanda who had six heads. The stellar equivalents of *nakṣatra*s cannot be identified from the Vedic texts, this is discussed later. An exception may be the Kṛttikās; the mythology, the number of stars, the position in the list of *nakṣatra*s and the similarity with other cultures (Frazer 1933; Aveni 2001) suggests an unambiguous identification of the Kṛttikās with the Pleiades. In the following text, the Kṛttikās will always be identified with the Pleiades. The myth above thus has obvious calendric interpretation; when the sun (Agni) is in conjunction with the Kṛttikās/Pleiades, the year (Skanda) with six seasons (six heads) is born (O'Flaherty 1975). In the northern hemisphere, this conjunction of the sun with the Kṛttikās/Pleiades would occur in spring (around the vernal equinox). Further support for this calendric interpretation is provided by the names of three of the six Kṛttikās, these are Abhrayantī, Meghayantī, Varṣayantī and there is an implication of rain in these three names. The interpretation of this myth is in agreement with one of the two possible times for the start of the Vedic year discussed in Chapter III. The unfortunate situation in which the six Kṛttikās found themselves was avoided by the seventh, Abhrayantī (or Arundhatī) wife of Vasiṣṭha, who was of particularly good character. For her good fortune or moral sense, she was rewarded by being placed in the sky as the asterism Arundhatī (Alcor a 4th magnitude star in the constellation of *sapta ṛṣis* or Ursa Major).

Not many *nakṣatra*s listed in the later Saṃhitās and Brāhmaṇas can be identified in the *Ṛgveda Saṃhitā*. However, statements like *soma* is stationed in the 'lap' of *nakṣatra*s (A.IV.13) does suggest that the position of the moon defined with reference to the *nakṣatra*s was important to the Āryas of the Ṛgvedic period. The *nakṣatra*s in the lists of post-Ṛgvedic texts are often identified by their presiding deities. Narahari Achar (2000a) has suggested that to identify the *nakṣatra*s in the Ṛgvedic period, it would be more appropriate to attempt to identify the *devatā*s of the *nakṣatra*s in *Ṛgveda Saṃhitā* rather than the *nakṣatra*s themselves. In the 15 *ṛcā*s of *sūkta Ṛ*V.V.51, 17 *devatā*s of *nakṣatra*s are mentioned explicitly, and Narahari Achar has suggested that the remaining ten are (probably) included in the *devatā sarvegaṇaḥ*. It is always difficult to determine the exact import of Ṛgvedic verses but Narahari Achar is of the opinion that, in this *sūkta*, the prayers are addressed to the *devatā* of the *nakṣatra*s because the last *ṛc* of this *sūkta* refers to 'the path of the sun and the moon' that is, to the ecliptic. Further study is required along this and other lines, to trace the origin and evolution of the Vedic *nakṣatra*s.

2. *NAKṢATRA* – STARS AND ASTERISMS

A complete list of 27 or 28 *nakṣatras* is given only in the post-Ṛgvedic Saṃhitās, Brāhmaṇas and Sūtras. These appear as complete lists in these texts, without a preamble or any trace of evolution. These lists along with the list of *nakṣatras* in *Vedāṅga Jyotiṣa* (Chapter V) are reproduced in Table IV.1. The Vedic texts in this table are given in the chronological order of Table I.5. In some lists in these texts both the *nakṣatras* and their *davatās* are given, in others only the *davatās* are given. These lists are broadly similar but differ in detail. The number of *nakṣatras* in each list is different; in *AV*.XIX.7.25, *KS*.39.13, *TS*.IV.4.10 and *VJ* there are 27 *nakṣatras*; in *MS*.II.13.20, *TB*.1.5.1, *TB*.3.1.1-2 and *SGS*.I.26 there are 28. Twenty-seven *nakṣatras* are also mentioned in *VS*.IX.7, *PB*.XXIII.23, *KB*.v.1 and *ŚB*.X.5.4.5 but the names of the *nakṣatras* are not given in these passages. In *AV*.XIX.7.2-5, 27 *nakṣatras* are listed but 28 are mentioned in the following passage (AV.XIX.8.2), the missing *nakṣatra* in this list is Uttara-Phālgunīs. In all other lists of 27 *nakṣatras*, the missing *nakṣatra* is Abhijit. Twenty-eight *nakṣatras* are mentioned in *TB*.3.1.1-2 (Dumont 1954), but the position of Abhijit is anomalous. This can be noted from *TB*.3.1.5 (Dumont 1954); this passage enjoins an *iṣṭi* (a rite) to the *davatā* (and therefore to the *nakṣatra*) on each successive day of a (synodic) month. On the nineteenth day, having offered an *iṣṭi* to *Viśvedevāḥ*, the *davatā* of Uttarāṣāḍhās, a second offering (on the same day) is made to *Brahman* the *davatā* of Abhijit. Unlike other *nakṣatras*, there is no *iṣṭi* for Abhijit on a separate day.

These *iṣṭis* to the *nakṣatras* start on the new moon day and the first is to Agni, the *davatā* of the Kṛttikās. This is followed by 27 *iṣṭis* to the *davatās* of the *nakṣatras* given in Table IV.1. On the fourteenth day, an additional *iṣṭi* is offered to the full moon, and on the twenty-seventh day, an additional *iṣṭi* is offered to the new moon. The 28 *nakṣatras* of *TB*.3.1.1-2 are divided into two groups of 14 *nakṣatras* each; the first group, from Kṛttikās to Viśākhās (Table IV.1), is called *deva-nakṣatras* and the second, from Anūrādhā to Bharaṇīs is called *yama-nakṣatras*. Thus, the *deva-nakṣatras* 'trace' the path of the sun from spring (roughly vernal equinox) to autumn (roughly autumnal equinox) and the *yama-nakṣatras* 'trace' the path of the sun from autumn to spring of the following year. A seasonal calendar is implicit in this division of the *nakṣatras*.

The daily *iṣṭis* for *nakṣatras* and the division of the *nakṣatras* into the two 'seasonal groups' sheds light on the anomaly of the number of *nakṣatras* in the lists given in the Vedic texts. The primary purpose of the *nakṣatras* was to identify the days of a synodic month. The day is identified by the *nakṣatra* that is in conjunction with the moon on that day. Since the moon is visible on 27 days of a synodic month, only 27 *nakṣatras* are required. However, the Āryan theologians liked symmetry and they wanted equal number of *nakṣatras* in *Devāyana* and *Pitrāyana* and therefore they had to have 28 *nakṣatras*. The theologians had divided a year into six seasons to achieve similar symmetry. Although two months for each season, is not compatible with the climate of South Asia. The theologians appear to have attempted to reconcile the requirements of identifying the days of the synodic month and the symmetry they desired by performing the *iṣṭi* for Abhijit on the same day as the *iṣṭi* for Uttarāṣāḍhās. Why they should have selected Abhijit cannot be established as any other *nakṣatra* could have been used. In *Vedāṅga Jyotiṣa*, *nakṣatras* mean sectors of the ecliptic. The location of these sectors is determined from the position of the moon during a *yuga* (a period of 62 synodic months) as described in Section 3 below. The moon (unfortunately) does not take notice of the symmetry of the *nakṣatras* with seasons that the Āryan theologians desired, and only 27 *nakṣatra*-sectors are required to completely describe the positions of the moon in a *yuga*. *Vedāṅga Jyotiṣa*, therefore, has 27 *nakṣatra*-sectors.

The lists of *nakṣatras* in all post-Ṛgvedic texts start with the Kṛttikās and the preferred position of Kṛttikās in the list of *nakṣatras* is echoed in the Vedic texts, e.g. in *TB*.1.1.2.1, 'Kṛttikās are the mouth of the *nakṣatras*'. Although all lists start with Kṛttikās, the names of other *nakṣatras* are not alike in all lists. In some lists, there are two *nakṣatras* of Rohiṇī and in some Jyeṣṭhā replaces the second Rohiṇī. The two Rohiṇīs are 13 *nakṣatras* apart (Table IV.1) that is they are (almost) on the opposite side of the ecliptic. Associated with each *nakṣatra* is a presiding deity but the same *nakṣatra* in two lists sometimes has different deities. In Table IV.1, two *nakṣatras* with the same name are differentiated in some lists by the addition of the prefix *pūrva* and *uttara* to the name of the *nakṣatra*. It has been suggested that the *nakṣatras* (and the Vedic calendar in general) were imported into South Asia from Mesopotamia in the fifth and fourth century BCE (Pingree 1989). One proposed 'evidence' of this import is that the Kṛttikās/Pleiades head the list of stars and constellations, along the path of the moon, in Mesopotamian astronomy texts (N.IV.1) just as they do in the Vedic texts. These proponents do not seem to have realized that the prominent role of the Pleiades is not unique to Mesopotamian astronomy; the Pleiades had a prominent and crucial role in the calendars of most ancient cultures (N.IV.3). The differences in the numbers, names and the *devatās* of the *nakṣatras* in different texts argue against a common and single source for the scheme of *nakṣatras*. These differences also suggest differences of opinion between different Vedic *śākhās* and within the same *śākhā* at perhaps different times.

It is possible that during the Ṛgvedic period, various stars and asterisms had been identified and these or the position of the moon relative to these had specific religious, social, agricultural, or calendric purposes. The post-Ṛgvedic texts mention astronomy (*nakṣatravidyā*, *TB*.3.4.4.1) as a discipline of study and the astronomer (*prajñānāya nakṣatradarśnaṃ*, *VS*.XXX.10) as a profession. The discipline of monitoring and calibrating the position of the moon with reference to a series of background stars may have been an entirely post-Ṛgvedic development and possibly during various iterations and redactions of the post-Ṛgvedic texts, all traces of formulation and evolution of *nakṣatras* were lost. The lists in Table IV.1 are only from the final redactions.

The Vedic texts do not indicate whether the lists of *nakṣatras* refer to sectors of the ecliptic or to stars and asterisms that mark the progress of the moon across the celestial sphere. However, the names in the lists of *nakṣatras* are singular, dual, or plural; this is the only (and perhaps the strongest) indicator that there was a time when the composers of these lists thought of the *nakṣatras* as stars and asterisms and not as sectors of the ecliptic. More often than not, *nakṣatras* have been interpreted as stars or asterisms with which the moon makes 'contact' each night. This is because, it was felt that only stars can provide the fixed (and absolute) frame of reference against which the motion of the moon can be observed and calibrated. Apart from a few exceptions (discussed above) the identification of *nakṣatra*s with stars or asterisms in modern catalogues of stars is almost impossible. The Vedic texts generally provide few if any clues to allow identification. Neither relative nor absolute coordinates nor the relative brightness of stars in the *nakṣatra*s are given in the Saṃhitās or Brāhmaṇas. The names of *nakṣatra*s can be of some help in identification, i.e. a singular name suggests a single star, a dual name suggests two stars, and a plural name more than two stars. Some Vedic texts do specify the number of stars in a *nakṣatra*, for example, seven stars in the Kṛttikās, four stars in Śraviṣṭhās (*TB*.3.1.2) and in Uttara-Proṣṭhapadas (*TB*.3.1.2). It is possible, that apart from the Kṛttikās there may not have been more than four stars in any Vedic *nakṣatra* (*ŚB*.II.1.2.2; Eggeling 1882). The number of stars in each *nakṣatra* is only given in the post-Vedic texts (Weber 1861). In South Asia, coordinates of stars are only given in astronomical texts produced after the fifth century. Perhaps the oldest catalogue of coordinates of one (principle) star of a *nakṣatra*, called the *yogatārā* of the *nakṣatra*, is that in Paitāmahasiddhānta of *Viṣṇudharmottara Purāṇa*. However, the coordinates (polar longitude and latitude) of *yogatārā*s given in this text are very inaccurate; barring a few exceptions, they are integer degrees. The provenance of this text is also disputed; Billard (1971) and others maintain that it was plagiarized from the *Brāhmasphuṭasiddhānta*, a text of the seventh century CE, but Pingree and Morrissey (1989) claim that this text was produced in the fifth century CE. Suffice it to say that there is a gap of more than thousand years between the Vedic texts and this catalogue of coordinates. This gap should be borne in mind in interpreting conclusions based on the coordinates of *yogatārā*s. Attempts to identify the *yogatārā*s from (modern) star catalogues were made in the late nineteenth century. More recently Pingree and Morrissey (1989) have undertaken somewhat involved reanalysis of this list of *yogatārā*s. They have taken a list of (bright) stars from modern star catalogues, and precessed these to what they believe to be the putative epoch (between CE 400 and 500) of Paitāmahasiddhānta and selected stars for which the difference between the precessed coordinates and the coordinates of Paitāmahasiddhānta is minimum. Their final list of stars with J2000.0 (N.IV.5) right ascension and declination is given in Table IV.2.

Abhyankar (1991) has also addressed the question of identification of the *yogatārās*. He maintains that the *nakṣatra* Abhijit was dropped from the list of *nakṣatras* sometime around 1500 BCE (he does not tell us how he arrived at this date!) and the correspondence between a number of *nakṣatras* and their *yogatārās* was lost in the following period. He has re-identified a number of *yogatārās*; his criteria for a match appear to be that the star or asterism should be within 30° of the ecliptic, it should be bright, and it should fall within the boundaries of the *nakṣatra*-sectors. He identifies the *nakṣatra*-sectors by defining a correspondence between the twelve *rāśis* (a Mesopotamian/Greek division of the ecliptic into 30° sectors adopted into South Asian astronomy during the medieval period) and twelve *nakṣatras*. The coordinates of his *yogatārās* are also given in Table IV.2. In Figure IV.1 is shown the distribution, around the ecliptic, of *yogatārās* identified by Pingree and Morrissey (1989) and by Abhyankar (1991). The *yogatārās* have been precessed to 1400 BCE to bring the *yogatārā* of Kṛttikās close to the vernal equinox i.e. right ascension and declination equal to 0° (the choice of this epoch is justified in Section 5). The

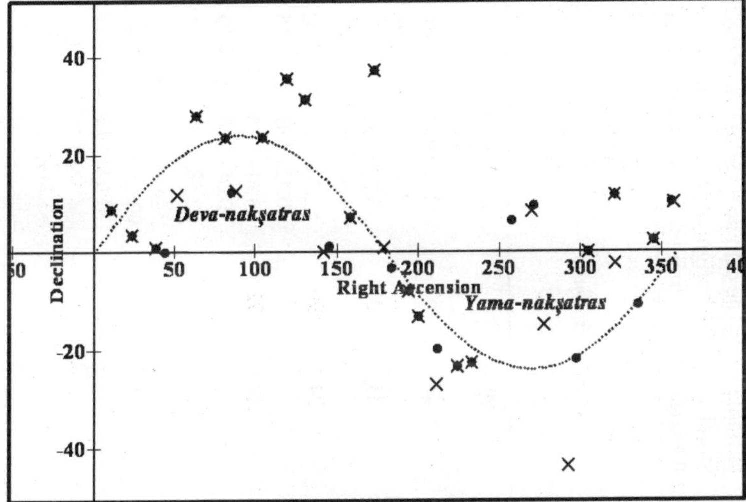

Figure IV.1: Distribution of *yogatārās* (precessed to 1400 BCE) around the ecliptic (the 'wavy' line). The regions of *deva-nakṣatras* and *yama-nakṣatras* are indicated. The filled circles are the *yogatārās* of Pingree and Morrissey (1989), and the crosses are those of Abhyankar (1991).

Table IV.2: The ecliptic coordinates of the first 27 full-moons of a *yuga*. The equatorial coordinates of the origin of the *nakṣatra*-sectors are given and for comparison the coordinates of the *yogatārā*s determined by Pingree and Morrissey (1989) and Abhyankar (1991) are included.

	Nakṣatra	FM	*Bhāṃśa*	Full-moon coordinates L B	O N-S α δ	P & M (1989) α δ J(2000)	P & M (1989) mag	A (1991) α δ J(2000)	A (1991) mag
1	Kṛttikās	10	3	3.22±0.72 2.45±2.24	00 12 14 + 1 19 31	03 47 29 +24 06 18	2.87	03 47 29 +24 06 18	2.87
2	Rohiṇī	23	55	22.20±0.61 1.21±3.16	01 01 26 + 06 33 02	04 35 55 +16 30 33	0.85	04 35 55 +16 30 33	0.85
3	Mṛgaśīrṣa	11	37	33.28±0.87 1.50±3.03	01 51 38 + 11 28 19	05 35 08 +09 56 03	3.66	05 35 08 +09 56 03	3.66
4	Ārdrā	24	90	52.37±0.61 -0.43±3.41	02 43 31 + 15 50 28	05 55 10 +07 24 25	0.50	06 37 42 +16 23 57	1.93
5	Punarvasus	12	72	63.61±0.70 -0.01±3.40	03 37 27 + 19 24 38	07 45 19 +28 01 34	1.15	07 45 19 +28 01 34	1.15
6	Puṣya	25	1	82.58±0.75 -2.01±2.90	04 33 26 + 21 56 58	08 44 41 +18 09 15	3.94	08 44 41 +18 09 15	3.94
7	Āśreṣās	13	104	93.65±0.82 -1.67±3.12	05 30 57 + 23 16 17	08 46 47 +06 25 08	3.38	08 55 23 +05 56 44	3.11
8	Maghās	1	81	104.72±0.75 -1.31±3.26	06 29 02 + 23 16 17	10 08 23 +11 58 02	1.35	10 08 23 +11 58 02	1.35

9	Pūrva-Phālgunīs	14	7	123.42±0.90 -2.93±2.32	07 26 33 + 21 56 58	11 14 06 +20 31 25	2.56	11 14 06 +20 31 25	2.56
10	Uttara-Phālgunīs	2	104	134.14±0.82 -2.62±2.57	08 22 32 + 19 24 38	11 49 04 +14 34 19	2.14	11 49 04 +14 34 19	2.14
11	Hasta	27	55	141.92±0.97 -3.45±1.96	09 16 28 + 15 50 28	12 29 52 -16 30 56	2.95	12 15 48 -17 32 31	2.59
12	Citrā	3	1	163.16±0.93 -3.38±1.91	10 08 21 + 11 28 19	13 25 12 -11 10 57	0.98	13 25 12 -11 10 57	0.98
13	Svātī	28	73	170.71±1.05 -2.90±2.62	10 58 33 + 06 33 02	14 15 40 +19 12 37	-0.04	14 15 40 +19 12 37	-0.04
14	Viśākhās	16	45	181.30± 0.94 -3.15±2.44	11 47 45 + 01 19 31	15 12 13 -19 47 30	4.54	14 50 53 -16 02 30	2.75
15	Anūrādhā	4	12	191.56±1.02 -3.20±2.24	12 36 45 -03 57 38	16 00 20 -22 37 17	2.32	16 00 20 -22 37 17	2.32
16	Jyeṣṭhā	17	55	209.38±0.95 -2.05±3.14	13 26 22 - 09 03 53	16 29 24 -26 25 53	0.96	16 29 24 -26 25 53	0.96
17	Mūla	5	27	219.75±0.96 -2.44±2.91	14 17 20 -13 44 28	17 27 21 -29 52 01	4.29	17 33 36 -37 06 14	1.63
18	Pūrvāṣāḍhās	18	66	237.36±0.95 -0.64±3.37	15 10 13 -17 44 27	18 20 59 -29 49 40	2.70	18 20 59 -29 49 40	2.70
19	Uttrāṣāḍhās	6	41	247.88±0.74 -1.03±3.37	16 05 13 -20 49 19	18 55 16 -26 17 46	2.02	18 55 16 -26 17 46	2.02
20	Śravaṇa	19	81	265.52±0.68 0.92±3.07	17 02 06 -22 46 20	19 50 47 +08 52 06	0.77	20 37 33 +14 35 43	3.63

(contd.)

Table IV.2. *(contd.)*

	Nakṣatra	FM	*Bhāṃśa*	Full-moon coordinates L B	O N-S α δ	P & M (1989) α δ J(2000)	P & M (1989) mag	A (1991) α δ J(2000)	A (1991) mag
21	Śraviṣṭhās	7	53	276.00±0.78 0.48±3.23	18 00 00 -23 26 21	20 39 38 +15 54 43	3.77	21 31 33 -05 34 16	2.91
22	Śatabhiṣaj	20	96	293.96±0.72 2.09±2.31	18 57 56 -22 46 17	22 52 37 -07 34 47	3.74	22 57 39 -29 37 20	1.16
23	Pūrva-Proṣṭapadas	8	74	304.52±0.57 1.81±2.55	19 54 46 -20 49 19	23 04 45 +15 12 19	2.49	23 04 45 +15 12 19	2.49
24	Uttara-Proṣṭapadas	33	16	312.05±0.82 2.60±1.97	20 49 46 -17 44 27	00 08 23 +29 05 26	2.06	00 13 14 +15 11 01	2.83
25	Revatī	21	117	322.73±0.74 2.61±1.90	21 42 39 -13 44 28	01 13 44 +07 34 31	5.24	00 08 23 +29 05 26	2.06
26	Aśvayujau	9	97	333.61±0.72 2.55±1.93	22 33 37 -09 03 53	01 54 38 20 48 29	2.64	01 54 38 20 48 29	2.64
27	Bharaṇīs	22	21	352.26±0.61 2.30±2.44	23 23 14 -03 57 38	02 43 27 +27 42 26	4.66	02 49 59 +27 15 38	3.63

P & M (1989) : Pingree and Morrissey (1989); A (1991) : Abhyankar (1991); J (2000) : reference equator and equinox;
α, δ : right ascension and declination; L,B : ecliptic longitude and latitude; FM: The full-moon number (Table IV.4; Col.# 1 & # 8);
O N-S: Coordinates of the origin of the *nakṣatra*-sectors

criteria the Āryas might have used to select *nakṣatra*s are not known and they may not have been the brightest stars (or asterism) or those close to the ecliptic. A number of *yogatārā*s of Paitāmahasiddhānta identified by Pingree and Morrissey (1989) and by Abhyankar (1991) are at a considerable distance from the ecliptic whereas the moon, in its monthly passage, will deviate only by ±5° from the ecliptic. This by itself would not preclude these *yogatārā*s being the markers of *nakṣatra*s along the path of the moon. A *yogatārā* has to be visible some time during the passage of the moon. The *yogatārā*s, like the *nakṣatra*s they represent, should divide into *deva-nakṣatra*s and *yama-nakṣatra*s (*TB*.3.1.1-2 above). The *yogatārā*s of *deva-nakṣatra*s should be in the northern hemisphere (i.e. positive declination), and the *yogatārā*s of *yama-nakṣatra*s should be in the southern hemisphere (i.e. negative declination). This is (almost) true of the *yogatārā*s that represent *deva-nakṣatra*s but a number of *yogatārā*s that should represent *yama-nakṣatra*s are in the northern hemisphere and far from the ecliptic. These *yogatārā*s cannot possibly represent *yama-nakṣatra*s and these *yogatārā*s (in Paitāmahasiddhānta) and their identification must be wrong (but see Section 7 below). In addition, the Āryas identified days by the *nakṣatra* close to the moon. If the *nakṣatra*s were as unevenly distributed as their identifying *yogatārā*s, then precise time reckoning would have been difficult, and this would have been problematic for the ritualists because they were very particular about the timing of their rituals. Certain rituals were forbidden 'under' some *nakṣatra*s. This should cast doubt on the identity of some *yogatārā*s in Paitāmahasiddhānta and it is possible that not all stars in this text represent the Vedic *nakṣatra*s.

The *nakṣatra*s as sectors of the ecliptic are believed to be a late development. On the development of this concept also, the Vedic texts are silent. The analytical derivation of the position and the boundaries of these sectors can only be inferred from *Vedāṅga Jyotiṣa*, and this is described in the next section. The understanding up to now has been that the *nakṣatra*-sectors are 27 divisions of the ecliptic and these can be identified by the *yogatārā*s of Paitāmahasiddhānta. In Figure IV.2, 27 equal '*nakṣatra*-sectors', or 13.33° wide bands of the ecliptic, are plotted. The origin of the '*nakṣatra*-sectors' that the Āryas might have assumed is not known (but see Section 3.2) and so an origin at Kṛttikās was assumed, that is, only a relative scale for the sectors is considered in this discussion. If the identification of Pingree and Morrissey is accepted, then five '*nakṣatra*-sectors' have no identifying *yogatārā* and further five sectors have two *yogatārās*. Abhyankar's identification results in a *yogatārā* for every *nakṣatra*-sector (because of the selection imposed by him) but for some sectors there can be two *yogatārās*. With the origin of the sectors at Kṛttikās, 14 (from identification of Pingree and Morrissey) or 15 (from identification of Abhyankar) stars are on the boundary of two sectors, the rest of the stars are within the sectors. The positions of the moon on successive days after the new moon at Kṛttikās (see *iṣṭis* in *TB*.3.1.5 described above) are plotted in Figure IV.2, and by a judicious choice of the origin of the *nakṣatra*-sectors, it is possible to locate the moon in each *nakṣatra*-sector. However, the moon will not be at the same location in each sector every night, and it will not return to the same sector each month (Section 7). This suggests that the currently accepted scheme of dividing the ecliptic into '*nakṣatra*-sectors' and identifying these by the *yogatārās* of Paitāmahasiddhānta (or the revisions that have been made to these identifications) is unsatisfactory in a number of ways. Principally it does not define the

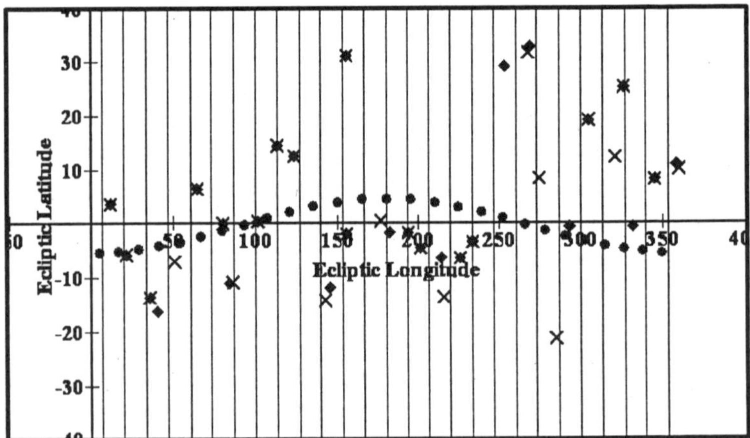

Figure IV.2: Ecliptic longitude and latitude of *yogatārās* of Pingree and Morrissey (1989) shown as filled diamonds, and those of Abhyankar (1991) shown as crosses. The filled circles are the daily positions of the moon, and the vertical lines are the pseudo *nakṣatra*-sectors, 13.33° wide and normalized to the position of Kṛttikās (see text for details).

origin of the sectors or the boundaries of the sectors and it does not define the criteria to match a sector to a *yogatārā* or a *nakṣatra*.

3. *NAKṢATRAS* – LUNAR MANSIONS

For over a hundred years, the discussions have focused on *nakṣatra*s as stars with which the moon conjoins each day of the month. Although *nakṣatra*s have always been accepted as lunar mansions, *nakṣatra*s as independent (of stars) entities have received almost no attention. This discussion has been one-sided perhaps because the stars were seen to define the only viable fixed (and absolute) frame of reference against which the motion of the moon could be observed and calibrated. The Vedic texts have aided and abetted this one-sided discussion by not distinguishing between the wide lunar mansions and their stellar markers. An invariant '*nakṣatra*-coordinate system' is essential to the calendar of *Vedāṅga Jyotiṣa*. The algorithms of *Vedāṅga Jyotiṣa* that define the position of the moon and the sun (discussed in Chapter V) would be meaningless without an invariant coordinate system. The absolute *nakṣatra*-coordinate system of *Vedāṅga Jyotiṣa*, described below, is independent of the background stellar frame of reference; instead, it is anchored to the ecliptic.

3.1. The *Jāvādi* Arrangement of *Nakṣatras*

In the Vedāṅga Jyotiṣa a list of *nakṣatra*s, identified by their presiding deity, is given in *RJ*.25-28 (verses 25-28 in the Ṛgvedic recension, Chapter V) and *YJ*.32-35 (verses 32-35 in the *Yājuṣa* recension, Chapter V), these lists are headed by (the *devatā* of) the Kṛttikās and are reproduced in Table IV.1. Allowing for minor differences in names, this list of *nakṣatra*s in *Vedāṅga Jyotiṣa* is similar to that in the Saṃhitās and the Brāhmaṇas and the Sūtras. In *RJ*.14 and *YJ*.18, *Vedāṅga Jyotiṣa* introduces a separate list of *nakṣatra*s identified by either a syllable abbreviating the name of the *nakṣatra* or a syllable abbreviating the name of its *devatā* (presiding deity). This arrangement is called the '*jāvādi* (*jau ādi* – beginning with *jau*, abbreviation for *Aśvayu**jau***) arrangement'. In *RJ*.18 and *YJ*.39, *Vedāṅga Jyotiṣa* states that the sun stays in each *nakṣatra* 13 and $^{5}/_{9}$ (*sāvana*) days. This verse unambiguously states that, in *Vedāṅga*

Jyotiṣa, a *nakṣatra* is not considered to be a star or an asterism but a sector of the ecliptic and this sector is 13.33° wide. There is also an implicit assumption here that all sectors are of equal width. The *jāvādi* arrangement of the *nakṣatra*(-sectors) is given in Table IV.3. The abbreviations of *Vedāṅga Jyotiṣa* are given in columns 2 and 5 and the *nakṣatra* or deity in columns 3 and 6. The *Vedāṅga Jyotiṣa* abbreviations are highlighted in these names. The scheme or the algorithm used to obtain this arrangement is not given in *Vedāṅga Jyotiṣa*. An inspection of the list of *nakṣatra*s in the Vedic texts (Table IV.1) and the list in the *jāvādi* arrangement (Table IV.3) suggests that the *nakṣatra*-sectors in the arrangement are every fifth *nakṣatra* from Table IV.1 starting from Śraviṣṭhās (*nakṣatra* number 21 in Table IV.1). That is, the Śraviṣṭhās are assumed to be *nakṣatra*-sector number 0 (in the *jāvādi* arrangement) and the fifth *nakṣatra* is

Table IV.3: The *jāvādi* arrangement of *nakṣatra*s. The abbreviations given in *Vedāṅga Jyotiṣa* are highlighted. (Kuppanna Sastry 1984)

Ser. No.	*VJ* Abbreviations	*Devatā* or *nakṣatra*	Ser. No.	*VJ* Abbreviations	*Devatā* or *nakṣatra*
1	*jau*	*Aśvayu**jau*** (*Aśvinī*)	14	*mā*	*Arya**mā*** *Uttraphālguṇī*
2	*drā*	*Ā**rdrā***	15	*dhāḥ*	*Anurā**dhāḥ***
3	*gah*	*Bha**gaḥ*** (*Pūrvaphālgunīs*)	16	*ṇaḥ*	*Śrava**ṇaḥ***
4	*khe*	*Viśā**khe***	17	*re*	***Re**vatī*
5	*śve*	*Vi**śve**devāḥ* *Uttarāṣāḍhā*	18	*mṛ*	***Mṛ**gaśīrsam*
6	*hiḥ*	*A**hir**budhnyaḥ* (*Uttraproṣṭapadas*)	19	*ghāḥ*	*Ma**ghāḥ***
7	*ro*	***Ro**hiṇī*	20	*svā*	***Svā**tī*
8	*ṣā*	*Āśre**ṣā***	21	*paḥ*	*Ā**paḥ*** (*Pūrvāṣāḍhās*)
9	*cit*	***Cit**rā*	22	*jaḥ*	*A**ja**ejapāt* (*Pūrvaproṣṭapada*)
10	*mū*	***Mū**lā*	23	*kṛ*	***Kṛ**ttikāḥ*
11	*ṣa*	*Śatabhiṣaj*	24	*ṣyaḥ*	*Pu**ṣyaḥ***
12	*ṇyaḥ*	*Bhara**ṇyaḥ***	25	*ha*	***Ha**staḥ*
13	*sū*	*Punarva**su***	26	*jye*	***Jye**ṣṭhā*
			27	*ṣṭhāḥ*	*Śravi**ṣṭhāḥ***

Aśvayujau, the first *nakṣatra* of the *jāvādi* arrangement (and hence the name) and the fifth after that is Ārdrā, and so on to choose 27 *nakṣatra*-sectors. This of course prompts the question, why choose every fifth *nakṣatra*? A pedestrian answer could be 'the Āryan predilection for number five (the Vedic lustrum)'. Thibaut (1877) has presented and discussed a rational explanation for the choice of *nakṣatras* of the *jāvādi* arrangement.

In a *yuga* there are
62 lunations/lunar months and
67 sidereal months (both these numbers are given in the *Vedāṅga Jyotiṣa*; Chapter V)
Therefore, in 1 lunation there are $67 \div 62$ sidereal months
Or 1 lunation = $1^{5}/_{62}$ sidereal months
In a sidereal month the moon passes by (or resides in) 27 *nakṣatras*
Therefore, in 1 lunation the moon passes by (or resides in) $27 \times 1^{5}/_{62}$ *nakṣatras* = $29^{22}/_{124}$ *nakṣatras*
Thus the separation of successive new (or full) moons is $29^{22}/_{124}$ *nakṣatras*
And the separation between a new and a full (or a full and a new) moon (or a *pakṣa*) is $14^{73}/_{124}$ *nakṣatras*

To obtain the *jāvādi* arrangement of *nakṣatra*-sectors a 'zero point' or an 'origin' is required from which the *nakṣatra*-sector of a full or a new moon in a *yuga* (the five-year intercalation period of *Vedāṅga Jyotiṣa*) can be counted. In Vedāṅga Jyotiṣa, the start of a *yuga* is on the new moon at the winter solstice 'when the sun and the moon occupy the same region of the sky together with (the *nakṣatra*) Śraviṣṭhās' (*RJ*.5 and *YJ*.6; Kuppanna Sastry 1984). Thus, starting with the new moon in the *nakṣatra* Śraviṣṭhās, it is possible to obtain the *nakṣatra*-sectors at successive full moon and new moon in a *yuga* using the scheme of Thibaut (1877) given above. The *nakṣatra*-sectors are counted from Śraviṣṭhās (i.e. this is the *nakṣatra*-sector number 0 or this *nakṣatra*-sector is the origin of the *jāvādi* arrangement of *nakṣatra*-sectors) from the list of *nakṣatras*. For example, the *nakṣatra*-sector of the first full moon after the start of a *yuga* will be the fourteenth *nakṣatra*-sector counted from Śraviṣṭhās that is Maghās. If the separation between adjacent *nakṣatra*-sectors is 124 parts (or each *nakṣatra*-sectors is divided into 124 parts) then the first full moon will be in the 73rd part of *nakṣatra*-sector Maghās. Similarly, the first new moon after the start of a *yuga* will be in the 22nd part of the twenty-ninth *nakṣatra*-sector. However, there are only 27 *nakṣatra*-sectors, therefore this new moon will be in the (27 plus 2, that is) second *nakṣatra*-sector after *nakṣatra* Śraviṣṭhās or Pūrva-Proṣṭhapadas. The division of the *nakṣatra*-sectors into 124 parts is given in *Vedāṅga Jyotiṣa* and each part is called a *bhāṃśa* (Chapter V). The *nakṣatra*-sectors of the new moon and full moon (and the location of the moon within a sector, in terms of the number of *bhāṃśas* from the origin of the sector) of the 62 lunations of a *yuga* are given in Table IV.4. The *nakṣatra*-sectors of the *jāvādi* arrangement are selected from this list of new moon and full moon *nakṣatra*-sectors. In the arrangement, the *nakṣatra*-sectors appear to be 'selected' when the moon is in the first 27 *bhāṃśa* (the sequence repeats after the twenty-seventh *nakṣatra*). These *nakṣatra*-sectors are highlighted in Table IV.4 and given in Table IV.5. Equivalently the position of a *nakṣatra*-sector in the *jāvādi* arrangement is given by X where,

$$N \equiv X \text{ modulo } (27) \quad (\text{N.IV.6})$$

and N is the *bhāṃśa* of the moon in a *nakṣatra*-sector in Table IV.4. For example, the first full moon in a *yuga* is in the *nakṣatra*-sector Maghās (Table IV.4.) at *bhāṃśa* 73 and

$$73 \div 27 = 2 \text{ and remainder } 19 \text{ or}$$

$$73 \equiv 19 \text{ modulo } (27)$$

and *nakṣatra*-sector Maghās is in the 19th position in the *jāvādi* arrangement as given in Tables IV.3 and IV.5. The *bhāṃśa* of all full and new moons in Table IV.4 will indicate the position of that *nakṣatra*-sector in the *jāvādi* arrangement when reduced in this manner, this is given in Table IV.5. It can be seen from this table that the new and full moons in a *yuga*, arranged by their ('reduced') *bhāṃśa*, reproduce the *jāvādi* arrangement. It is not clear why the composer(s) of *Vedāṅga Jyotiṣa* chose to arrange the *nakṣatra*-sectors of the *jāvādi* arrangement in this ('reduced') *bhāṃśa* sequence. But the *jāvādi* arrangement or the derivation of the *jāvādi* arrangement emphasizes the fundamental aspect of *nakṣatra*s in the calendar of

Table IV.4: The *bhāṃśa*s and the *nakṣatra*s at new moon and full moon of sixty-two synodic months of a *yuga*. The *jāvādi* arrangement of *nakṣatra*s is highlighted. The serial numbers of the new moons and full moons in a *yuga* are given in columns 1 and 8. The *nakṣatra* numbers (from Table IV.1) of these moons are given in column N, the *bhāṃśa*s of these moons are given in column B, and the *nakṣatra*s of these moons are given in column Nk.

No.	New moon			Full moon			No.	New moon			Full moon		
	N	B	Nk	N	B	Nk		N	B	Nk	N	B	Nk
1	0	0	Śraviṣṭhās	14	73	Maghās	32	13	62	Āśreṣās	**1**	**11**	**Śatabhiṣaj**
2	**2**	**22**	**P. Proṣṭapadas**	16	95	U. Phālguṇīs	33	15	84	P. Phālguṇīs	3	33	U. Proṣṭapadas
3	4	44	Revatī	18	117	Citrā	34	17	106	Hasta	5	55	Aśvayujau
4	6	66	Bharaṇīs	**21**	**15**	**Anurādhā**	35	**20**	**4**	**Viśākhās**	7	77	Kṛttikās
5	8	88	Rohiṇī	23	37	Mūla	36	**22**	**26**	**Jyeṣṭhā**	9	99	Mṛgaśīrṣa
6	10	110	Ārdrā	25	59	U. Aṣāḍhās	37	24	48	P. Aṣāḍhās	11	121	Punarvasus
7	**13**	**8**	**Āśreṣās**	0	81	Śraviṣṭhās	38	26	70	Śravaṇa	**14**	**19**	**Maghās**
8	15	30	P. Phālguṇīs	2	103	P. Proṣṭapadas	39	1	92	Śatabhiṣaj	16	41	U. Phālguṇīs
9	17	52	Hasta	**5**	**1**	**Aśvayujau**	40	3	114	U. Proṣṭapadā	18	63	Citrā
10	19	74	Svātī	**7**	**23**	**Kṛttikās**	41	**6**	**12**	**Bharaṇīs**	20	85	Viśākhās
11	21	96	Anurādhā	9	45	Mṛgaśīrṣa	42	8	34	Rohiṇī	22	107	Jyeṣṭhā
12	23	118	Mūla	11	67	Punarvasus	43	10	56	Ārdrā	**25**	**5**	**U. Aṣāḍhās**
13	**26**	**16**	**Śravaṇa**	13	89	Āśreṣās	44	12	78	Puṣya	**0**	**27**	**Śraviṣṭhās**
14	1	38	Śatabhiṣaj	15	111	P. Phālguṇīs	45	14	100	Maghās	2	49	P. Proṣṭapadas
15	3	60	U. Proṣṭapadās	**18**	**9**	**Citrā**	46	16	122	U. Phālguṇīs	4	71	Revatī
16	5	82	Aśvayujau	20	31	Viśākhās	47	**19**	**20**	**Svātī**	6	93	Bharaṇīs

(contd.)

Table IV.4 *(contd.)*

No.	New moon			Full moon			No.	New moon			Full moon		
	N	B	Nk	N	B	Nk		N	B	Nk	N	B	Nk
17	7	104	Kṛttikās	22	53	Jyeṣṭhā	48	21	42	Anurādhā	8	115	Rohiṇī
18	**10**	**2**	**Ārdrā**	24	75	P. Aṣāḍhās	49	23	64	Mūla	**11**	**13**	**Punarvasus**
19	**12**	**24**	**Puṣya**	26	97	Śravaṇa	50	25	86	U. Aṣāḍhās	13	35	Āśreṣās
20	14	46	Maghās	1	119	Śatabhiṣaj	51	0	108	Śraviṣṭhās	15	57	P. Phālgunīs
21	16	68	U. Phālgunīs	**4**	**17**	**Revatī**	52	**3**	**6**	**U. Proṣṭapadas**	17	79	Hasta
22	18	90	Citrā	6	39	Bharaṇīs	53	5	28	Aśvayujau	19	101	Svātī
23	20	112	Viśākhās	8	61	Rohiṇī	54	7	50	Kṛttikās	21	123	Anurādhā
24	**23**	**10**	**Mūlā**	10	83	Ārdrā	55	9	72	Mṛgaśīrṣa	**24**	**21**	**P. Aṣāḍhās**
25	25	32	U. Aṣāḍhās	12	105	Puṣya	56	11	94	Punarvasus	26	43	Śravaṇa
26	0	54	Śraviṣṭhās	**15**	**3**	**P. Phālguṇīs**	57	13	116	Āśreṣās	1	65	Śatabhiṣaj
27	2	76	P. Proṣṭapadas	**17**	**25**	**Hasta**	58	**16**	**14**	**U. Phālguṇīs**	3	87	U. Proṣṭapadas
28	4	98	Revatī	19	47	Svāti	59	18	36	Citrā	5	109	Aśvayujau
29	6	120	Bharaṇīs	21	69	Anurādhā	60	20	58	Viśākhās	**8**	**7**	**Rohiṇī**
30	**9**	**18**	**Mṛgaśīṛsa**	23	91	Mūlā	61	22	80	Jyeṣṭhā	10	29	Ārdrā
31	11	40	Punarvasus	25	113	U. Aṣāḍhās	62	24	102	P. Aṣāḍhās	12	51	Puṣya

Table IV.5: The *jāvādi* arrangement of *nakṣatra*s arranged by *bhāṃśa*, and differentiated by the new and full moons in a *yuga*. The numbers in the columns below the 'new moon' and 'full moon' are the new and full moon '*nakṣatra* numbers' from Table IV.4, column 1 and column 8.

Bhāṃśa	New Moon #			Full Moon #			Nakṣatra
1	16	53		9	34	59	Aśvayuja
2	6	18	43	24	61		Ārdrā
3	8	33		14	26	51	P. Phālguṇīs
4	23	35	60	16	41		Viśākhās
5	25	50		6	31	43	U. Aṣāḍhās
6	15	40	52	33	58		U. Proṣṭapadas
7	5	42		23	48	60	Rohiṇī
8	7	32	57	13	50		Āśreṣās
9	22	59		3	15	40	Citrā
10	12	24	49	5	30		Mūla
11	14	39		20	32	57	Śatabhiṣaj
12	4	29	41	22	47		Bharaṇīs
13	31	56		12	37	49	Punarvasus
14	21	46	58	2	39		U. Phālguṇīs
15	11	48		4	29	54	Anurādhā
16	13	38		19	56		Śravaṇa
17	3	28		21	46		Revatī
18	30	55		11	36		Mṛgaśīrṣa
19	20	45		1	38		Maghās
20	10	47		28	53		Svātī
21	37	62		18	55		P. Aṣāḍhās
22	2	27		8	45		P. Proṣṭapadas
23	17	54		10	35		Kṛttikās
24	19	44		25	62		Puṣya
25	9	34		27	52		Hasta
26	36	61		17	42		Jyeṣṭhā
27	1	26	51	7	44		Śraviṣṭhās

Vedāṅga Jyotiṣa — the *nakṣatra*s are equal sectors of the ecliptic and are evenly distributed along it. These sectors are anchored to the ecliptic and are independent of the background stars. The *jāvādi* arrangement thus defines an invariant coordinate system with origin at new moon around the winter solstice, which by definition (*RJ*.5 and *YJ*.6) is at the beginning of the *nakṣatra*-sector Śraviṣṭhās. The coordinates of this reference-frame are the *nakṣatra*-sector and the *bhāṃśa* within this sector. This frame of reference is very similar to the ecliptic frame of reference of modern astronomy, with origin at the First Point of Aries (N.IV.7). In the ecliptic frame of reference, the coordinates are ecliptic longitude and latitude. The *nakṣatra*-sector and *bhāṃśa* are equivalent to the ecliptic longitude. *Vedāṅga Jyotiṣa* does not have an equivalent of ecliptic latitude as it is interested only in the position of the sun and the moon and the sun does not deviate from the ecliptic (it defines the ecliptic) and the moon deviates by only ±5° from the ecliptic.

A note of caution should be introduced here. The *jāvādi* arrangement explicitly assumes that the length of the synodic month is fixed. This is an approximation (N.III.3) as the length of the synodic month will vary in a quasi-random manner. In every *yuga*, the *nakṣatra*-sector of the full and new moon will be same as that given in Table IV.4, the *bhāṃśa* of the moon, however, will have values different from those given in Table IV.4. This is discussed in detail in Chapter V, Section 1.5.

3.2. *Nakṣatras* – Ecliptic and Equatorial Coordinates of Lunar Mansions

Vedāṅga Jyotiṣa implicitly assumes that *nakṣatras* are wide sectors of the ecliptic and the moon moves through these sectors. As shown above, the *jāvādi* arrangement gives the location of the moon (new and full) on the ecliptic in terms of a *nakṣatra*-sector and a *bhāṃśa*. If the (ecliptic or equatorial) coordinates of this new or full moon can be determined, then the ecliptic or the equatorial coordinates of a point in the *nakṣatra*-sector can be established. Thus, the position on the ecliptic of the *nakṣatra*-sectors can be given in astronomical units used at present that is in ecliptic and equatorial coordinate units.

The procedure for determining the (equatorial and ecliptic) coordinates of the full moons in a *yuga* has been described in Gondhalekar (2009) and is as follows:

- Determine the date(s) of the new moon at or around the winter solstice.
- Determine the separation in days between the new moon at winter solstice and the following full moon(s) in a *yuga*.
- Convert the separation to (equatorial or ecliptic) coordinates. The

reference frame for this conversion is the standard frame of modern astronomy, with the origin at the First Point of Aries (N.IV.7).

- The *nakṣatra* of this full moon is obtained by the scheme described above and is given in Table IV.4.

Two procedures have been followed to determine the coordinates of the full moon. In the first method, the dates of the new moon at the winter solstice between CE 1900 and 2000 were obtained from the available calendars. The dates of the 62 full moons following each of these new moons were also obtained from these calendars. In the second method, the dates of the new moon at or within one day of the winter solstice between 1500 BCE and 500 BCE were obtained by first calculating the dates of the winter solstice, that is, dates of minimum declination of the sun each year. The separation between the sun and the moon was calculated for each winter solstice and the dates when this separation was within five degree (i.e. the sun and the moon were in conjunction) were retained. These dates were considered to be the dates of the new moon at the winter solstice. For each new moon date, the following 62 dates when the sun and the moon were in opposition (i.e. about 180° apart) were determined, these were considered the dates of the full moons of a *yuga*. For computational convenience, the dates, for both methods, were determined in Julian days (N.IV.8). The dates of winter solstice and the full moon for the period from 1500 BCE to 500 BCE were computed with the currently available orbital parameters of the earth and the moon. The secular changes in the orbital parameters are included in these and all computations in this book, which require these parameters. The separations between the new moon at the winter solstice and the following full moons were converted to equatorial and ecliptic coordinates. For the two periods, the coordinates of the same full moon in every *yuga* were averaged to obtain the spread (root mean square deviation) in the mean value of the full moon coordinates. The full-moon number, the corresponding *nakṣatra*-sector (from Table IV.4), the mean ecliptic coordinates of the full moon and the equatorial coordinates of the origin of the *nakṣatra*-sectors are given in Table IV.2. Only the mean coordinates determined for the period 1500 to 500 BCE are given in this table as the mean coordinates for the CE 1900 to 2000 period are within the spread of the values obtained for the earlier period. The coordinates for the first twelve full moons of a *yuga* for both periods are shown in Figure IV.3. The computed *nakṣatra* (as counted from Śraviṣṭhās at the winter solstice) and the *bhāṃśa* at each full moon in Table IV.2 can be compared with the *nakṣatra* and *bhāṃśa* in Table IV.4. The match between the *nakṣatra*s is prefect but the *bhāṃśa*s differ. This is because of the difference between the (fixed) mean length of the synodic month (used in Table IV.4) and the varying length (due to moon's position in its orbit) of the synodic month in the two methods described. It is important to emphasize that the procedures to obtain the (equatorial and ecliptic) coordinates and the *nakṣatra*-sector of the full moon are independent. The coordinates of the (full) moon are computed from the currently accepted orbital parameters of the earth and the moon and the origin at the First Point of Aries on the ecliptic but the *nakṣatra*-sector and the *bhāṃśa* of this moon are obtained from the procedure followed to obtain the *jāvādi* arrangement of *nakṣatra*s with origin at Śraviṣṭhās. These positions and the *nakṣatra*-sectors of the moon do not change with epoch (as demonstrated by the similarity of values obtained for the two periods considered here and shown in Figure IV.3) because they are sectors of

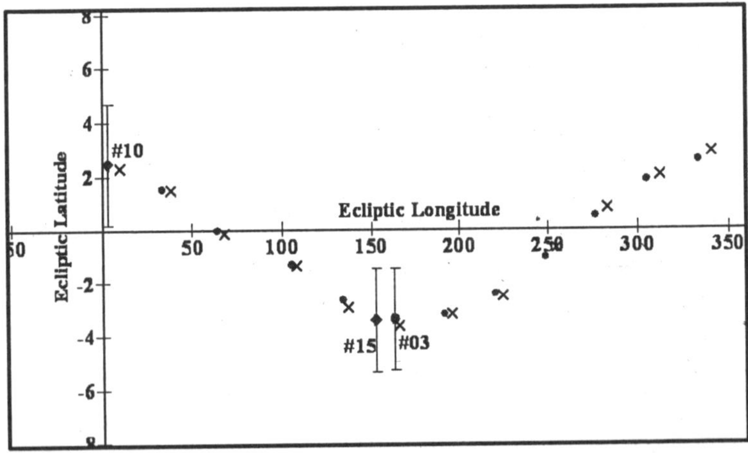

Figure IV.3: The ecliptic longitude and latitude of first twelve full moons of a *yuga*. The 'lay-out' of the positions of all subsequent full moons will be similar but slightly shifted to the left. The data obtained for the CE 1900 to 2000 period (see Section 4) are shown by dots, and the data obtained for the 1500 BCE to 500 BCE period are shown by crosses. To avoid crowding only representative spread bars are shown. The *nakṣatra* of the full moon 10 is Kṛttikās and the *nakṣatra* of full moons 3 and 15 is Citrā.

the ecliptic (independent of the background stars) and are not affected by precession of the equinox (N.IV.9). If background stars are used to identify these *nakṣatra*-sectors, the position or coordinates of these stellar markers will be epoch-dependent because of the precession of the equinox; the same set of stars will not identify the same *nakṣatra*-sector at different epochs.

In the calendar of *Vedāṅga Jyotiṣa* the division of the ecliptic into (wide) *nakṣatra*-sectors is not a hypothetical concept and these sectors have a fundamental role in various algorithms of *Vedāṅga Jyotiṣa*. A full moon (or a new moon) can occur in the same *nakṣatras*-sector at different times in a *yuga*, i.e. in a *yuga*, the moon can be full in the same *nakṣatra*-sector but at different *bhāṃśa*s, for example, full moons 3, 15 and 40 occur in the *nakṣatra*-sector of Citrā but at *bhāṃśa* 117, 9 and 63 respectively (positions of the full moon 3 and 15 are shown in Figure IV.3). In Table IV.6, are collated all *nakṣatra*-sectors in which the moon is full more than once during a *yuga*. The full moon numbers (FM), the *bhāṃśa* of the full moon and the ecliptic longitude (L; obtained for the CE 1900 to 2000 period) of the full moon are given in this table. The width of the *nakṣatra*-sectors (WNS) calculated from these pairs of ecliptic longitudes of the full moons, in the same *nakṣatra*-sector, is also given in this table. The mean width of the *nakṣatra*-sector is 12.65±3.28°. The mean width obtained from the ecliptic longitude of the full moon determined for the 1500 BCE to 500 BCE period is 11.38±5.07°. The (root mean) spread in these mean values suggests that all *nakṣatra*-sectors are of equal width. These analytically derived widths of the *nakṣatra*-sector are consistent with the width given in *Vedāṅga Jyotiṣa* (in terms of transit time of the sun across a *nakṣatra*-sector, Chapter V, Section 1.6). *Vedāṅga Jyotiṣa* emphasizes the importance of *bhāṃśa* and gives algorithms for calculating the *bhāṃśa*s of the sun and the moon in a *nakṣatra*-sector. However, it does not indicate whether the *bhāṃśa* was determined empirically. Since *Vedāṅga Jyotiṣa* describes the use of (and the units used with) a clepsydra (water-clock, Chapter V) it is highly likely that the the *bhāṃśa* of the moon and the sun within a *nakṣatra*-sector was determined by timing the passage of these celestial objects across a sector as suggested by *RJ*.18 and *YJ*.39 (Chapter V, Section 1.6).

The tenth full moon of a *yuga*, which is in the *nakṣatra*-sector of Kṛttikās (Tables IV.4 and IV.5), will be in autumn (*śarad*); this is

Table IV.6: The width of the *nakṣatra*-sector (WNS, in degrees) obtained from the ecliptic longitude of the full moon when it is in the same *nakṣatra*-sector (but different *bhāṃśa*) during a *yuga*.

Nakṣatra	FM	*Bhāṃśa*	L	FM	*Bhāṃśa*	L	WNS
Kṛttikās	10	23	10.17	35	77	15.98	13.34
Rohiṇī	23	61	27.49	48	115	33.45	13.69
Rohiṇī	48	115	33.45	60	7	21.23	14.03
Mṛgaśīrṣa	11	45	38.32	36	99	44.53	14.26
Ārdrā	24	83	55.91	61	29	50.48	12.47
Punarvasus	12	67	67.90	37	121	73.74	13.41
Punarvasus	37	121	73.74	49	13	61.69	13.84
Puṣya	25	105	85.34	62	51	79.22	14.05
Āśreṣās	13	89	96.13	50	35	91.45	10.75
Maghās	1	73	107.91	38	19	102.49	12.45
Pūrva Phālgunīs	14	111	125.68	26	3	113.58	13.89
Pūrva Phālgunīs	26	3	113.58	51	57	119.71	14.08
Uttara Phālgunīs	2	95	137.33	39	41	131.45	13.50
Hasta	27	25	143.26	52	79	149.37	14.03
Citrā	3	117	166.12	15	9	154.32	13.55
Citrā	15	9	154.32	40	63	160.53	14.26
Svātī	28	47	171.62	53	101	177.76	14.10
Viśākhās	16	31	183.87	41	85	189.36	12.61
Anurādhā	4	15	195.82	29	69	201.47	12.97
Anurādhā	29	69	201.47	54	123	207.27	13.32
Jyeṣṭhā	17	53	212.80	42	107	218.70	13.55
Mūla	5	37	224.49	30	91	229.90	12.42
Pūrva Aṣāḍhā	18	75	242.20	55	21	235.78	14.74
Uttara Aṣāḍhās	6	59	254.28	31	113	259.81	12.70
Uttara Aṣāḍhās	31	113	259.81	43	5	247.40	14.25
Śravaṇa	19	97	271.29	56	43	265.19	14.01
Śraviṣṭhās	7	81	282.63	44	27	277.04	12.84
Śatabhiṣəja	20	119	300.29	32	11	288.25	13.82
Śatabhiṣaja	32	11	288.25	57	65	294.02	13.25
Pūrva Proṣṭapadas	8	103	312.37	45	49	305.56	15.64
Uttara Proṣṭapadas	33	33	317.99	58	87	323.20	11.96
Revatī	21	17	329.53	46	71	335.34	13.34
Aśvayujau	9	1	340.56	34	55	346.52	13.69
Aśvayujau	34	55	346.52	59	109	352.28	13.23
Bharaṇīs	22	39	358.21	47	93	3.71	12.62

Mean Width of the *nakṣatra*-sector in degrees 12.65±3.28

FM: Full Moon Number L: Ecliptic Longitude

WNS: Width of the *nakṣatra*-sector in degrees

ascertained by counting the number of lunations from the first full moon after the new moon at the winter solstice (the zero point of a *yuga*). The sun will occupy this position six months later, that is, the sun will be in the *nakṣatra*-sector of Kṛttikās in spring (*vasanta*). The *jāvādi* arrangement thus states that, in spring (*vasanta*), the sun is in the *nakṣatra*-sector of Kṛttikās. In the calculation of the coordinates of the full moons in a *yuga* the origin was assumed to be the First Point of Aries and a *nakṣatra*-sector would be expected to be close to this origin or close to the vernal equinox. That this *nakṣatra*-sector should be the Kṛttikās is perhaps not entirely fortuitous. The start of the seasonal year when the sun is in the *nakṣatra* Kṛttikās suggests that the Āryas may have setup their scheme of *nakṣatra*s with the origin in *vasanta* or the *nakṣatra* close to the sun at the start of the Vedic seasonal year was considered the first *nakṣatra*. This is why all lists of *nakṣatra*s in Vedic texts start with Kṛttikās.

4. VEDIC NEW YEAR – REVISITED

The post-Ṛgvedic Saṃhitās and Brāhmaṇas indicate that the start of the Vedic year was either in spring (*vasanta*) or in winter (*śiśira*) (Chapter III, Section 2.5.1). In this section, only the start of the year in spring (*vasanta*) is discussed. The start of the year in winter (*śiśira*) is deferred to Chapter V, Section 1.5. The month of spring (*vasanta*) in which the Vedic seasonal new year commenced can be identified from the list of names of seasonal months given in the Saṃhitās and Brāhmaṇas (Chapter III, Table III.5). This list begins with Madhu in all Vedic texts in which the names of the seasonal months are given. It is also stated in these texts that Madhu is the first month of spring (*vasanta*); it is therefore, reasonable to assume that the start of the Vedic seasonal year was in the month of Madhu. When the Āryan calendar-makers named or identified the months by the *nakṣatra* that conjoined the monthly full moon (the *nakṣatra*-names of the months are discussed in Section 6), Madhu was renamed or became part of the month of Caitra. The start of the seasonal year would then have been in the month of Caitra. This is echoed in the statement 'the full moon in Citrā, is the beginning of the year' (*TS*.VII.4.8).

As noted above, the mythology the Kṛttikās/Pleiades and Skanda suggests that, in spring (*vasanta*), the new year commenced when 'the sun (*agni*) is in the Kṛttikās'. This prompts the question how the Āryas could have determined that the sun was in the Kṛttikās/Pleiades. Two simple observations would have made this possible. There is evidence in the Vedic text (*AB*.viii.28 (xl.5); Keith 1998; *ŚB*.I.6.4.18-20; Eggeling 1882; Chapter III, Section 1) that the Āryas were aware that at full moon the sun and the moon are in opposition (at a great distance). Thus, the Āryas would have known that the sun 'would be in the *nakṣatra* Kṛttikās/Pleiades' six months after or before the full moon in autumn conjoins *nakṣatra* Kṛttikās/Pleiades. The second way to ascertain when the sun is 'in' a specific *nakṣatra* is to see if the sun and the *nakṣatra* are visibly close in the sky. This is not possible during the day because the *nakṣatra*s are lost in the glare of the sun, and this is not possible at night as the sun is not visible. However, there is a narrow window just before sunrise and just after sunset when a *nakṣatra* can be visible close to the location on the horizon where the sun rises or sets. That is, the *nakṣatra* is visible just before sunrise or just after sunset. This is known as the heliacal rising or setting of a *nakṣatra*. There is considerable historical evidence (Frazer 1933; Aveni 2001; N.IV.3) that the heliacal rising or setting of prominent stars and in particular the Pleiades were observed and noted by many ancient civilizations. The Āryas were familiar with observations of stars around sunrise and sunset (*TB*.1.5.2.1, A.V.3) and there is no reason to believe that they would not have observed the heliacal rising and setting of the Kṛttikās/Pleiades.

The *nakṣatra* Citrā is 12 *nakṣatra*s from the Kṛttikās/Pleiades, almost on the opposite side of the ecliptic. Thus on the evening of the full moon day of the first month of *vasanta* (i.e. Caitra), the Āryan calendar-makers would have seen the Kṛttikās/Pleiades in the western sky shortly after sunset. A little while later, they would have seen the moon rise in the eastern sky, in the *nakṣatra* Citrā. Thus the two 'requirements' that the new year in *vasanta* should begin when the full moon is in *nakṣatra* Citrā and the sun is in the Kṛttikās are satisfied.

Vedic texts, however, do have a propensity to be ambiguous and the texts recommend that 'the full moon in Phālgunīs, is the beginning of the year' (*TS*.VII.4.8; *ŚB*.VI.2.2.18; *TB*.1.1.2.8). The texts (*TS*.VII.4.8) state that commencement of the seasonal new year should be moved from the full moon in Phālgunīs' to 'the full moon in Citrā' to avoid *viṣuvat* day (roughly mid-year) falling in a cloudy season. Dīkṣita (1896) has conjectured that the start of the Vedic seasonal year was moved from the full moon in Caitra to full moon in Phālgunīs because precession of the equinox would have 'moved' the full moon of spring (*vasanta*) from Citrā to Uttara-Phālgunīs. This would take about 1,000 years. This may be so, but cannot be established unambiguously from the Vedic texts. Moreover, the texts cite inclement weather for this change. However, the Vedic texts may not be telling the full story. The spring (*vasanta*) ceremony of the *cāturmāsya* (or seasonal) sacrifices was performed when the full moon was in Uttara-Phālgunīs (Section 5). These sacrifices were performed four synodic months apart as the name of the sacrifices (*cāturmāsya*) implies. The *nakṣatra*s in which the seasonal full moon should be for the performance of these sacrifices are given in the Vedic texts. This order of the *nakṣatra*s and the interval of four synodic months between the sacrifices can only be maintained if the spring (*vasanta*) ceremony is performed when the full moon is in *nakṣatra* Uttara-Phālgunīs and not when it is in *nakṣatra* Citrā. This suggests that the spring (*vasanta*) sacrifice may not have been performed in spring (*vasanta*), but a month earlier, perhaps in anticipation of spring. The start of the seasonal year was probably moved to the full moon in Uttara-Phālgunīs to make it consistent with the *cāturmāsya* sacrifices. The year will then 'not be born' when the sun is in Kṛttikās/Pleiades.

5. SEASONS – REVISITED

The three *cāturmāsya* (or seasonal) sacrifices have been described in Chapter II. In the post-Ṛgvedic texts, the 'dates' of the performance of these sacrifices were specified by the coincidence of the full moon with three prescribed *nakṣatra*s. The spring sacrifice of *Vaiśvadeva* was performed when the full moon was in the *nakṣatra* (Uttara-)Phālgunīs (*KŚS* 5.1.1; Ranade 1978), the summer sacrifice, *Varuṇapraghāsa* was performed when the full moon was in (Uttara-) Āṣāḍhās (*KŚS* 5.3.1; Ranade 1978) and the autumn sacrifice, *Sākamedha* was performed when the full moon was in Kṛttikās (*KŚS* 5.6.1; Ranade 1978). Sometimes a fourth sacrifice, *Śunāsīrīya* was performed. The Vedic texts (e.g. *AB*. iv.26(xix.4); Chapter III, Section 2.7) stipulate that the ritual year (or the *sattra* of *Gavām ayana*) should commence in the cool season (*śiśira*) in the months of Māgha or Phālguna. The *cāturmāsya* sacrifices must have been a feature of the Vedic ritual year. For a start of the ritual year at new moon of the month of Māgha, the specified coincidence of the full moon and the *nakṣatra*s suggests that the *cāturmāsya* sacrifices were performed as follows:

- *Vaiśvadeva* in the second half of *śiśira* or first half of *vasanta* (approximately second half of January or first half of February).
- *Varuṇapraghāsa* in the second half of *grīṣma* (approximately second half of May).
- *Sākamedha* in the second half of *śarad* or first half of *hemanta* (approximately second half of September or the first half of October).
- *Śunāsīrīya*, if performed on the fifth full moon after *Sākamedha*, would have coincided with the performance of *Vaiśvadeva* of the following year. It was therefore performed on the day before the full moon (*ŚB* II.6.3.13; Eggeling 1882). However, this sacrifice could also be performed on any day after the *Sākamedha* sacrifice (*KŚS* 5.11.2; Ranade 1978).

The 'dates' imply that the sacrifices were performed at an interval of four synodic months in agreement with the name (*cāturmāsya*) of the sacrifices. The specification of the 'date' of performance of the sacrifices by the coincidence of the full moon and the *nakṣatra*s was probably a later development because the *nakṣatra*s were fully incorporated in the calendar only in the post-Ṛgvedic period. Vogel (1971) has speculated that the performance of *Vaiśvadeva* (spring) sacrifice when the full moon was in the *nakṣatra* Phālgunīs was an earlier practice and at a later date it was performed when the full moon was in the *nakṣatra* of Citrā. He does not provide textual evidence for this proposal nor does he discuss the performance of the summer and autumn sacrifices, these would have had to be shifted to the next *nakṣatra* if the spring sacrifice was shifted to Citrā. This shift would have been necessary to maintain the four (synodic) month interval between the sacrifices. Unless the Āryas had simply abandoned the rationale of the *cāturmāsya* sacrifices and performed these sacrifices at any time. This seems wholly unlikely. To perform the *cāturmāsya* sacrifices every year, when the full moon was in coincidence with these three specific *nakṣatras*, a correction of the synodic calendar by intercalation, would have been necessary.

The sacrifice of *Sākamedha* was performed when the seasonal full moon was in Kṛttikās/Pleiades (*KŚS* 5.6.1). This is the autumn sacrifice and the ecliptic longitude and latitude of the full moon in autumn are shown in Figure IV.4. The coordinates of the full moon obtained for the 1500-500 BCE have been used here. For *nakṣatra* Kṛittikās/Pleiades to conjoin the full moon of the first autumn of a *yuga*, it is necessary to precess the stars of this asterism to about 1400 BCE. The *yogatārā* (η Tau) of *nakṣatra* Kṛttikās precessed to 1400 BCE is shown in Figure IV.4. Consistency requires that the *yogatārā*s of seasonal *nakṣatras* Uttara-Phālgunīs and Uttara-Āṣāḍhās should also be precessed to this epoch and these are also shown in Figure IV.4. In this figure, the spring, summer, and autumn full moons or the second, sixth and, tenth full moon in the first year of a *yuga* are shown as black dots. The coincidence between the full moon of the first autumn of a *yuga* and *nakṣatra* Kṛttikās (or the *yogatārā* of this *nakṣatra*) is good, as expected. However, the coincidence between the full moon of the first spring and the first summer of a *yuga* and the *nakṣatras* (or the *yogatārā*s of their *nakṣatras*) prescribed for these seasons is poor. However, the lack of coincidence between the *yogatārā*s and the full moon for these two seasons is equivalent to an error of one day in the day of the full moon. In this analysis there is thus an implicit assumption that the coincidence between the seasonal full moon and the *yogatārā*s must have been determined from observations over an extended period and an error of a day (or two) in the full moon day would have been corrected. As the identity of *nakṣatra* Kṛttikās is secure, the epoch determined above is retained. The position of these seasonal full moons in the years following the first year of a *yuga* are also shown in Figure IV.4 (black dots). Because the synodic year is shorter than the tropical/seasonal year, the full moon 'drifts' to the left or to lower ecliptic longitude each year. If this is not corrected after first couple of years of a *yuga*, the *cāturmāsya* sacrifices will be performed when the seasonal full moon is in *nakṣatras* different from those prescribed in the Vedic texts. The correction required is the synchronization of the seasonal/tropical year (that fixes the position of the seasons) and the synodic year (that determines the position of the full moon). The seasonal full moons were corrected by the

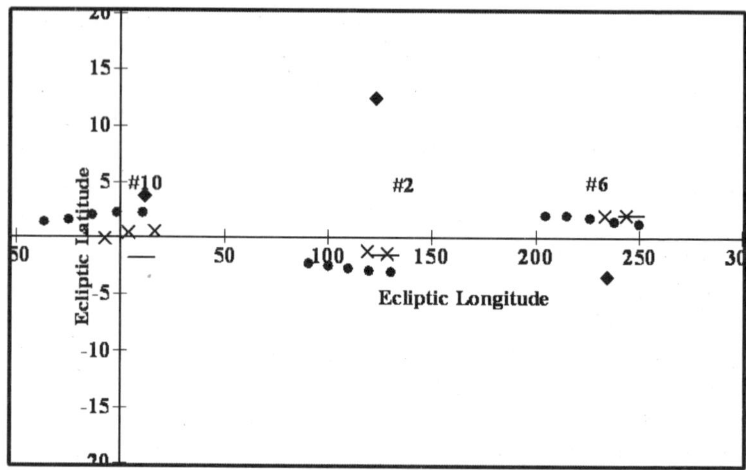

Figure IV.4: The ecliptic longitude and latitude of *cāturmāsya nakṣatras* (dimonds), *nakṣatra*-sectors (lines) and the second (spring), sixth (summer) and tenth (autumn) full moons (dots and crosses) in a *yuga*. The three sacrifices are; *Vaiśvadeva* at 2, *Varuṇapraghāsas* at 6 and *Sākamedha* at 10. The dots indicate the positions of the seasonal full moons without intercalation. The crosses indicate the positions of the seasonal full moons with intercalation. Only the positions of the full moons that have been corrected are shown, that is, the positions in the fourth and fifth year of the full moon 2 and 6 and positions in the third, fourth and fifth year of the full moon 10. The *nakṣatras* have been precessed to 1400 BCE.

intercalation scheme proposed in the *Vedāṅga Jyotiṣa;* A similar scheme is given in *MS*.I.10.8 (Chapter III, Section 2.7). That is, a synodic month was intercalated after 30 synodic months of a *yuga*, and an additional synodic month was intercalated after 61 synodic months of a *yuga*. The positions of the spring (full moon 2) and summer (full moon 6) full moons are not affected in the first three years of a *yuga* because the first intercalation is after the first 30 synodic months of a *yuga*. The position of full moons of these two seasons in the fourth and the fifth years of a *yuga* are affected, and these are shown as crosses in Figure IV.4. Similarly, the position of the autumn (full moon 10) full moon is not affected in the first two years of a *yuga* but the position in the last three years of a *yuga* is affected, and this is shown in Figure IV.4. After intercalation, all five seasonal full moons of a *yuga* cluster close to their prescribed *nakṣatras* and thus the *cāturmāsya* sacrifices would have been performed at their proper times every year of a *yuga*. This cycle repeats in the following *yuga*s.

The three *cāturmāsya-nakṣatra*s considered here will continuously precess away from their respective seasons (because of precession of the equinox). It is difficult to determine the period after which the three seasonal *nakṣatra*s prescribed in the Vedic texts would have been considered not to conjoin the seasonal full moons. This is because the accuracy with which observations may have been made during the Vedic period is not known nor is it possible to guess the separation between a seasonal full moon and a seasonal-*nakṣatra* that the Āryas would have accepted as coincidence. For lack of more data, it is assumed that the spread, after intercalation, in the ecliptic longitude of the five seasonal full moons of a *yuga* is the maximum separation between a full moon and a *nakṣatra* that would have been considered as coincidence. This suggests that a recalibration of the *nakṣatra*s and the seasons would have been required after about 500 years. However, there is no evidence in the Vedic texts that the *cāturmāsya* sacrifices were ever re-calibrated.

In the calendar of *Vedāṅga Jyotiṣa* (Chapter V), a *nakṣatra* does not mean a star or an asterism (as it does in the earlier Vedic texts) but, as shown above, it means a sector of the ecliptic. The width and the (invariant) coordinates of these *nakṣatra*-sectors have been discussed in Section 3.5. The three 'seasonal' *nakṣatra*-sectors are shown in Figure IV.4 as short black lines. As expected, the three 'seasonal full moons' in the first year of a *yuga* coincide with their respective *nakṣatra*-sectors. However, in the following years the 'seasonal full moons' drift away from their *nakṣatra*-sectors if not corrected for the lack of synchronism of the tropical/seasonal year and the synodic year. After intercalation the three 'seasonal full moons' in the five years of a *yuga* (and all the subsequent *yuga*s) cluster by their respective *nakṣatra*-sectors. No recalibration would have been necessary if the 'dates' of the *cāturmāsya* sacrifices had been specified by the coincidence of the seasonal full moon and the *nakṣatra*-sector as the sectors are not affected by precession of equinox. In the Vedic texts, there is no evidence that the *nakṣatra*s were ever 'recalibrated' to establish a 'new' match with the seasons. It is possible that the 'dates' of the *cāturmāsya* sacrifices were not always determined by the coincidence of the seasonal full moons with the *nakṣatra*s but by the coincidence of these full moons with the appropriate *nakṣatra*-sectors. It should be emphasized that even if the dates of the seasonal sacrifices had been determined by the coincidence of the full moon and a *nakṣatra*-sector, the correction by intercalation would have been necessary as the solar cycle (that determines the seasons) and the synodic cycle (that determines the phases of the moon) are not congruent.

6. VEDIC MONTHS REVISITED

The earliest names of Vedic months were probably seasonal names (Chapter III, Section 2.3) in the sense that a name described the weather and environment of the season. At some stage during the evolution of the Vedic calendar, the Āryas began to identify the months by *nakṣatra*s. A month was identified by the *nakṣatra* with which the full moon of that month had conjoined. The *nakṣatra*-names of the months (along with the seasonal names) are given in Table IV.7. A list of *nakṣatra* names of the months is not given in the Saṃhitās and the Brāhmaṇas and these names are relatively infrequent in these

Table IV.7: The Vedic seasons, the seasonal names and the *nakṣatra* names of the months. The *nakṣatra*s of the first 12 full moons of a *yuga* are given in the last column

Seasons	Vedic Seasons	Seasonal names of months	*Nakṣatra*-names of months	*Nakṣatra*s of first 12 full moons of a *yuga*
Cool	*Śiśira*	Tapas	Māgha	Maghās
		Tapasya	Phālguna	U. Phālguṇīs
Spring	*Vasanta*	Madhu	Caitra	Citrā
		Mādhava	Vaiśākha	Anurādhā
Summer	*Grīṣma*	Śukra	Jyaiṣṭha	Mūla
		Śuci	Āṣāḍha	U. Aṣāḍhās
Rains	*Varṣā*	Nabhas	Śrāvaṇa	Śraviṣṭhās
		Nabhasya	Bhādrapada	P. Proṣṭapadas
Autumn	*Śarad*	Iṣa	Aśvina	Aśvayujau
		Ūrja	Kārttika	Kṛttikās
Winter	*Hemanta*	Sahas	Mārgaśīrṣa	Mṛgaśīrṣa
		Sahasya	Pauṣa	Punarvasus

texts. This suggests that the *nakṣatra* names of months were a late development. These names have to be inferred from passages in the Saṃhitās, Brāhmaṇas and Sūtras that specify the times of rites or ceremonies, for example,

- One should get consecrated on the Phālgunī full moon day because Phālguna full moon is the mouth of the year (*TS*.VII.4.8; Ketih 1914).
- Perform the preliminary ceremony on the full moon of Caitra (*TB*.3.8.1; Dumont 1948).
- Lay down fire on the new moon of the month of Vaiśākha (*ŚB*. XI.1.1.7; Eggeling 1900).
- Consecrate themselves on one day after the new moon of Taiṣa/ Pauṣa or Māga (*KB*. xix. 2; Keith 1998).
- Perform the *pañcaśāradīya* in the sixth *śarad* season in the Kārttika month (*LŚS*.IX.12.13; Ranade 1998).
- When the herbs appear . . . in (the month of) Śrāvaṇa (*ĀGS*.III.5.2; Oldenberg 1964).
- On the full moon of Mārgaśīrṣa . . . (*ĀGS*.II.3.1; Oldenberg 1964).

The *nakṣatra*-names of the months suggest evolution in the precision (and reproducibility) with which months (and by inference seasons) were determined in the Vedic period. This method of timing ceremonies demands an ability to calculate the day when the full moon will be in conjunction with a specific *nakṣatra* (because the month itself starts on the previous new moon day). This suggests that the Āryas had 'calibrated' the *nakṣatra*s with respect to both the seasons and times of the full moon. The names of the months are derived from the names of the *nakṣatras* in which the moon is full. The Sanskrit grammarian Pāṇini (between 600 BCE and 400 BCE) has given rules for obtaining the name of the full-moon day and the month from the name of the appropriate *nakṣatra* (*Aṣṭādhyāyī* iv.2.22; *sāsmin paurṇamāsīti*). These names have survived for over 2,000 years and are in use in most parts of India today.

The *nakṣatra*s selected to name the months were those with which the full moon is in conjunction. In a year, the moon passes 12 times through the 27 *nakṣatra*s. Thus in a year the moon will be full when close to 12 *nakṣatra*s. However, the *nakṣatra*s are not separated by equal distances, the moon's orbital motion is not uniform, and the moon's sidereal period is not equal to its synodic period, so the moon will not be full 'besides' the same *nakṣatra* every year. Dīkṣita (1896) suggested that this nomenclature might have been introduced when the vernal equinox occurred in the month of Caitra. However, he does not give a rationale for the names of the months. To get round the confusion that would be caused in identifying a month from a *nakṣatra*, Sewell and Dīkṣita (1896) suggested that the months were named after groups of *nakṣatra*s and they allocated sometimes two and sometimes three *nakṣatra*s to a month for this purpose. However, they do not explain why they chose these groups of *nakṣatra*s to identify the months, and why the months were named after only one particular *nakṣatra* from this group. Vogel (1971) has proposed a similar scheme for grouping (and omitting) *nakṣatra*s to obtain the names of the months; however, his groupings and omissions are different. Saha and Lahiri (1992) suggested that the 12 *nakṣatra*s to name the months were selected because they are 'approximately at equal interval'. These authors have selected and grouped the *nakṣatra*s to match the known *nakṣatra* names of the months. To put it differently, they have 'retrofitted' the *nakṣatra*s to the names and have not identified the 'causal link' between a *nakṣatra* and the *nakṣatra* name of a month. Narahari Achar (2000b) has suggested that the months were named after *nakṣatra*s whose presiding deities can be identified with Agni, Prajāpati or Yajña. These *nakṣatra*s just happen to be distributed almost evenly along the path of the sun (ecliptic). The suggested groupings of *nakṣatra*s to obtain the *nakṣatra* names of the months are also illogical, as will be shown below. These authors have also ignored a fundamental aspect of the Vedic calendar, namely intercalation. Gondhalekar (2010) has shown that the schemes to synchronize the synodic year with the seasons, described in the Vedic texts, must be included in any discussion of the *nakṣatra* names of the months.

The Vedic texts suggest both spring (*vasanta*) and winter (*śiśira*) as likely seasons for the start of the year. The evidence for start of the year in winter is more extensive and will be considered first. The start of the Vedic year in winter (*śiśira*) is suggested in the Saṃhitās and the Brāhmaṇas (*AB*.iv.26 (xix.4); *KB*.xix.3; Keith 1998; *TS*.VII.4.8, Keith 1914; *PB*.V.9; Caland 1931; Chapter III, Section 2.7) and in the late Vedic text *Vedāṅga Jyotiṣa* (Chapter V). The calendar of *Vedāṅga Jyotiṣa* has an intercalation period of five years (*yuga*) and this period or *yuga* starts at new moon at (or near) the winter solstice when the moon and the sun are in the *nakṣatra* Śraviṣṭhās (Kuppanna Sastry 1984). Starting with the new moon in the *nakṣatra* Śraviṣṭhās; the *nakṣatra*s of the 62 full moons in a *yuga* are given in Table IV.4 and the *nakṣatra*s of the 12 full moons in the first year of a *yuga* are reproduced in column 5 of Table IV.7.

The positions (ecliptic longitude and latitude) of these full moons were calculated with the currently available orbital parameters of the earth and the moon, and the origin of the coordinate system was assumed to be at the First Point of Aries. The epoch of the *nakṣatra*s when the monthly (or seasonal) full moon conjoins its respective *nakṣatra* can be determined from the *cāturmāsya* sacrifices as has been described in Section 5 (above). The *yogatārā*s of the twelve *nakṣatras* that conjoin the first twelve full moons of a *yuga* were precessed to 1400 BCE. The proper motion (N.IV.10) of the *yogatārā*s has not been included in the precessed positions, as the small change that this will cause will not affect the conclusions reached here. As expected, the precession to 1400 BCE brings the spring, summer and autumn full moons in conjunction with the *nakṣatra*s prescribed for the *cāturmāsya* sacrifices. In the montage of Figure IV.5 are shown the positions of the *yogatārā*s of the *nakṣatra*s Bharaṇīs, Kṛttikāṣ and Rohiṇī and the positions of the tenth (autumn) full moon in the five years of a *yuga*; these positions are shown with and without intercalation (crosses and dots in Figure IV.5). The intercalation scheme used is that proposed in *Vedāṅga Jyotiṣa* that is, one synodic month is intercalated after 30 synodic months of a *yuga* and a second synodic month is intercalated after 61 synodic months of a *yuga*. The tenth full moon in the first year of a *yuga* is in the *nakṣatra* Kṛttikās. In successive years, the position of this full moon will 'shift to the left' or its ecliptic longitude will decrease (if not corrected by intercalation) because the synodic year is shorter (by about eleven days) than the seasonal/tropical year. After intercalation, the positions of the full moon in the third, fourth, and the fifth year of a *yuga* will be 'shifted to the right' or to the higher ecliptic longitude relative to the uncorrected positions. Judging from the (uncorrected) positions of the full moon on successive years of a *yuga* it would have been more logical to group Bharaṇīs and Kṛttikās rather then Kṛttikās and Rohiṇī as suggested by Sewell and Dīkṣita (1896). However, as can be seen from Figure IV.5, there is really no need to group Kṛttikās (from which the name of the tenth month is derived) with the adjacent *nakṣatra*s as the tenth (autumn) full moon of every year of a *yuga* will 'cluster' by this *nakṣatra* when intercalation is included. Similarly, there is no need to group the *nakṣatra*s Revatī, Aśvayujau and Bharaṇīs because the ninth full moon of every year of a *yuga*, after intercalation, clusters around the *nakṣatra* Aśvayujau as shown in Figure IV.5. It can be seen from Table IV.7 that the *nakṣatra*-names of eight months of a year can be derived unambiguously from the

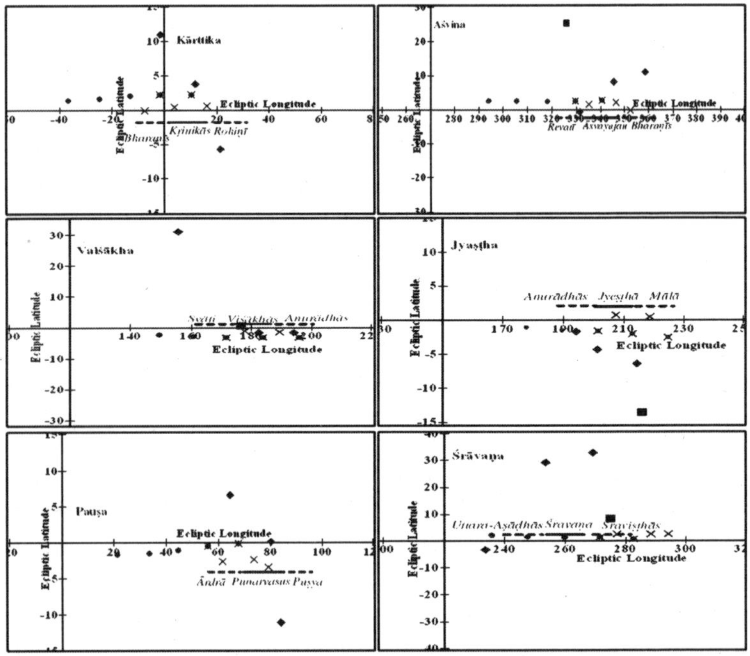

Figure IV.5: Examples of coincidence of full moons in a *yuga* with *nakṣatra*s and *nakṣatra*-sectors. The full moons without intercalation are shown as full dots and those with intercalation are shown as crosses. The *nakṣatra*s from the list of Pingree and Morrissey (1989) are shown as diamonds and those from the list of Abhyankar (1991) are shown as squares. If the *nakṣatra*s from the two lists coincide then the *nakṣatra* from Abhyankar (1991) is omitted. The *nakṣatra*s have been precessed to 1400 BCE. The *nakṣatra*-sector that identifies the month is shown by full line and the adjoining sectors are shown by dotted line. For clarity the *nakṣatra*-sectors have been off-set from the ecliptic.

*nakṣatra*s in which the moon is full during the first year of a *yuga*. With intercalation, at least three full moons for these months will cluster around these *nakṣatra*s in successive years of a *yuga*. Thus, the same *nakṣatra* names can be used for these months every year, as indeed they have been. However, the *nakṣatra* names of the remaining four months of a year cannot be derived from the *nakṣatra*s in which the moon is full in the first year of a *yuga* (Table IV.7) and it is worth looking at these in detail.

In Figure IV.5 are shown the ecliptic longitude and latitude of the *yogatārā*s of the *nakṣatra*s

- Svātī, Viśākhās, and Anurādhā
- Anurādhās, Jyeṣṭhā, and Mūlā
- Uttarāṣāḍhās, Śravaṇa, and Śraviṣṭhās
- Ārdrā, Punarvasus, and Puṣya

These are the *nakṣatra*s of the four 'discrepant' months in Table IV.7 and the adjacent *nakṣatra*s. The fourth full moon in the first year of a *yuga* is in the *nakṣatra* Anurādhā, but in the following years, the full moon for this month moves into Viśākhās and Svātī. After intercalation, three of the five full moons of a *yuga* for this month cluster by Viśākhās. It is because of this preferred clustering of full moons that this month is named after the *nakṣatra* Viśākhās and not Anurādhā. This is also true of the month of Jyeṣṭha as three of the five full moons of a *yuga* for this month (that is the fifth full moon) cluster by the *nakṣatra* Jyeṣṭhā after intercalation. However, this is not true of the seventh month of a *yuga*, this month is Śrāvaṇa but the intercalated full moons for this month cluster by Śraviṣṭhās (Figure IV.5) and the *nakṣatra* name of this month should have been derived from this *nakṣatra*. Similarly, the name of the twelfth month is derived from the *nakṣatra* Puṣya. However, after intercalation, in a *yuga* three full moons for this month cluster by Punarvasus and this month could have been better named after this *nakṣatra*. The Vedic texts provide no clues for alternative names for the months of Śrāvaṇa and Pauṣa, and it has not been possible to determine the reasons for the discrepancy in the *nakṣatra*-names of these two months when the *nakṣatra*-names of ten months of a year can be derived so unambiguously by the scheme proposed here.

The Vedic texts also suggest a start of the year when the spring full moon is in the *nakṣatra* (Uttara-)Phālgunīs (*TS*.VII.4.8; Keith 1914; *KB*.iv.4; *KB*.v.1; Keith 1920; *ŚB*.VI.2.2.18; Eggeling 1894; *PB*.V.9; Caland 1931; Chapter III and Section 4 above). The procedure described above can be repeated to obtain the *nakṣatras* of the full moons in a year starting with the spring full moon in Uttara-Phālgunīs. From this sequence of *nakṣatras* of the full moons, the names of only three months in a year can be derived; the remaining nine months would have *nakṣatra* names different from those known at present.

In the description of the coincidence of the *nakṣatras* and the full moons in a *yuga* (Figure IV.5), the *yogatārās* of the *nakṣatras* have been precessed to 1400 BCE. As discussed (Section 5), this coincidence between (*yogatārās* of the) *nakṣatras* and the 'monthly full moons' was probably acceptable for about 500 years either side of 1400 BCE. Beyond this limit, the (*yogatārās* of the) *nakṣatras* would have precessed far enough not to be considered to conjoin the 'monthly full moons'. It therefore, would have been necessary to assign new names to the months. In the Vedic texts, there is no evidence for renaming the months.

In the calendar of *Vedāṅga Jyotiṣa*, a *nakṣatra* does not mean a star or an asterism (as it does in the earlier Vedic texts). It means a sector of the ecliptic as discussed above (Section 3 above and Chapter V). In Figure IV.5, the full black bars denote the *nakṣatra*-sectors in which the moon is full in the first year of a *yuga*. The dashed bars denote the adjacent *nakṣatra*-sectors. After intercalation, the monthly full moon will be within the same *nakṣatra*-sector every year and a month therefore, will have the same *nakṣatra*-name. The months will retain their *nakṣatra*-names indefinitely, as the *nakṣatra*-sectors are not affected by precession of the equinox. The longevity of the *nakṣatra*-names of the months suggests that after the *nakṣatras* (stars and asterisms) had precessed away from their respective months (and seasons), a month was identified by the *nakṣatra*-sector in which the moon was full in a *yuga*. It should be stressed that intercalation has to be considered even if the *nakṣatra*-names of the months are derived from the names of *nakṣatra*-sectors

7. EPOCH OF THE *NAKṢATRAS*

The annual appearance of bright stars constitutes chronological markers in nature's calendar. This was noted by a number of ancient cultures (Frazer 1933; Aveni 2001) to fix important civil, religious and agricultural (or food gathering) dates in the year. There is considerable historical evidence (N.IV.3.) that the heliacal rising or setting of the Pleiades was observed by many ancient civilizations, including that of South Asia. Like the Kṛttikās/Pleiades, different stars and asterisms were probably added to the list of chronological markers as the civil and ecclesiastic ceremonies grew in number and complexity. It seems very likely that the Āryas, in order to continue their tradition of marking significant ceremonies by lunar phase, identified stars and asterisms along and around the ecliptic as chronological markers, which eventually became the 27 (or 28) *nakṣatra*s. It is impossible to establish the epoch(s) when the stars or asterisms were added to the list of *nakṣatra*s. However, when the *nakṣatra*s were reformulated as (invariant) sectors of the ecliptic, the formulators would have required a correspondence between the *nakṣatra*-sectors and the *nakṣatra*s that were already in use as chronological markers. To address the question, when was the list of *nakṣatra*s to represent the *nakṣatra*-sectors formulated, it is worth asking, how many *nakṣatra*s match their respective sectors in any particular epoch. The period between 2500-500 BCE was selected for this analysis. The date of 2500 BCE was selected arbitrarily. The final date of 500 BCE was selected as the putative end of the Vedic era. This period was divided into 21 epochs, each 100 years apart. Unfortunately, the only available identifiers of *nakṣatra*s are the *yogatārā*s, and the coordinates of the *yogatārā*s available at present are those based on Paitāmahasiddhānta. Thus, the analysis presented here will have all the caveats (described above) that go with the data from this text. Since there is no viable alternative to these data, the *yogatārā*s from Pingree and Morrissey (1989) and Abhyankar (1991) were used in this analysis. The J2000 coordinates (N.IV.5) of *yogatārā*s of the *nakṣatra*s were precessed to one of the epochs between 2500 and 500 BCE and the *nakṣatra*s whose ecliptic longitudes were within the width of the *nakṣatra*-sector they represent were identified. This was repeated for all 21 epochs in the period between 2500 BCE and 500

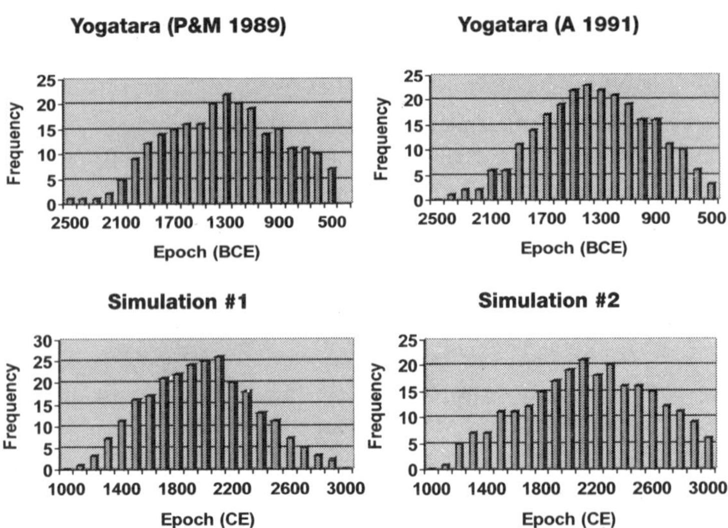

Figure IV.6: The cross-correlation of *yogatārā*s and *nakṣatra*-sectors. The frequency distribution obtained with *yogatārā*s identified by Pingree and Morrissey (1989) is shown in the top left panel and that with *yogatārā*s identified by Abhyankar (1991) is shown in the top right panel. The results with pesudo-*yogatārā*s selected at epoch CE 2000 are shown in the bottom two panels. The frequency distribution for pesudo-*yogatārā*s around the position of the full moons in a *yuga* are shown in Simulation 1 and that around the position of the moon for 27 nights of a month is shown in Simulation 2. See text for detail.

BCE. The results of this cross-correlation are shown in Figure IV.6. For both sets of data, 80 per cent of the *yogatārā*s fall in their respective *nakṣatra*-sectors at 1300±300 BCE (N.IV.11). The (surprisingly) large number of coincidences between the *yogatārā*s and their respective *nakṣatra*-sectors within this 'window' around 1300 BCE suggests that this may well have been the epoch when the list of *nakṣatra*s was created. It is possible that *nakṣatra*s to identify *nakṣatra*-sectors were selected from a list of *nakṣatra*s that were already in use as chronology markers. From this list, the *nakṣatra*s that corresponded to the *nakṣatra*-sector were (probably) retained, those that did not were discarded, and new *nakṣatra*s were added for *nakṣatra*-sectors that had no corresponding *nakṣatra*. It is highly likely that Kṛttikās is one of these 'old' *nakṣatra*s. The large number of coincidences also suggests that the *nakṣatra*s and their identifying *yogatārā*s have been transmitted with considerable fidelity over the gap of more than a thousand years between the Vedic texts and Paitāmahasiddhānta.

To verify the results of cross-correlation of the *yogatārā*s and the *nakṣatra*-sectors, a list of 27 *nakṣatra*s and their *yogatārā*s was created for the current epoch, i.e. CE 2000. The criteria the Āryas may have used to select the *nakṣatra*s and their *yogatārā*s are not known but as discussed above, *nakṣatra*s that conjoin the full moon have a central role in the Vedic calendar. They identify the months and the date of the three seasonal sacrifices. In addition, in the Vedic texts, the *nakṣatra*s close to the moon were used to identify the days of a (synodic) month. Two simulations were performed; in the first, the dates of all full moons in a *yuga* starting with the new moon near the winter solstice (on 22 December 1995) were identified and the coordinates (ecliptic) of the full moons were computed with the current orbital parameters for the moon and the earth. The *cāturmāsya* (or seasonal) sacrifices define the *nakṣatra*s of the full moon in spring, summer, and autumn. These three *nakṣatra*s establish the 'absolute' *nakṣatra*-grid (at epoch 2000) for the full moons of this *yuga*. That is, the *nakṣatra* of the full moon in spring (in year 2000) was designated (Uttara-)Phalgunīs, that of the full moon in summer was (Uttara-) Āṣāḍhās and that of the full moon in autumn was Kṛttikās. Since each full moon is separated by $29^{22}/_{124}$ *nakṣatra*s (see Section 3.1 above), the *nakṣatra*s of all other full moons in the *yuga* can be determined from Table IV.1. If two full moons in a *yuga* have the same *nakṣatra*, only one was selected to create a list of 27 *nakṣatra*s (at epoch 2000) and their associated full moons. A *yogatārā* for each *nakṣatra* was selected from SIMBAD Astronomical Database J2000.0 (N.IV.5) catalogue of stars.

The following criteria were used to select these *yogatārā*s:

- Only stars brighter than sixth magnitude were selected
- Only stars with ecliptic latitude within ±20° of the ecliptic were selected.
- Only stars with ecliptic longitude within ±5° of the corresponding full moon were selected.
- If there was a choice of stars, then the brightest was selected.

The J2000 coordinates of these pseudo-*yogatārā*s (or J2000-*yogatārā*s) were cross-correlated with the coordinates of the *nakṣatra*-

sectors, i.e. the cross-correlation described above was replicated for these pseudo-*yogatārā*s. The frequency distribution for this cross-correlation is shown as Simulation 1 in Figure IV.6. The profile of this frequency distribution is similar to that of the frequency distribution for the cross-correlation of the *yogatārā*s and the *nakṣatra*-sectors, but the peak of the distribution is around CE 2000, the epoch of the psudo-*yogatārā*s.

For the second simulation, the *nakṣatra*s close to the moon on 27 nights of a month were determined. The *nakṣatra* on the full moon night in autumn was designated Kṛttikās (as prescribed for the *cāturmāsya* sacrifices, Section 6) and the *nakṣatra*s of the thirteen nights on either side of this night were obtained from Table IV.1. A *yogatārā* for each *nakṣatra* was selected from SIMBAD Astronomical Database J2000.0 catalogue of stars, using the criteria given above. The J2000 coordinates of these pseudo-*yogatārā*s (or J2000-*yogatārā*s) were cross-correlated with the coordinates of the *nakṣatra*-sectors, and the frequency distribution for this cross-correlation is shown as Simulation 2 in Figure IV.6. The peak of this distribution is also around CE 2000, the epoch of these psudo-*yogatārā*s.

The distribution of *yogatārā*s at epochs other than those around 1300±300 BCE (or CE 2000, in the simulations in Figure IV.6) is to be expected because the *nakṣatra*-sectors are wide and will straddle more than one epoch. In addition, misidentification of *yogatārā*s is likely and should be expected when data have been transmitted over a considerable period. These uncertainties will also be included in the frequency distribution of the simulated *yogatārā*s at the current epoch. Each night the moon does not move through a complete *nakṣatra*-sector, and depending on the choice of (pseudo-)*yogatārā*s, some *nakṣatra*-sectors will have two *yogatārā*s and this will influence the frequency distribution of the cross-correlation of the 'nightly' pseudo-*yogatārā*s and the *nakṣatra*-sector (Simulation 2). Notwithstanding these deficiencies, these simulations demonstrate that the frequency distribution of the cross-correlation of the *yogatārā*s and the *nakṣatra*-sectors peaks around the epoch at which the *nakṣatra*s were selected. There is, therefore, a high probability that *nakṣatra*s listed in the Vedic texts were selected at 1300±300 BCE (N.IV.11).

In the montage of Figure IV.7 are shown the 27 *nakṣatra*-sectors and the *yogatārā*s from the identifications of Pingree and Morrissey (1989; black dot) and Abhyankar (1991; cross) precessed to 1300 BCE. Five *yogatārā*s from the list of Pingree and Morrissey (1989)

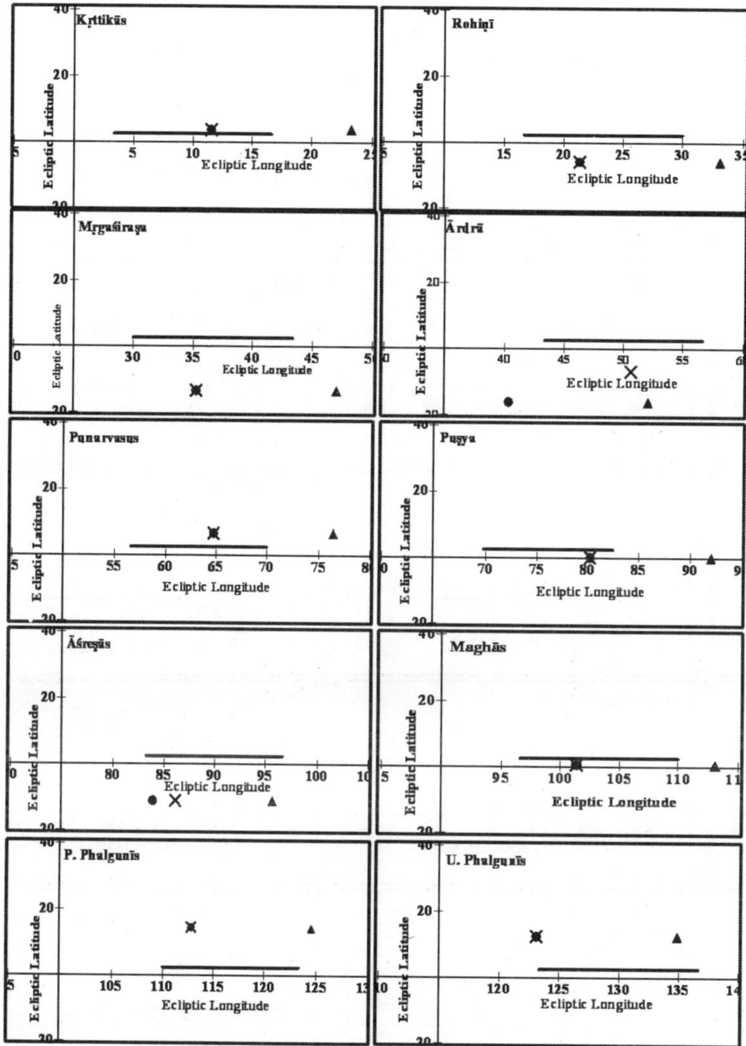

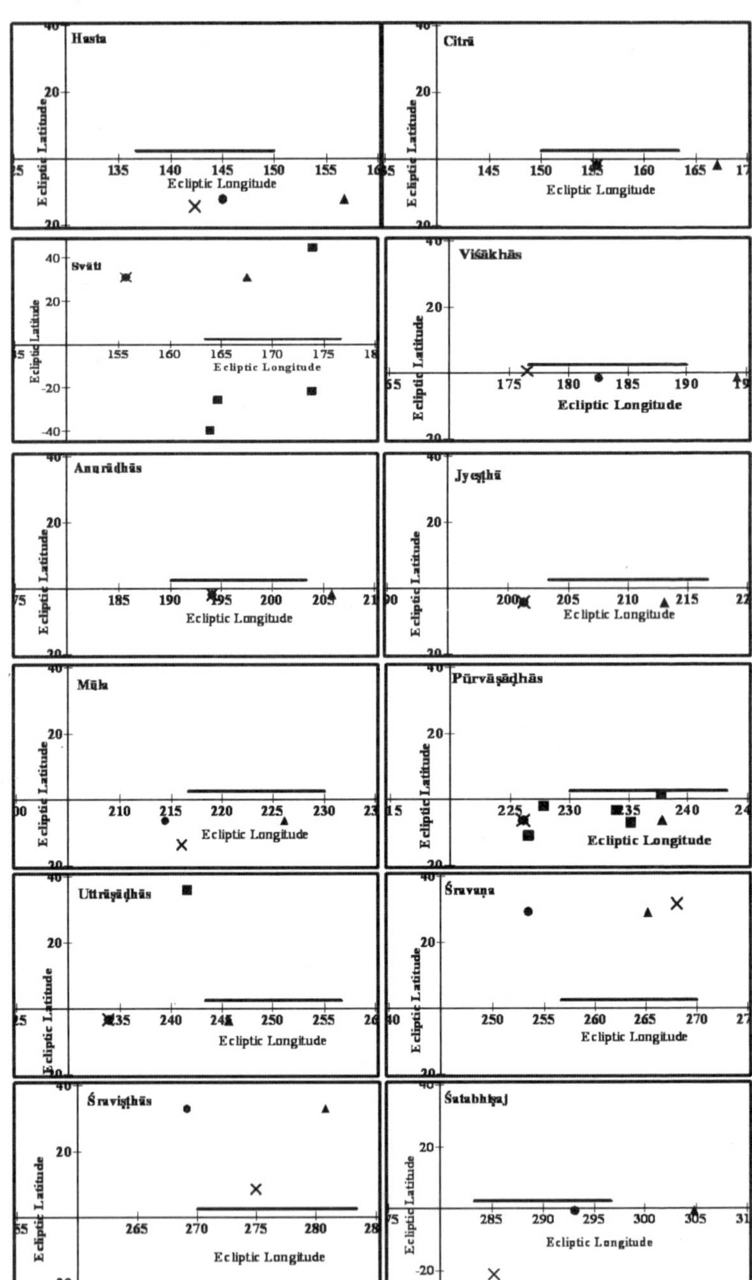

Hasta
Citrā
Svāti
Viśākhās
Anurādhās
Jyeṣṭhā
Mūla
Pūrvāṣāḍhās
Uttrāṣāḍhās
Śravaṇa
Śraviṣṭhās
Śatabhiṣaj
Ecliptic Latitude
Ecliptic Longitude

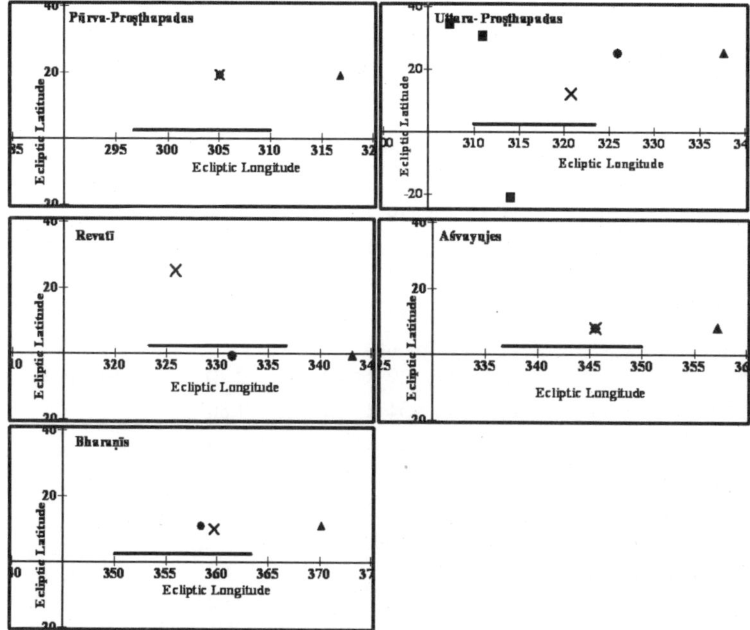

Figure IV.7: The 27 *nakṣatra*s and *nakṣatra*-sectors. The *yogatārā*s from Pingree and Morrissey (1989) are shown as dots, and those from Abhyankar (1991) are shown as crosses. These *yogatārā*s have been precessed to 1300 BCE. The *nakṣatra*-sectors are shown as straight lines and have been off-set from the ecliptic for clarity. Filled squares are stars (precessed to 1300 BCE) which match the *nakṣatra*-sectors better than the proposed *yogatārā*s (see text and Table IV.8 for detail). The triangles are *yogatārā*s from Pingree and Morrissey (1989) precessed to 500 BCE.

and five from Abhyankar (1991) do not fall in their own *nakṣatra*-sector around 1300 BCE. Four *nakṣatra*-sectors that do not have a 'matching' *yogatārā* in the list of either Pingree and Morrissey (1989) or Abhyankar (1991) are given in Table IV.8. Also, given in Table IV.8 are possible alternatives to these *yogatārā*s, stars that do fall in the appropriate *nakṣatra*-sector at this epoch. Only stars brighter than 3rd magnitude are given. The criteria that the Āryas may have used to select the *nakṣatra*s and their *yogatārā*s are not known. These may not have been the criteria used by Pingree and Morrissey (1989) and Abhyankar (1991) that is, brightness and proximity to the ecliptic, and

Table IV.8: *Yogatārās* from the identifications of Pingree and Morrissey (1989; P&M) and Abhyankar (1991; A) that do not match their *nakṣatra*-sectors. Alternative stars (brighter than 3rd mag.) that do match the *nakṣatra*-sectors are also given.

Star ID	α, δ (J2000)	mag	Ref
	Svātī		
α Boo (Arcturus)	14 15 40; 19 12 37	-0.04	P&M; A
γ Cen	12 41 31; -48 57 36	2.18	
ι Cen	13 20 36; -36 42 44	2.70	
θ Cen	14 6 41; -36 22 12	2.06	
	Pūrvāṣāḍhās		
δ Sgr	18 20 59; -29 49 40	2.70	P&M; A
ε Sgr	18 24 10; -34 23 5	1.80	
λ Sgr	18 27 58; -25 25 18	2.83	
ζ Sgr	19 2 37; -29 52 48	2.61	
ϖ Sgr	19 9 46; -21 01 25	2.89	
	Uttrāṣāḍhās		
σ Sgr	18 55 16; -26 17 46	2.02	P&M; A
ζ Aql	19 05 25; 13 51 49	2.99	
	Uttara-Proṣṭapadās		
α And	00 08 23; 29 05 26	2.06	P&M
γ Peg	00 13 14; 15 11 01	2.83	A
η Peg	22 43 00; 30 13 16	2.95	
β Peg	23 03 46; 28 04 58	2.42	
β Cet	00 43 35; -17 59 12	2.04	

it is these criteria that have been used here to suggest the alternative identifications for *yogatārās*. The *yogatārā* α Boo (Arcturus) appears to have been selected as the thirteenth *nakṣatra* (Svātī) because it is a bright star. It is not within the *nakṣatra*-sector *Svātī* and a better identifier for this sector would be one of the alternatives suggested in Table IV.8. The *yogatārā* δ Sgr is a poor identifier of the eighteenth *nakṣatra*-sector Pūrvāṣāḍhās as brighter stars around the centre of the *nakaṣtra*-sector could have been selected. Similarly, a brighter star could have been selected to identify the twenty-fourth *nakṣatra*-sector (Uttara-Proṣṭapadas) rather than α And or γ Peg the current *yogatārās* of this *nakṣatra*. Similar choice of a brighter star is not possible for the identification of the nineteenth *nakṣatra*-sector (Uttarāṣāḍhās).

Some Western scholars have claimed that the scheme of *nakṣatra*s and mathematical astronomy generally were imported into South Asia from Mesopotamian sources during the Achaemenid occupation of parts of South Asia (513-326 BCE). That is, 500 BCE was the epoch of both the Vedic *nakṣatra*s and *Vedāṅga Jyotiṣa* (specially the Ṛgvedic recension; see Chapter V). In the montage of Figure IV.7 all *yogatārā*s precessed to 500 BCE are shown by black triangles. At this epoch, only six (22 per cent) *yogatārā*s 'fall in' their respective *nakṣatra*-sectors. This shows that these *nakṣatra*s and their respective *yogatārā*s could not have been used to represent these *nakṣatra*-sectors at or after 500 BCE. The scheme of dividing the ecliptic into 27 *nakṣatra*-sectors is unique to *Vedāṅga Jyotiṣa* (and is not mentioned in the Mesopotamian astronomy texts) and must have been developed before or during the formulation of this text. If it is assumed that *Vedāṅga Jyotiṣa* was composed (borrowed) around 500 BCE, then the very few coincidences between the *nakṣatra*-sectors and the *yogatārā*s suggests that the composers of *Vedāṅga Jyotiṣa* ignored the list of *nakṣatra*s from the Saṃhitās and the Brāhmaṇas. The composers created an entirely new list of *nakṣatra*s to identify the *nakṣatra*-sectors, and this list is now lost. Considering the innate conservatism of cultures generally and the Āryan culture particularly, it is extremely unlikely that the Āryas, at the time of composition of *Vedāṅga Jyotiṣa*, would have abandoned the list of *nakṣatra*s from earlier texts. Timing the passage of a *yuga* and keeping track of the new and full moons from the start of the *yuga* to identify *nakṣatra*-sectors is possible, but prone to error. It is likely that the *nakṣatra*-sectors and their identifying list of *nakṣatra*s were defined at the time of formulation of *nakṣatra*-sectors. As shown earlier (Figure IV.6), the most likely epoch for this is 1300±300 BCE because of the maximum number of coincidences between the *nakṣatra*-sectors and their representing *nakṣatra*s (or *yogatārā*s) occur at this epoch. This list of *nakṣatra*s would have included *nakṣatra*s from earlier epochs that were important to the religious and social life of the Āryas.

Corroboration for this as the epoch of the *nakṣatra*s is difficult to find in the Vedic texts. However, the coincidence of the (*yogatārā*s of the) *nakṣatra*s prescribed for the performance of the *cāturmāsya* sacrifices and the full moon of the appropriate seasons is only possible around this epoch. This epoch is independent of the *yogatārā*s identified from Paitāmahasiddhānta, and is based on the robust identification of *nakṣatra* Kṛttikās with the Pleiades. The description, in *Vedāṅga Jyotiṣa*, of the season, the position of the sun and the *nakṣatra* that conjoin the sun, and the moon at the start of *yuga*, provide supporting evidence (Gondhalekar 2008a) for the proposed epoch of the *nakṣatra*s (this will be discussed further in Chapter V). The references in Vedic texts to the 'cask or barrel turned over by various gods to pour riches on the earth', discussed above, suggests that *sapta ṛṣi*s was a circumpolar asterism throughout the year, when these texts were composed. Viewed from Madhyadeśa (Delhi 77°12′E; 28°35′N) all stars of *sapta ṛṣis* are not circumpolar at present. Gondhalekar (2011) has shown that the constellation would have been 'high in the firmament/night (*uccā nakaṃ*)' and circumpolar before 1500 BCE. Whilst this is only circumstantial evidence for the epoch discussed above it does suggest that around 1500 BCE, the night sky and the relative position of the stars were under scrutiny of the Āryas.

NOTES

1. It has been claimed (Pingree 1989) that the *nakṣatra*s are of Mesopotamian origin and were 'copied', along with most of mathematical astronomy, by South Asians sometime in the fifth century BCE. The Mesopotamian certainly had star lists of their own, and it is worth comparing these with the *nakṣatra*s to access the validity of this claim. The primary source of information on Mesopotamian astronomy is the Assyrian astronomical compendium *MUL. APIN* (The Plough), which in its present form may be dated to the century around 686 BCE (Hunger and Pingree 1989). The general catalogue lists 60 stars or constellations which later included five planets. The stars and constellations are distributed along the 'paths of Enlil, Anu, and Ea', which correspomd to sectors of the eastern horizon over which the constellations rose. The path of Enlil extended north from about +17°, of Anu between +17° and -17° and Ea south from about -17°. All other Mesopotamian star lists are derived from this general catalogue. Like the *nakṣatra*s, the stars and constellations of *MUL.APIN* are distributed along the ecliptic. However, there are fundamental differences. The star catalogue of *MUL.APIN* has a number of applications and provides information on calendric parameters that were of importance and relevance to the Mesopotamians. The *nakṣatra*s have a limited application; to define the position of the moon along the ecliptic. The *nakṣatra*s were always 27 (sometimes 28) and never included the planets. The Kṛttikās/Pleiades is the first constellation in the lists of *nakṣatra*s in all

Vedic texts. The Pleiades are the 31st constellation in the general catalogue of *MUL.APIN* and the fifth constellation in the path or list of Anu (Hunger and Pingree 1989). The Pleiades are the first constellation in the *MUL.APIN* list of simultaneously rising and setting stars or constellations. The Pleiades are also the first constellation in the *MUL.APIN* list of 17 constellations along the path of the moon. A number of stars and constellations are common to the list of *nakṣatra*s and the Mesopotamian list of 17 stars and constellations along the path of the moon. This is hardly surprising (and proves nothing) as both people looked at the same sky and identified stars along the ecliptic. Āryas had 27 *nakṣatra*s because they identified a (synodic) day by the *nakṣatra* that was in the vicinity of the moon on that day. In *MUL.APIN* the days are never identified by stars in the vicinity of the moon. In *Vedāṇga Jyotiṣa*, which, it is also claimed, was copied from Mesopotamian sources, the *nakṣatra*s are not stars (or constellations) but evenly spaced equal sectors of the ecliptic (Chapter IV, Section 3). This concept is not found in *MUL.APIN*.

2. As the earth rotates on its axis, the stars appear to revolve around the celestial poles. In the northern hemisphere the stars will appear to revolve around the northern celestial pole and in the southern hemisphere, the stars will appear to revolve around the southern celestial pole. Depending on an observer's latitude (no star is circumpolar from the equator), stars that are close to the celestial pole will remain above the horizon, that is, they will never set. These stars are called 'the circumpolar (circle around the pole) stars' because they are visible for the entire night on every night of the year (and would be continuously visible throughout the day but for the glare of the sun). Because of the precession of the equinox, the same group of stars will not be circumpolar at all epochs.
3. The Mesopotamian and South Asian were not the only cultures to give a position of prominence to the Pleiades. Scholars who see the first place of Pleiades in the Mesopotamian and South Asian star lists as 'proof' that the South Asian astronomy was borrowed from the Mesopotamians would do well to broaden their knowledge-base. In his monumental study of myths, *The Golden Bough*, Frazer (1933: 307-19) has shown that the Pleiades have a central role in the calendar of cultures of Americas, Africa, Polynesia, (indigenous) Australia, etc. The star group accorded the place of prominence throughout Mesoamerica is the Pleiades (Aveni, 2001: 29-32). In Japanese astronomy, the only constellation with a Japanese name is the Pleiades (*Subaru*); all other constellations have Chinese names. This suggests that the Pleiades were important to the Japanese before they encountered Chinese astronomy. All these cultures observed the appearance and disappearance of this conspicuous group of stars to identify the time for food gathering or to hunt for certain foods or time of commencement of agricultural activities like planting or harvesting. Consider two examples, one from the 'old' world and one from the 'new' world. In Hesiod's *Works and Days* 'When the Pleiades, daughters of Atlas, are rising, begin your harvest, and your ploughing when

they are going to set, ll. 383-404'. The New Year Day of the Inca calendar was on June 8/9, coinciding with the heliacal rising (the rising just before sunrise) of the Pleiades. The importance of the Pleiades for intercalation has been noted above. It is unlikely that all these 'under-developed' cultures 'borrowed' the Pleiades as chronological marker from the Mesopotamians. The Pleiades are not difficult to observe, although individual stars of this cluster are faint, their combined light and the light emitted by the gas ionized by the stars, spreads over a considerable area of the sky and is easily recognizable. The importance of this constellation to (almost) all cultures is due to a unique 'accident' in astronomy; its conspicuousness in the sky, its proximity to the ecliptic and its heliacal rising and setting during periods that are significant in the calendar of food and fodder gathering and production (i.e. spring and autumn). The importance of the Pleiades in the calendars of different cultures lies in the Pleiades and not with the Mesopotamians.

5. The precession of earth's axis (N.IV.9) results in a steady increase of about 50 arc seconds per year (or 1° in 71.6 years) in the ecliptic longitude of a star. It is, therefore, essential, when reporting the coordinates of an astronomical object, to qualify the coordinates with a date or an epoch. Specifying the epoch identifies the equator and equinox that define the coordinate system that is being used. The coordinates of astronomical objects in catalogues are defined with respect to a single mean equator and equinox. Thus, J2000 coordinates are coordinates with respect to the mean equator and equinox of CE 2000.

6. Modular arithmetic (or clock arithmetic) is a system of arithmetic for integers in which the numbers 'wrap around' after reaching a certain value called the modulus. Consider an analogue 12-hour clock; if it is 6:00 a.m. now, 8 hours later it should be 14:00 but because of the limiting value or modulus 12, the time will be 2:00 p.m. In the application of this arithmetic to the *nakṣatras*, the modulus is 27. Thus the tenth *nakṣatra* counted from Śraviṣṭhās (*nakṣatra* 21, Table IV.1) will be *nakṣatra* 31, but there are only 27 *nakṣatras* and by modular arithmetic, the required *nakṣatra* will be *nakṣatra* 4. Symbolically the simplest operation of modular arithmetic is expressed as

$$Y \equiv X \text{ modulo } (Z)$$

which means Y ÷ Z remainder X. The use of the simplest form of modular arithmetic is implicit in the formulation of the *jāvādi* arrangement of *nakṣatras* and a number of algorithms given in *Vedāṅga Jyotiṣa*. Carl Friedrich Gauss (1777-1855) introduced modular arithmetic to modern mathematics, in his book *Disquisitiones Arithmeticae*.

7. Astronomers generally use two coordinate systems to identify the position of a celestial object (planets, stars, etc.) in the sky; these are the equatorial coordinate system and the ecliptic coordinate system. The equatorial system is analogous to the terrestrial longitude and latitude, i.e. the equatorial plane is projected on to the sky and an object's position is given by right ascension

and declination. The declination is an angular measure along an arc from the equator to the pole, and the right ascension is measured eastwards along the equator and is given in 'hours' rather than 'degrees' to emphasize the clock like behaviour of the celestial sphere. In the second coordinate system, the position of a celestial object is defined with respect to the ecliptic and the ecliptic pole. The ecliptic latitude is measured along an arc from the ecliptic to the ecliptic pole. The ecliptic longitude is measured eastwards along the ecliptic. The ecliptic is inclined (by about 23.5°) to the equator and intersects the equator at two points known as the vernal/spring equinox and the autumnal equinox (Chapter III). The origin of both celestial coordinate systems is known as the First Point of Aries and is the point where the equator and ecliptic intersect in spring or vernal/spring equinox. The First Point is not fixed in space but moves along the ecliptic due to precession of earth's axis. When the equinox was first recognized (in the first millennium BCE), the point of intersection of the celestial equator and the ecliptic was in the constellation of Aries hence the name. Today this point is in the constellation of Pisces.

8. The Julian date is the interval in days and fraction of a day measured from 1 January 4713 BCE. The Julian date system was introduced by the French astronomer Joseph Scaliger in 1583, at the time of Gregorian calendar reform. Apart from the name, this Julian date is not related to the Julian calendar introduced by Julius Caesar in 46 BCE. The Julian date system or the continuous reckoning of days (Julian days) was introduced to provide a single system of dates that could be used when working with different calendars. In South Asia, the continuous reckoning of days was introduced by Āryabhaṭa I (CE 476-525) in his *Āryabhaṭīya*. He called it the *ahargaṇa* (heaps of days) 'system of time reckoning'.
9. The precession of the equinox is the gradual shift in the earth's axis of rotation similar to the conical path traced by the spin axis of a spinning top. This precession is caused by gravitational drag of the moon and the sun on earth's equatorial bulge. The period of precession is 25,771.5 years or 1° every 71.6 years. Earth's precession is historically called 'precession of the equinox' because the equinoxes move westwards along the ecliptic relative to the stars. This motion is opposite to the motion of the sun along the ecliptic. The precession of earth's axis has a number of observable effects. The north and south celestial poles appear to move in a circle against the background stars. The position of the earth in its orbit around the sun, defined relative to the seasons, gradually changes. Therefore, a star or a group of stars prominent in one season will gradually move into the next season.
10. The proper motion of a star is its movement on the celestial sphere. This motion is caused by the true movement of the star relative to the solar system. Proper motion reflects only transverse motion, i.e. the component of motion

across the line of sight to the star; the component of motion towards or away from the sun is called the radial motion of the star. The proper motion of stars is very small, and over historical times, the shape of constellations does not appear to change. However, precise long-term observations show that the shape of constellations does change as each star has an independent movement in the constellation. Amongst the 27 *nakṣatra*s, Svātī (α Boo) has the highest proper motion, but over 4,000 years this causes a change of only 0.3° in ecliptic longitude and 2.5° in ecliptic latitude.

11. It is impossible to calculate the error in the date of 1300 BCE obtained for the maximum in the frequency distribution of correlations between the *yogatārā*s and the *nakṣatra*-sectors. The estimated spread (±300 years) is the width of the distribution (Figure IV.7) at the point where the distribution has dropped by $\sqrt{27} \approx 5$ below the peak value. The implicit (and inaccurate) assumption here is that the *yogatārā*s are random events.

V

Vedāṅga Jyotiṣa

Vedāṅga Jyotiṣa – the astronomical 'arm' or auxiliary (N.V.1) of the Veda is the earliest mathematically codified calendric text of South Asia. The text is a manual for determining the appropriate times of Vedic ceremonies and rituals. The Vedic period has left us a rich store of religious, sacerdotal and liturgical texts but little on secular matters. *Vedāṅga Jyotiṣa* and *Śulbasūtras* (N.I.2.) are the only two known texts of this period that do not discuss liturgy or ritual matters and are the only texts that can be considered 'scientific and mathematical' texts.

Vedāṅga Jyotiṣa has survived in two recensions – a *Ṛgveda* recension called Ārca-Jyotiṣa and a *Yajurveda* recension called Yājuṣa-Jyotiṣa (N.V.2). There are minor differences between the two recensions; the Ṛgvedic recension has 36 verses, 30 of these are repeated in the Yajurvedic recension, and there are 13 additional verses in this recension. The Ṛgvedic recension is attributed to Lagadha but, unfortunately, nothing else is known about him. The author of the *Yajurveda* recension is not known. This recension does have a (rather unsatisfactory) *bhāṣya* or commentary by Somākara. The *Ṛgveda* recension is considered the older of the two. *Vedāṅga Jyotiṣa* is not a treatise on the calendar but a hand-book for ritual officiants (probably) educated orally, to make the necessary observations and calculations to determine the correct time for a ceremony. It is possible that the text was used mostly to jog the memory of an officiant, who was familiar with the detail. The contents of the available recensions are not in a thematic or logical order and topics on the same subject are distributed in different parts of text. This suggests that the original text has not come down to us. The language of Ārca-Jyotiṣa is post-Vedic, but not yet classical. Pingree (1972, 1981) has dated this recension to fourth or fifth century BCE. This date is based on his belief that Mesopotamian calendric methods and parameters were transmitted to India during the Achaemenid occupation of the Indus valley between 513 and 326 BCE. There is no internal or external evidence to support this belief. Pingree (1972) dates the Yājuṣa-Jyotiṣa to third or fifth century CE.

The early Indologists like William Jones and Henry Thomas Colebrooke had noticed *Vedāṅga Jyotiṣa* and the prospects it offered of determining Vedic chronology. The Ṛgvedic recension was first published in 1834 by Captain Jarvis, who was investigating the measurement of time in India. *Vedāṅga Jyotiṣa*, written in terse Sanskrit verse is not for the faint hearted and attempts to interpret it, by the nineteenth-century Sanskritist William Dwight Whitney were unsuccessful. He dismissed it as 'delusive phantom' (Whitney 1874). This, however, did not deter other nineteenth-century scholars. In 1862, the German Indologist Weber (1862) published both recensions of the *Vedāṅga Jyotiṣa* from the manuscripts available to him. He too was unable to interpret the text, apart from a few easy verses. Thibaut (1877) took a second stab at the *Yajur*-recension; he was able to interpret some of the difficult verses but left out several. Dīkṣita (1896) presented and discussed the interpretation and translation (in Marathi) of 36 of the total of 49 verses. Tilak (1925) has summarized attempts made in the first quarter of the twentieth century to 'decipher' *Vedāṅga Jyotiṣa*. Further attempts to translate and interpret it culminated in the Sanskrit commentary and English translation by Shamasastry (1936). In 1979, a translation and interpretation of *Vedāṅga Jyotiṣa* was produced by Kuppanna Sastry (1984); this, along with critical editions of both the Ṛgvedic and Yajurvedic recensions (from 20 manuscripts) produced by K.V. Sarma, were published in 1984 by the Indian National Science Academy. The discussion in this chapter is based on this edition and translation of *Vedāṅga Jyotiṣa*.

A full exposition of *Vedāṅga Jyotiṣa* is not attempted here. Here a number of algorithms of *Vedāṅga Jyotiṣa* are discussed and described to demonstrate the continuity with the calendric tradition of the Saṃhitās and the Brāhmaṇas. The verses of *Vedāṅga Jyotiṣa* are not reproduced; these have taxed the ingenuity and scholarship of a number of eminent Sanskritists of last 150 years and would be incomprehensible to most lay persons.

1. PREAMBLE

The introductory verses of *Vedāṅga Jyotiṣa* follow the standard format of Vedic texts with salutations and obeisance to the gods and the teacher (*RJ*.1, 2, 3, verse 1, 2 and 3 of the Ṛgvedic recension; *YJ*.1, 2, verse 1 and 2 of the Yajurvedic recension) and descriptions of the benefits (eternal life, progeny, etc.) to be accrued from the knowledge of the text (*RJ*.30; *YJ*.43). *Vedāṅga Jyotiṣa* is a text on 'the effect on time of movement of the luminaries (the sun, the moon and the stars)'. The principal segment of time is the five-year *yuga* (*RJ*.1,3; *YJ*.1,2) which consists of days, months, seasons, and *ayana*, the (apparent) northern and southern course of the sun. *Vedāṅga Jyotiṣa* notes (*YJ*.3) that the purpose of the Vedas is the performance of sacrifices and these have to be performed at appropriate times. Only he who has knowledge of the science-of-time (*kālavidhānaśāstra*) understands the performance of sacrifice. The contents list of the *Vedāṅga Jyotiṣa* (*RJ*.1, 3; *YJ*.1, 2) gives only the wide subject areas covered by the text, and the detail included in each area is not given. The broad subject areas addressed in the *Vedāṅga Jyotiṣa* can all be traced to the Saṃhitās and the Brāhmaṇas.

2. THE CALENDAR OF *VEDĀṄGA JYOTIṢA*

The calendar described in *Vedāṅga Jyotiṣa* is the culmination of a process that started in the Vedic Saṃhitās and Brāhmaṇas. These texts mention the division of the day, into *muhūrtas* and perhaps also time-keeping, or the measurement of time defined by the *muhūrta*. Stray references in these texts (see Chapter III) suggest that the passage of time may have been measured by the duration of chants. *Vedāṅga Jyotiṣa* ends this uncertainty by introducing a mechanical means of measuring time – a clepsydra (a water-clock), discussed in Section 2.1. *Vedāṅga Jyotiṣa* also introduces three additional units of time; the four units of time in *Vedāṅga Jyotiṣa* define respectively the solar time or the passage of the day (*muhūrta*), mechanical time (*nāḍikā*), synodic time (*kalā*) or the passage of the moon, and the 'local time'. The 'local time' is most intriguing; it is 124th part of a day measured from sunrise or the east and thus defines the position of the sun in the sky. A 'unit' for this time is not given and there is no indication of how this time was measured.

Vedāṅga Jyotiṣa follows the Saṃhitās and Brāhmaṇas in its list of *nakṣatra*s, and these define the positions of the moon in the sky. However, unlike the earlier texts, this text never mentions stars (and asterisms) as markers of lunar positions. In *Vedāṅga Jyotiṣa* the *nakṣatra*s are wide sectors of the ecliptic and by implication (no proofs or measurements are given) are of equal width (Chapter IV, Section 3.1). In the earlier Vedic texts, only the position of the moon was described or defined by the *nakṣatra*s. In a significant departure, *Vedāṅga Jyotiṣa* defines the position of both the moon and the sun by the *nakṣatra*s. However, it should be noted that the *nakṣatra*s with respect to which the apparent motion of the sun and the moon is described, are not the 27 lunar *nakṣatra*s of the Saṃhitās and the Brāhmaṇas, but 27 *nakṣatra*-sectors. The position of the sun and the moon are defined by a *nakṣatra*-sector and the separation of these bodies from the leading-edge of this sector. The twelve 'solar *nakṣatras*' of the later South Asian astronomical texts have not made their appearance in *Vedāṅga Jyotiṣa*.

In *Vedāṅga Jyotiṣa* there is no evidence of systematic day-by-day observations and the algorithms used to obtain various parameters of a calendar are based on the mean motion of the sun and the moon, that is, the motion of these objects was observed over a long period and an average set of parameters computed. A number of new astronomical terms of calendric significance are introduced in *Vedāṅga Jyotiṣa*; these have not previously been mentioned in the Saṃhitās or Brāhmaṇas. These terms are not defined in *Vedāṅga Jyotiṣa* and it is worth looking at the definitions of these terms before proceeding to the discussion of the principle topics.

Sāvana (day) – day from sunrise to sunrise or a civil day. A *sāvana* year is 366 *sāvana* days. A *sāvana* day is sub-divided into 124 parts and a quarter of a day or 31 parts is called a *pāda* (step).

Tithi – In the Saṃhitās and the Brāhmaṇas the synodic month of 29.53 days is rounded-up to 30 days. In *Vedāṅga Jyotiṣa* this discrepancy is confronted by retaining the number of days in a synodic month but by redefining the day. A synodic day or a *tithi* is 1/30th of a synodic (lunar) month, that is a *tithi* is 29.53 ÷ 30 = 0.984 *sāvana* day (or 122 parts of a *sāvana* day). A synodic year is 360 *tithi*s. Thus, the *Vedāṅga Jyotiṣa* calendar-makers abandoned the 360 day ritual year and replaced it with the 360 *tithi* (synodic) year. The *tithi* as a 'lunar day' is not encountered in the Saṃhitās and the Brāhmaṇas. The word '*tithi*' occurs once, in *Aitareya Brāhmaṇa* (*AB*.vii.11 (xxxii.10); A.V.1), and it is not clear if it means 'a lunar day'. The *tithi* is not defined in *Vedāṅga Jyotiṣa* but is a crucial parameter of this calendar. Since the moon's motion around the earth is not uniform (N.III.3), the *tithi*s in a year will not be of equal length. The average length of a *tithi* is 23.62 hours but the actual length can vary from 20.0 hours to 26.8 hours. Since the *tithi* varies in length, sometimes a *tithi* had to be skipped and sometimes two days have the same *tithi*. By defining a *tithi*, the ritual requirement of a synodic month of 30 days and a ritual year of 360 days are retained but the days are 'lunar days' or *tithes*. The synodic (lunar) month is divided into two *pakṣa* or fortnights. The bright fortnight or the *śukla pakṣa* is from new moon to the first fifteenth *tithi*s and the dark fortnight or the *kṛṣna pakṣa* is the following fifteen *tithi*s. Within a *pakṣa*, *tithi*s are identified by their ordinal number; the 15th *tithi* of the bright fortnight is the full moon or *pūrṇimā*, and the 15th *tithi* of the dark fortnight is the new moon or *amāvāsyā*. The *tithi* of the moon at sunrise is taken to be the *tithi* for the day.

Yuga – this is the fundamental unit of the calendar of *Vedāṅga Jyotiṣa* and equals 62 lunations or 1830(.86) *sāvana* days. Each *yuga* starts in the month of Māgha at new moon around the winter solstice (*RJ*.32; *YJ*.5; see Section 1.6). As discussed later (Section 1.5), the lunar calendar is harmonized with seasons over this period by intercalation of two synodic months. *Vedāṅga Jyotiṣa* does not explain the derivation of the *yuga* and takes it as given. This period is implicit in the Ṛgvedic recension but is given in Yajurvedic recension (*YJ*.28) rather indirectly by the length of the seasonal (*sāvana*) year, which is given as 366 *sāvana* days (the *yuga* is five *sāvana* years). This period of 1830 *sāvana* days (during which the synodic year is harmonized with seasons) follows from the mathematical interpretation of *TS*.VII.4.8 (see Chapter III, Section 2.7). The *yuga* is discussed in detail in Section 1.6.

Parvan – 124th part of a *yuga*. *Parvan* means a 'joint' and there are 124 joints in a *yuga*. The 'value' of a *parvan* is equal to that of a *pakṣa* (both are equal to 15 *tithi*) but whereas the *pakṣa* refers to a single lunation, the *parvan* is used to count the number of half lunations in a *yuga*. *Vedāṅga Jyotiṣa* does not provide the rationale for dividing a *yuga* into 124 parts but in the Vedic texts a *pakṣa* – a fortnight – is considered *kevala* (complete, not connected with anything else; *TB*.3.12.8.7, 8) and it is possible that a *parvan* was considered a fundamental unit. The total number of *parvans* (*parvan-rāśi* – a heap of *parvans*) from the start of a *yuga* can be obtained from the number of elapsed years, months, and *parvans* (*RJ*.4; *YJ*.13). In each year, there are 12 synodic months, and in each synodic month there are 2 *parvans*, thus:

$$\textit{parvan-rāśi} = (12\times(y-1)+m)\times 2+p+\mathrm{K}$$

where y is the (current) year of a *yuga*, m is the number of elapsed months in the current year and p is the number of elapsed *parvans* in the current month. The correction K is 2 *parvans* for every 60 *parvans* in the *parvan-rāśi*. This is the intercalation of one synodic month (2 *parvans*) after 30 synodic months (60 *parvans*) in a *yuga*. This intercalation is required to synchronize five synodic years with five tropical years (see Section 1.5). Since a *yuga* starts at new moon (Section 1.6) all odd numbered *parvans* end a bright fortnight (that is at full moon) and all even numbered *parvans* end a dark fortnight (that is at new moon).

Bhāṃśa (also *aṃśa*) – 124th part of a *nakṣatra*-sector (Chapter IV, Section 1.7) and is used to measure the position of the sun and the moon in a *nakṣatra*-sector. A *yuga* is 1830 *sāvana* days or 1,860 *tithi* (see Section 1.6). In a *yuga* the sun makes five revolutions of the circle of 27 *nakṣatra*-sector. Thus in one *tithi* the number of *bhāṃśas* the sun crosses are

$$5\times27\times124\div1860=9$$

If the *parvan* from the start of a *yuga* is known then the *bhāṃśa* of the sun can be calculated:

1 *yuga* = 124 *parvans*
the sun traverses through 27 *nakṣatras* in 1 year, so
in 1 *yuga* the sun traverses through 5×27 = 135 *nakṣatras*

the number of *nakṣatra*s traversed in
1 *parvan* = 135÷124 = 1 (complete) *nakṣatra* + 11 *bhāṃśa*s
the *bhāṃśa* of the sun after *p parvan* = $11 \times p$

The complete *nakṣatra*s-sector is ignored as only the current *bhāṃśa* of the sun is required. For the same reason the final number of *bhāṃśa* (i.e. $11 \times p$), is factored by 124 to determine the complete number of *nakṣatra*s-sectors, and these are ignored because only the *bhāṃśa* of the sun and not its *nakṣatra*s-sector is required. To put it differently, the *bhāṃśa* of the sun is X, where $11 \times p \equiv X$ modulo (124) (N.IV.5). If the value of the *bhāṃśa* of the sun is odd, then the *parvan* has ended a bright fortnight. That is, the moon is full and it will be 13½ *nakṣatra*s-sectors from the sun (because the moon will be in opposition). Thus, the *bhāṃśa* of the moon is the *bhāṃśa* of the sun plus 62 (again ignoring the complete number of *nakṣatra*s-sectors). Similarly, if the value of the *bhāṃśa* of the sun is even, the *parvan* has ended a dark fortnight or it is a new moon and the *bhāṃśa* of the moon is same the as that of the sun, as the sun and the moon will be in conjunction.

For reasons not entirely obvious, *Vedāṅga Jyotiṣa* goes about this in a somewhat involved way (*RJ*.10; *YJ*.15). The reasoning appears to be as follows:

Suppose the sun has traversed *p parvan*
In a synodic month, there are 2 *parvans* and in a synodic year there are 12×2 *parvans*
Therefore *p parvans* can be represented as $p = 12 \times 2y + r$
Were y is the number of synodic years and r is the remainder of *parvans* or $p = 12q + r$ $(q = 2y)$
In 124 *parvans* (*yuga*) the sun traverses through 135 *nakṣatra*-sectors
Therefore, the number of *nakṣatra*-sectors traversed in *p parvan* is
$p \times 135 \div 124$
Substituting for *p*, the *nakṣatra*-sectors traversed are
$135\,(12q + r) \div 124$
Now $(135 \times 12q) \div 124 = (13^{8}/_{124})q$
And $(135 \times r) \div 124 = (1^{11}/_{124})\,r$
Since only the *nakṣatra*-sector and *bhāṃśa* at the end of the *parvan* are required

The complete number of *nakṣatra*-sectors can be ignored and the the *bhāṃśa* of the sun is $b = q \times 8 + r \times 11$
(where *parvan* $p \div 12 = q$ (quotient) + r (remainder))

The *bhāṃśa* obtained is factored by 124 (or modulo (124)) to eliminate the complete number of *nakṣatra*s-sectors traversed. The *bhāṃśa* of the moon is obtained as described above.

Kalā – 603rd part of a (*sāvana*) day or $2^m\ 23^s$ (see Table V.1). This unit is used to time the motion of the moon across a *nakṣatra*-sector. The origins of this (rather cumbersome) unit may be in the desire to obtain an integer number for the 'time' taken by the moon to traverse a *nakṣatra*-sector. In a sidereal month (interval between successive passages of the moon by the same star) the moon traverses past 27 (*nakṣatra*s or) *nakṣatra*-sector and in a *yuga* of 1830 (*sāvana*) days there are 67 sidereal months (see Section 1.6). Thus, the moon crosses every *nakṣatra*-sector in (1830÷27×67) or $^{610}/_{603}$ days (since all *nakṣatra*-sectors are of equal angular width). If a day is divided into 603 units (*kalā*s) then the moon will cross one *nakṣatra*-sector in 610 units (*kalā*s).

Ṛtuśeṣa – The cycle of six seasons (*ṛtu*) is completed in (12 months, two months per season) 365.24 (*sāvana*) days, or 371.18 *tithi*s. The synodic year is 360 *tithi*s. Thus for the moon to complete the cycle of seasons, a synodic year plus 11.18 *tithi*s are required. These additional *tithi*s are the *ṛtuśeṣa* (*RJ*.23; *YJ*.41).

2.1. Measurement of Time

In *Vedāṅga Jyotiṣa* time is measured by a clepsydra (a water-clock) but the clock itself is not described. There has been much

Table V.1: Units of time in *Vedāṅga Jyotiṣa*

1 *ahorātra* (day)	30 *muhūrttas*	24^h
1 *muhūrta*	2 *nāḍikās*	48^m
1 *nāḍikā*	$10^1/_{20}$ *kalās*	24^m
1 *kalā*	124 *kāṣṭās*	$2^m\ 23^s$
1 *kāṣṭā*	5 *akṣaras*	1.15^s
1 *akṣara*	2 *mātrās*	0.57^s

discussion on water-clocks in South Asia. Pingree (1972) contends that these are Mesopotamian inventions, imported into South Asia during the Achaemenid occupation of South Asia in the fifth and fourth century BCE. Ôhashi (1993) has discussed in detail the astronomical instruments and observations in the Vedic period of South Asia. He has shown that the water-clocks of the post-Vedic period (no descriptions exist of water-clocks that may have been in use during the earlier period) were significantly different from those described in the Mesopotamian texts, and there is no *a priori* reason to believe that these were not South Asian inventions. There may also be a reference to a water-clock in the Vedic text, *AV*.XIX.53, 54. These two *sūktas* are known as the *kālasūkta*s and are dedicated or addressed to Eternal Time (*kāla*) and are considered to have theological and cosmogonic character. Narahari Achar (1998a) has argued that a word division (*padavibhāga*) rather different from that given by Whitney (1962) can suggest that *AV*.XIX.53.3 refers to mundane time and to the measurement of time with a water-clock (A.V.2). He has also suggested that this verse may refer to a water-clock of the overflowing type, one of the two types mentioned in the post-Vedic texts (the other is the floating bowl clock).

In *Vedāṅga Jyotiṣa* the measured unit of time is the *nāḍikā* (*YJ*.24) and corresponds to the (out) flow of 50 *pala* of water. Unfortunately, nothing in the text allows us to relate a *pala* to a modern unit of weight or volume. However, the *nāḍikā* is linked to the absolute unit of time – a day – via the *muhūrta* (*RJ*.16, 17; *YJ*.38) and it would not have been difficult to calibrate the water-clock in 'absolute' units. The relation between *muhūrta* and *nāḍikā* leads to the relation between *nāḍikā* and *kalā* since a *sāvana* day of 30 *muhūrta* is also 603 *kalā* (see definition of *kalā*, above). Since a *kāṣṭā* is $^{1}/_{124}{}^{th}$ part of a *kalā* (by definition), it follows that $^{1}/_{124}{}^{th}$ part of a *sāvana* day is equal to 603 *kāṣṭā*s. Similarly, the moon will traverse one *nakṣatra*-sector in 610 *kalā*s (see definition of a *kalā* above) and since a *nakṣatra*-sector is 124 *bhāṃśa*s wide, the moon will traverse one *bhāṃśa* in 610 *kāṣṭā*s. The *nāḍikā* may also have been linked to the musical notes or speech via *akṣara* – each of two *mātrā*s (*YJ*.12b, 30d, 39d). A *mātrā* is the length of time required to pronounce a short vowel and a long vowel requires two *mātrā*s or an *akṣara*. The units of time in *Vedāṅga Jyotiṣa* converted to the modern units of time are shown in Table V.1. Apart from the *muhūrta* these units of time are different from the units of time described in the Saṃhitās and the Brāhmaṇas (Table III.4).

2.2 Days

Vedāṅga Jyotiṣa recognizes two (types) of days; the *sāvana* or civil day and the lunar day (see definitions above). Of these only *sāvana* (sun-rise to sun-rise) has physical reality, the lunar day is a mathematical entity constructed for a calendar. *Vedāṅga Jyotiṣa* describes the measurement of (fractions of) a (*sāvana*) day by mechanical means (water clock). *Vedāṅga Jyotiṣa* also recognizes that the time of the day can be defined by the position of the sun in the sky and the diurnal cycle (*nāḍīmaṇḍala*) is divided into 124 *bhāgas* (parts) with the origin at sunrise. The 'time' of the day (or the position of the sun on this scale) can be determined if the *parvan* and *tithi* are known. The time at the end of a *parvan* can be obtained as follows:

1 *yuge* = 1,830 (*sāvana*) days = 124 *parvan*

1 *parvan* = 1,830÷124 days

p (*parvan*) = $p \times 1{,}830 \div 124$ days

$= 14p + 0.758p$

$= q + r$

where q is the complete number of days (elapsed) and
r is the fraction (on a scale of 124) of the day elapsed at the end of the *parvan* p

Vedāṅga Jyotiṣa obtains this time in a slightly different (and involved) way (*YJ*.12).

1 *parvan* = 1,830÷124 days

= 15 days – 30 parts (on a scale of 124) of a day

= 15 days – (approx.) ¼ day

therefore p *parvans* = $(15 \times p)$ complete days – $(\tfrac{1}{4} \times p)$ days

ignore the complete days.

$p \div 4 = q$ (quotient) + r (remainder)

Thus, the fraction of the day (on a scale of 124) at the end of *parvan* p is

p if $r = 0$

$p + 93$ (3 *pāda*) if $r = 1$

$p + 62$ (2 *pāda*) if $r = 2$
$p + 31$ (1 *pāda*) if $r = 3$
If the sum is greater than 124 then it has to be factored by 124
The remainder is the part of the day (on a scale of 124) when the *parvans* p end

The complete number of days is ignored as only the time in the current day is required. Surprisingly, this algorithm is not given in the Ṛgvedic recension although it is required by *RJ*.20 as shown below. If both the *parvan* and the *tithi* (at the end of the *parvan*) are known, then the time at the end of the *parvan* (obtained above) has to be corrected for the elapsed *tithis* to obtain the time of the day or the position of the sun in the *nāḍīmaṇḍala* (*RJ*.20; *YJ*.22). Since a *tithi* is smaller than a *sāvana* day (of 124 parts) by two parts:

Time of the day (on a scale of 124) = time at the end of *parvans* – 2×*tithi* (of the day)

The sun will be at the end of the part (on a scale of 124) obtained above and the parts are counted from east. The division of the diurnal cycle into smaller intervals suggests a possible 'mechanical' means of measuring time; this could be by timing (with a water-clock) or by measuring the length of the shadow of a gnomon (*śaṅku*). *Vedāṅga Jyotiṣa* is silent on gnomon and timing may have been the preferred method of measuring fractions of a day.

2.3. Seasons and the Year

Vedāṅga Jyotiṣa divides the tropical year into 366 (*sāvana*) days (*YJ*.28) and in each year there are six *ṛtu* (seasons), two *ayana* (the apparent northern and southern course of the sun) and twelve months. There is here a significant departure from the Saṃhitās and the Brāhmaṇas, which never explicitly refer to the exact length of the tropical year. References to the tropical year can be traced back to the earliest Saṃhitās but the length of the year is always couched in uncertain terms; it is twelve months plus extra or it is longer than 364 days and shorter than 366 days (Chapter III, Section 2.7). A year of 366 days can be derived from the possible intercalation schemes in these early texts (Chapter III, Section 7). This value of the length of the year in *Vedāṅga Jyotiṣa* is obtained from the 1830 days in a five-year *yuga*. There is no reference in *Vedāṅga Jyotiṣa* of a measurement of the length of the tropical year. It is interesting to note that this value of the year is not mentioned in the Ṛgvedic recension and is only mentioned in the later Yajurvedic recension. This value of the length of a year has no significant role in the calendar of *Vedāṅga Jyotiṣa* and it is extremely unlikely that the value was of any significance to the Vedic ritualists. Five (such) years make a *yuga* (*YJ*. 28). Both the five-year period (with individually named years) and the word *yuga* signifying a long period, occur in the Saṃhitās and the Brāhmaṇas. In addition, a five-year period of intercalation occurs explicitly in the calendar discussed in *MS*.I.10.8 and implicitly in the calendar defined by *TS*.VII.4.8 (Chapter III, Section 2.7). However, *Vedāṅga Jyotiṣa* is the only Vedic text in which a five-year period is explicitly called a *yuga*.

In the Saṃhitās and the Brāhmaṇas a *ṛtu* (season) is two ritual months (each 30 days) long and the start of a season (celebrated by a *cāturmāsya* sacrifice) is on the full moon that conjoins a predefined *nakṣatra* (Chapter IV, Section 6). In these texts, the length of a season is never described in terms of the motion of the moon past the *nakṣatras*. However, *Vedāṅga Jyotiṣa* expresses a *ṛtu* in terms of both the synodic months and the *nakṣatra*-sectors:

- A *ṛtu* is two synodic months plus two *tithis* (*YJ*.11). A seasonal year has six *ṛtus*; since a *yuga* is five seasonal years, a *yuga* has 30 *ṛtus*. A *yuga* is also 62 synodic months long (see Section 1.5 below), thus a *ṛtu* is 62÷30 synodic months long or 2 synodic months plus 2 *tithis* (as each synodic month is 30 *tithis*). This also means that each *ṛtu* is 62 *tithis* (*YJ*.11).
- In a year of six *ṛtu* the sun traverses through 27 *nakṣatra*-sectors or each *ṛtu* is four and a half *nakṣatra*-sectors long (*RJ*.9; *YJ*.10).

2.4. Intercalation – Synodic and Sidereal

The intercalary month to synchronize the tropical and the ritual or quasi-synodic years is introduced in the Saṃhitās and the Brāhmaṇas, but the rationale for it is not stated. In the calendar of *TS*.VII.4.8, synodic months are intercalated through the *ekāṣṭakā* condition and in *MS*.I.10.8 an arithmetic rule is prescribed for intercalating two synodic months in a period of five seasonal years or 60 synodic months (Chapter III, Section 2.7). The approach in *Vedāṅga Jyotiṣa* is more formal, it explains that the synodic/lunar day (*tithi*) is shorter than the *sāvana* day. This difference is reconciled by adding (intercalating) one synodic month (*adhimāsa*) at the end of half the *yuga* and another synodic month at the end of the *yuga* (*YJ*.37). This verse unambiguously states that intercalary months are required to synchronize the synodic years with the seasons in a *yuga*. An intercalary month also follows from the definition of *ṛtuśeṣa* given above. One *ṛtu* is 62 *tithi*s long (*YJ*.11) and 15 *ṛtu*s pass in 930 *tithi*s or 30 synodic months (of 30 *tithi*s) plus 30 *tithi*s or one synodic month. Fifteen *ṛtu*s (seasons) are completed in about 30 tropical months. That is, after 30 synodic months, it is necessary to add one synodic month to keep the synodic year synchronized with the tropical year or the cycle of seasons.

The intercalation scheme of *Vedāṅga Jyotiṣa* is remarkably similar to that described in *MS*. I.10.8, the only difference being in the time when the first synodic month is intercalated. The intercalation scheme of *Vedāṅga Jyotiṣa* is modelled, and the accumulated difference (in days) between the tropical years and the intercalated years (see Chapter III, Section 2.7 for detail) is shown for a period of 500 years in Figure V.1. For comparison, the data for curves 1 and 2 of Figure III.4 (Chapter III, Section 2.7) are also plotted in this figure. To obtain Curve 3 of Figure V.1, the days in the synodic year were calculated by assuming each synodic month to be 29.53 days long. A month was added (intercalated) after the 30th and 61st synodic months respectively, and the length of the intercalated month was assumed 29.53 days. To obtain Curve 4 of Figure V.1, alternate months of a year were assumed to be respectively 29 and 30 days long, and the intercalated month was assumed to be 30 days long. The extra month was added after the 30th and 61st quasi-synodic months. Both these schemes are consistent with the scheme described in *Vedāṅga Jyotiṣa* and that described in *MS*.I.10.8. In both, the accumulated difference increases less rapidly than that for a scheme in which a month of 30 days is intercalated every fifth year (shown in curve #1 of Figure V.1 and Figure III.4). The increase is nevertheless quite fast and the schemes are far from the stability of the corrected five-

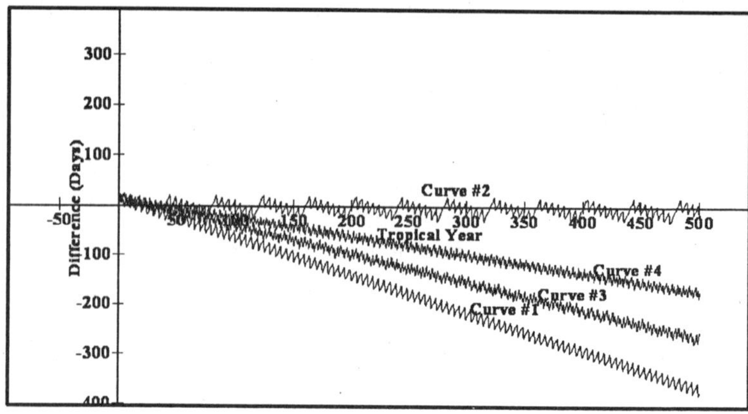

Figure V.1: The accumulated difference (in days) between the tropical year and the intercalated year. Curves 1 and 2 are reproduced from Figure III.4. To obtain curve 1 a month of 30 days was intercalated every fifth (ritual) year. To obtain curve 2 the algorithm of curve 1 was modified by not intercalating a month every fortieth year. Curve 3 was obtained by assuming a synodic month of 29.53 days and a month was intercalated every 30th and 61st synodic month in a *yuga* of 62 synodic months. Similarly, curve 4 was obtained by assuming alternate months of a year to be respectively 29 and 30 days long. A month of 30 days was intercalated after the 30th and 61st quasi-synodic months of a *yuga* of 62 quasi-synodic months.

year *yuga* scheme shown in Curve 2 (see Chapter III, Section 2.7). *Vedāṅga Jyotiṣa* does not mention any correction of the intercalation scheme it describes. However, the difference of 2.36 *tithi*s between 31 synodic months and thirty tropical months would have been obvious. Ritualists who were very particular about the timing of their rituals and ceremonies would not have tolerated the inaccurate calendar that results from this intercalation scheme. It is possible that the correction scheme is missing from the versions of the *Vedāṅga Jyotiṣa* that have come down to us. It is also possible that this intercalation scheme was included in *Vedāṅga Jyotiṣa* as an example of a method considered in earlier times and the calendar makers of *Vedāṅga Jyotiṣa* had developed a more stable intercalation scheme.

In *Vedāṇga Jyotiṣa* a *yuga* starts 'when the sun and the moon occupy the same region of the sky with the *nakṣatra* Śraviṣṭhās' (*RJ*.5-6; *YJ*.6-7), at this time also begins 'the (synodic) month of

Māgha, the (seasonal) month of *tapas* . . . the sun and the moon begin to move north . . .'. These verses have been interpreted as – the New *Yuga* Day is the day when the new moon at the winter solstice conjoins the *nakṣatra* Śraviṣṭhās (the sun will also conjoin *nakṣatra* Śraviṣṭhās as it will be in conjunction with the moon). The new moon, of course, cannot be at the winter solstice every year. A realistic interpretation of these verses would be that a *yuga* starts when the new moon conjoins *nakṣatra* Śraviṣṭhās around the winter solstice. This would also bring the sun at the winter solstice close to *nakṣatra* Śraviṣṭhās. *Vedāṅga Jyotiṣa* does not describe how 'when the sun (and the moon) occupy the same region of the sky with the *nakṣatra* Śraviṣṭhās' (*RJ*.5-6; *YJ*.6-7) was determined. Thus to interpret *RJ*.5-6 and *YJ*.6-7 it is necessary to examine the annual visibility of *nakṣatra* Śraviṣṭhās. The annual position of α *Delphini* (the *yogatārā* of Śraviṣṭhās identified by Pingree and Morrissey (1989)) at astronomical twilight (N.V.3) is shown in Figure V.2; these data are the altitude of Śraviṣṭhās at astronomical twilight in the morning and the evening, from the beginning of January to the end of December. The Śraviṣṭhās are close to the horizon on four days. At #1, the Śraviṣṭhās will be just above the eastern horizon at sunset and will be visible during the night. At #2, the Śraviṣṭhās are on the western horizon just before sunrise and will set soon afterwards. At #3, the Śraviṣṭhās are on the western horizon and will be visible after the sun sets and the stars will set soon afterwards (stars set heliacally). At #4, the Śraviṣṭhās are on the eastern horizon just before sunrise (stars rise heliacally) and will be lost in the glare of the sun as it rises. It is these last two days that are of interest in interpreting verses *RJ*.5-6 and *YJ*.6-7 of *Vedāṅga Jyotiṣa*. If it is assumed that the verses *RJ*.5-6 and *YJ*.6-7 mean that the sun is 'visually' close to *nakṣatra* Śraviṣṭhās, then this is only possible at heliacal rising and setting of this *nakṣatra*. The data in Figure V.2 are for 1500 BCE and a location in Madhyadeśa (Delhi 77°12'E; 28°35'N). Because of the precession of the equinox, the day of the year when the *nakṣatra* rises or sets heliacally will vary with the epoch (every 72 years the day of heliacal rising or setting will increase by one). The day of the year, between 2000 and 500 BCE, when α *Delphini*/Śraviṣṭhās rises and sets heliacally is shown in Figure V.3. Around 1550 BCE, the *nakṣatra* Śraviṣṭhās rises heliacally at the winter solstice (note, computational error of one day is possible

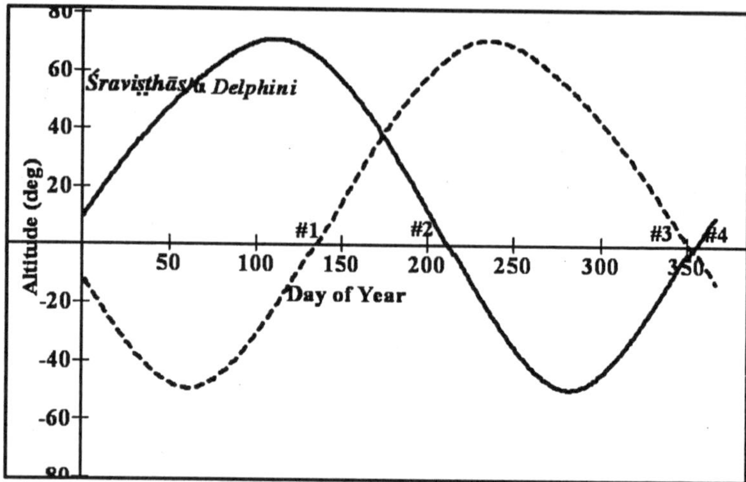

Figure V.2: The annual apparent path of *nakṣatra* Śraviṣṭhās. The data plotted are the altitude of the star α *Delphini* (*yogatārā* of Śraviṣṭhās identified by Pingree and Morrissey (1989)) at astronomical twilight (N.IV.4). The full line curve is the altitude of the star at morning twilight and the dashed line curve is the altitude of the star at evening twilight. The epoch of the data is 1500 BCE. The annual path of the star intersects the horizon on four days. At #1 the Śraviṣṭhās will be just above the eastern horizon at sunset and will be visible during the night. At #2 the Śraviṣṭhās will be on the western horizon just before sunrise and will set soon afterwards. At #3 the Śraviṣṭhās are on the western horizon and will be visible after the sun sets and the stars will set soon afterwards (stars set heliacally). At #4 the Śraviṣṭhās are on the eastern horizon just before sunrise (stars rise heliacally) and will be lost in the glare of the sun as it rises.

in the date of the winter solstice) but five days earlier the *nakṣatra* sets heliacally (Figure V.3). In 1150 BCE, the *nakṣatra* sets heliacally on the winter solstice and rises heliacally five days later (Figure V.3). Between these two epochs, the *nakṣatra* Śraviṣṭhās rises and sets heliacally at the winter solstice, or the sun and the *nakṣatra* 'occupy the same region of the sky', i.e. on the horizon (although they will be separated by about 30° in azimuth). Thus, the requirement of verses *RJ*.5-6 and *YJ*.6-7 is satisfied. The moon will, of course, be close to *nakṣatra* Śraviṣṭhās during this period when it is in conjunction with the sun, that is, at new moon. From about 1550 BCE to about 1150 BCE the *nakṣatra* Śraviṣṭhās/α *Delphini* and the sun would have been close visually at the winter solstice. Outside this (epoch-)window, the *nakṣatra* Śraviṣṭhās/α *Delphini* will rise and set heliacally but not at or around the winter solstice.

The star β *Aquarii* has also been identified as a possible *yogatārā* of *nakṣatra* Śraviṣṭhās (Abhyankar, 1991). The separation between the heliacal rising and setting of α *Delphini* is just over five days (Figure V.3). The separation between the heliacal rising and setting of β *Aquarius* is 30 days or a month. Thus, β *Aquarii* will not set and rise heliacally close to the winter solstice and this star (and the associated *nakṣatra*) cannot be considered to 'occupy the same region of the sky' as the sun at the winter solstice.

To the calendar makers of *Vedāṅga Jyotiṣa*, *nakṣatra*s were sectors of the ecliptic, and the stars and asterisms may only have been used to identify the position of the *nakṣatra*-sectors. Because the *nakṣatra*-sectors are wide and invariant, the sun and the new moon will be in

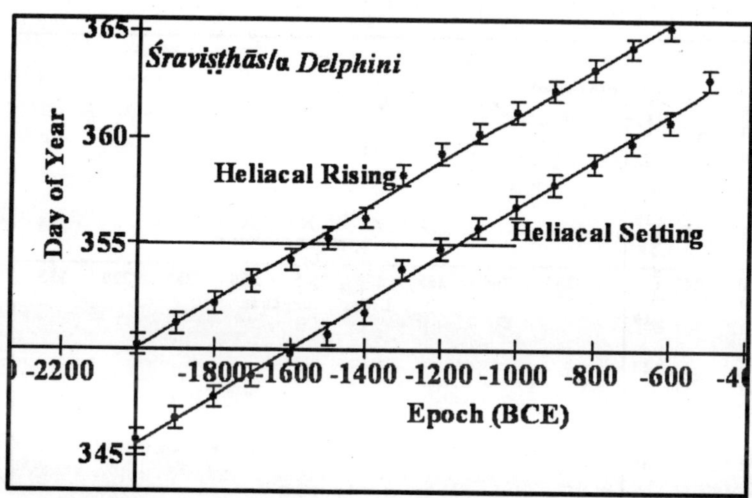

Figure V.3: The day of heliacal rising and setting of Śraviṣṭhās/α *Delphini* from 2000 – 500 BCE. The days of the (Gregorian calendar) year when the stars rise and set heliacally are shown. There can be a computational error of one day in this Day of the Year and this is represented by the error bars in the data. The lines through the data are straight line fits to the data. The winter solstice is at day 355 of the (Gregorian calendar) year. The line at day #355 intersects the heliacal rising line at about 1550 BCE and intersects the heliacal setting line at about 1150 BCE.

the *nakṣatra*-sector Śraviṣṭhās at the start of every *yuga*. The ecliptic longitude and latitude of the new moon at the start of a *yuga* over a period of 500 years are shown in Figure V.4. For the calculation of the coordinates of the new moon, the fiducial point was a date when the new moon was at the winter solstice. The coordinates of the new moon, every 62 lunations (i.e. at the start of a *yuga*), were calculated from this date. The new moon and therefore the sun are clearly 'in *nakṣatra*-sector Śraviṣṭhās' during these 500 years. Assuming that in the calendar of *Vedāṅga Jyotiṣa* every tropical/seasonal year starts on the winter solstice, the difference in the total number of days in a *yuga* (of 62 lunations) and the total number of days in the corresponding five tropical years, is plotted in Figure V.5. At the end of the first *yuga* the difference in days between the length of five tropical years and a *yuga* is

$$5 \times 365.24 - 62 \times 29.53 = -4.66 \text{ days}$$

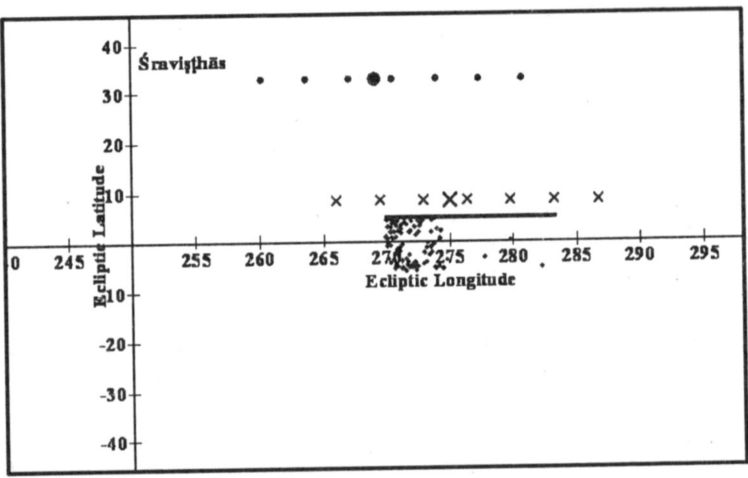

Figure V.4: Position of the new moon around winter solstice, every *yuga*, for 500 years (small diamonds). The *nakṣatra*-sector Śraviṣṭhās is shown by the horizontal line and is off-set from the ecliptic for clarity. The stars α *Delphini* and β *Aquarius* are shown by the black circle and the black cross respectively. These stars have been identified as *yogatārās* of *nakṣatra* Śraviṣṭhās by respectively Pingree and Morrissey (1989) and Abhyankar (1991). The coordinates of *yogatārās* from 2000 to 500 BCE, every 250 years apart, are shown by small dots and crosses respectively. The large dot and cross are coordinates at 1300 BCE.

This difference will accumulate with successive *yuga*s as shown in Figure V.5. After 30 years or six *yugas*, this difference will accumulate to about one synodic month. Because the beginning of a *yuga* is anchored to the new moon in the *nakṣatra*-sector Śraviṣṭhās (*RJ*.5-6 and *YJ*.6-7) around winter solstice, a lunation (or synodic month) in the sixth (sometimes seventh) *yuga* 'drops out'. This gives the 'saw-tooth' appearance to the plot in Figure V.5. By restricting the start of a *yuga* to the new moon in the *nakṣatra*-sector Śraviṣṭhās whose origin is at the ecliptic longitude of 270° (the ecliptic longitude of the sun at the winter solstice), a lunisolar calendar had developed that corrected itself for lack of synchronization of the synodic years and the seasons. Since the positions of the *nakṣatra*-sectors on the ecliptic are invariant, this intercalation scheme (and therefore the Vedic calendar) is stable over hundreds of years. This self-correcting aspect of the calendar of *Vedāṅga Jyotiṣa* was hinted at by Kuppanna Sastry (1984) and has been noted by Narahari Achar (1998b). Given the stability of this calendar, it is rather surprising that *YJ*.37 describes an intercalation scheme that is inherently inaccurate (see above). This inaccurate scheme is not mentioned in the *Ārca-Jyotiṣa* and the scheme is remarkably similar to that mentioned in *Maitrāyaṇī*

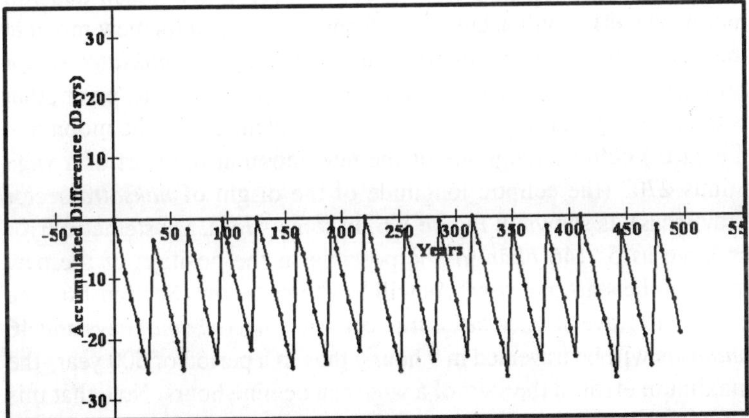

Figure V.5: Accumulated difference in the number of days in five tropical years and a *yuga* (62 synodic months). The fiducial point for these data is a new moon within a day of winter solstice. This and the constraint that every *yuga* should start when the new moon is in the *nakṣatra*-sector Śraviṣṭhās causes the 'saw-tooth' shape of the plot.

Saṃhitā (*MS*.I.10.8), a Vedic text that is earlier than the *Vedāṅga Jyotiṣa*. Was the composer of *Yājuṣa-Jyotiṣa* simply 'recalling' an earlier intercalation scheme? We will never know!

If the length of the synodic month is fixed, as *Vedāṅga Jyotiṣa* assumes, then every *yuga* will start at the beginning of the *nakṣatra*-sector Śraviṣṭhās at *bhāṃśa* zero. However, because of tidal interaction between the moon, the earth, and the sun (N.III.3) the length of the synodic month is not fixed. Consequently every *yuga* will not start at *bhāṃśa* zero at the beginning of *nakṣatra*-sector Śraviṣṭhās. Moreover, the positions of the new moon and full moon in every *yuga* will not correspond exactly to the positions given in Table IV.4. For example, the first new moon after the start of a *yuga* is in the *nakṣatra*-sector Pūrva Proṣṭapadās at *bhāṃśa* 22, for the mean length of the synodic month (Table IV.4). For the true length of the synodic month (calculated from the orbital elements of the moon and the earth), this new moon will be in the *nakṣatra*-sector Pūrva Proṣṭapadās but at *bhāṃśa* 36. There will be an error of 14 *bhāṃśa* (2 hours 45 min) in the time of the new moon (that is, in the time when the Vedic calendar makers think it is the new moon and the actual time of the new moon). Because the length of the synodic month varies, this error will not be the same at every new moon and full moon. The error will accumulate during a *yuga* and the new moon at the start of every *yuga* will not be at *bhāṃśa* zero of *nakṣatra*-sector Śraviṣṭhās. For a period of 500 years, the frequency of the 'error', that is the true (i.e. calculated from the orbital elements of the moon and the earth) ecliptic longitude of the new moon at the start of a *yuga* minus 270° (the ecliptic longitude of the origin of *nakṣatra*-sector Śraviṣṭhās), is shown in Figure V.6. In some *yugas*, a systematic error as large as 5° (46 *bhāṃśas*) is possible in the position of the new moon at the start of a *yuga*. As will be shown below (Section 1.6), the moon will traverse each *nakṣatra*-sector in $1^{7}/_{603}$ (*sāvana*) days and 46 *bhāṃśas* will be traversed in 9 hours; thus in a period of 500 years the maximum error in the start of a *yuga* can be nine hours. Note that this difference in the length of the synodic month used in *Vedāṅga Jyotiṣa* and the true length of the synodic month will introduce an error in all *bhāṃśas* calculated in the algorithms given in *Vedāṅga Jyotiṣa*. It is possible that the composer(s) of *Vedāṅga Jyotiṣa* were aware of this source of inaccuracy because the Saṃhitās and the Brāhmaṇas

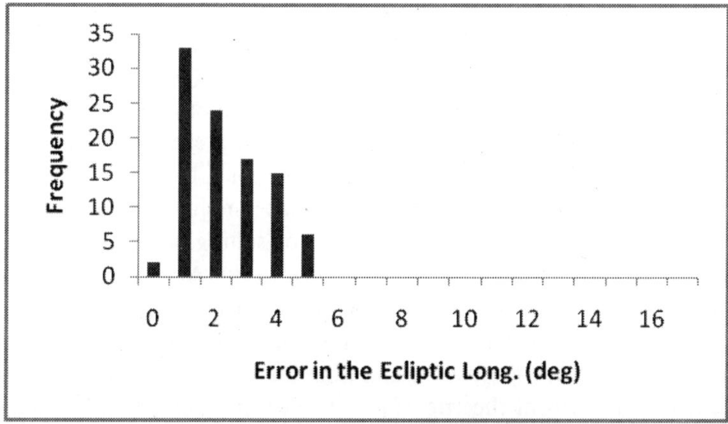

Figure V.6: Frequency of the error in the ecliptic longitude of the new moon at the start of a *yuga*. The error is defined as the ecliptic longitude of the new moon minus 270°, the ecliptic longitude of the origin of *nakṣatra*-sector Śraviṣṭhās. The data are over a period of 500 years.

caution sacrificers against reliance on just calculations in planning ceremonies (see Chapter III, Section 2.7).

As the earth orbits the sun, the apparent position of the sun against the stars appears to move gradually along the ecliptic. This motion is difficult to observe directly because the stars cannot be seen when the sun is in the sky. However, if one looks regularly at the sky before dawn, the annual motion is very noticeable. The last stars seen to rise, before sun rise, are not always the same and the motion of these stars to higher altitude can be observed. A sidereal (derived from the Latin *sidus* meaning a star) year is the time taken, by the earth, to orbit the sun once with respect to the background stars. Due to the precession of the equinox (N.IV.9) the sidereal year is longer than the (mean) tropical year and over a period of time the sidereal year will drift out of sync with the seasons at the rate of one day in 71 years. The author(s) of *Vedāṅga Jyotṣa* had recognized the need to synchronize the sidereal year with the seasons. To achieve this, *Vedāṅga Jyotṣa* proposes (*RJ*.12; *YJ*.27) addition of one-third of a day ($^{124}/_{366} = {}^{1}/_{2.952} \approx {}^{1}/_{3}$) to each *sāvana* day to obtain a sidereal day. This adds up to 122 parts in 366 *sāvana* days (or a *sāvana* year). The remaining discrepancy of two parts (each *sāvana* day is 124 parts) was corrected by adding one extra part at mid-year and another extra part at the end of the year. Note that for this correction *Vedāṅga Jyotiṣa* has approximated a sidereal day to a *sāvana* day (of 124 parts); the sidereal day is actually four minutes shorter than the mean solar day.

The discussion of the beginning of the *yuga* presented above (and the beginning of the seasonal year presented in Chapter IV, Section 4) is based on observations of the heliacal rising and setting of stars. Observations of heliacally rising or setting stars to time the performance of ceremonies not unknown to the Āryas. Ṛṣi Mātsya explains (*TB*.1.5.2.1) that if a *nakṣatra* prescribed for a rite rises after sunrise (and the *nakṣatra* is, therefore, lost in the glare of the sun), then it is necessary to observe and time the stars, visible just before sunrise and perform the rite after an interval when the prescribed *nakṣatra* will have risen (A.V.3). This passage suggests that the Āryas not only observed helically rising stars but also timed the transit of the stars.

It is possible that, at some early age, the Āryan New Year was in *vasanta* (spring) and was moved to *śiśira* (winter/dewy season) at a later date. Possible corroboration for this hypothesis is found in a myth in the later Sanskrit text, *Mahābhārata* (Vanaparva 230/8-10) that narrates the 'fall' of Abhijit from the list of *nakṣatras*. Briefly, Abhijit, the younger sister of Rohiṇī, desired the 'position of eldership' and undertook extreme *tapas* (meditation/penance) to achieve this. Skanda (year, see the myth of Kṛttikās in Chapter IV, Section 1) reported this to Brahmā who moved the beginning of time to Dhaniṣṭhās/Śraviṣṭhās (from Kṛttikās) to thwart Abhijit who was dropped from the *nakṣatras*. It is possible that this myth also refers to the period of beginning of composition of *Vedāṅga Jyotiṣa* because, as discussed above, in this calendric text the five-year (intercalation period) *yuga* starts in *śiśira* with the new moon in Dhaniṣṭhās/Śraviṣṭhās.

2.5. Yuga

A *yuga* of five years commences with the bright fortnight of the month of Māgha and ends in the dark fortnight of the month of Pauṣa (*RJ*.32; *YJ*.5). This period starts when the sun (and the moon) and *nakṣatra*-sector Śraviṣṭhās are in the same region of the sky at the winter solstice (*RJ*.5, *YJ*.6). This allows the date of the composition of *Vedāṅga Jyotṣa* (and certainly this verse) to be determined, this is discussed Section 3 below. *Vedāṅga Jyotiṣa* specifies that this is also the start of the (synodic) month of Māgha and the (tropical/ seasonal) month of *tapas*. The Vedic months start on the days after the new moon; however, a new moon on the winter solstice would be rather infrequent (see Section 2). These verses either define a very specific date, or by 'the start of the month' they mean 'the first half of the month', as indeed the later parts of the verses (and *RJ*.32; *YJ*.5) suggest. Allowing for this slight uncertainty, these verses define a calendar that is remarkably similar to the possible calendar suggested by *TS*.VII.4.8, (Chapter III, Section 2.7). *Vedāṅga Jyotiṣa* identifies the epoch when this calendar was formalized, it also recognizes the five-year period of intercalation (which occurs naturally in the scheme of *TS*.VII.4.8). A *yuga* (of five years) has ten *ayana* (solstices) in 1860 *tithi*s (lunar days). There are thus 186 *tithi*s in an *ayana* and

$$186 \equiv 6 \text{ modulo } (30) \quad \text{(N.IV.5)}$$

Therefore in a *yuga* the first five *ayana* occur (starting with the new moon in Śraviṣṭhās; see above) on the 1st, 7th, 13th, 19th and 25th lunar day or the 1st, 7th, and 13th *tithi* of the bright fortnight and the 4th and the 10th *tithi* of the dark fortnight. This sequence is repeated for the next five *ayana*s (*RJ*.8; *YJ*.9; Table V.2).

In 10 *ayanas*, i.e. one *yuga*, the moon traverses

$$27 \times 67 = 1809 \text{ } \textit{nakṣatra}\text{-sectors (see below)}$$

Therefore in 1 *ayana* the moon traverses $180^9/_{10}$ *nakṣatra*-sectors and in *n ayana* the moon will traverse

$$n \times 180^9/_{10} \text{ } \textit{nakṣatras}$$
$$= n \times (6 \times 27 + 18^9/_{10})$$
$$= n \times 6 \times 27 + n \times 18^9/_{10}$$

ignoring the complete cycles (of 27 *nakṣatra*-sector) traversed by the moon the position or the *nakṣatra*-sector of the moon at the beginning of successive *ayana* can be obtained for $n = 0, 1, 2, 3 \ldots 9$. Again ignoring the complete number of revolutions of the *nakṣatra*-sectors, the moon, at the start of successive *ayana*, will be in the 1st, 19th, 11th, 3rd, 22nd, 14th, 6th, 25th, 17th, and 9th *nakṣatra*-sector (see Table V.2.) counting from Śraviṣṭhās that is, considering this *nakṣatra*-sector as

Table V.2: The *tithes*, *pakṣa* and *nakṣatra* at the start of each *ayana* (solstice) in a *yuga*.

Year (Table II.7.)	Winter Solstice		Summer Solstice	
	Month, *pakṣa*, *tithi*	*Nakṣatra*	Month, *pakṣa*, *tithi*	*Nakṣatra*
saṃvatsara	Māgha SP.1st	Śraviṣṭhās	Śrāvaṇa SP.7th	Citrā
parivatsara	Māgha SP.13th	Ārdrā	Śrāvaṇa KP.4th	Pūrva-Proṣṭapadas
idūvatsara	Māgha KP.10th	Anūrādhā	Śrāvaṇa SP.1st	Āśreṣās
idvatsara	Māgha SP 7th	Aśvinīs	Śrāvaṇa KP 13th	Purvāṣāḍhās
vatsara	Māgha KP. 4th	Uttara-Phālgunīs	Śrāvaṇa KP.10th	Rohiṇī

SP: *śukla pakṣa* (bright fortnight)
KP: *kṛṣna pakṣa* (dark fortnight)

the *nakṣatra*-sector number 1 (*RJ*.9; *YJ*.10). The *pakṣa*, *tithi* and the *nakṣatra* at the start of each *ayana* of a *yuga* are shown in Table V.2. The names of the five years (as a group) found in the Saṃhitās and the Brāhmaṇas are given in this table although these names are not mentioned in *Vedāṅga Jyotṣa*.

Vedāṅga Jyotiṣa identifies various 'parameters' of a *yuga*, these parameters, apart from the width of the *nakṣatra*-sectors, are only given in the Yajurvedic recension. The interpretation of these parameters in modern astronomical terms and units is as follows:

- A *yuga* is composed of five years of 366 *sāvana* days each (*YJ*.28) or 1830 *sāvana* days (the actual number is 1,826.28 days, note that 62 synodic months equals 1,830.86 *sāvana* days). This is the first time the length of a year (that is not a ritual year) is explicitly stated in a Vedic text; this length of a year is not given in the *Ṛgveda* recension. *Vedāṅga Jyotiṣa* does not mention the canonical ritual year of 360 days (found in the Saṃhitās and the Brāhmaṇas). As noted above the fundamental period of the calendar of *Vedāṅga Jyotiṣa* is 1,830 *sāvana* days, that is, a *yuga* of five years and thus the length of a year is (1,830 ÷ 5 =) 366 days. *Vedāṅga Jyotiṣa* does not explain how the period of 1,830 days was obtained. It is unlikely that the composer(s) of *Vedāṅga Jyotiṣa* would have just taken 5×360+30=1,830 as the number of days in a *yuga* and then resolved these into the intercalation scheme (involving synodic months) described above. It is more likely that the composer(s) were aware of the calendar of *MS*.I.10.8 and that derived from *TS*.VII.4.8. An intercalation period of 1,830 days is a natural consequence of mathematical interpretation of *TS*.VII.4.8 (see Chapter III, Section 2.7).

- The number of risings of a *nakṣatra* in a *yuga* or the number of 'sidereal days' is the number of days (in a *yuga*) plus five or 1,830 + 5 = 1,835 days (*YJ*.29). A mean sidereal day is 0.99727 *sāvana* day; therefore there are (366÷0.99727=) 367 sidereal days in a year (of 366 *sāvana* days). Thus in a *yuga* of five *sāvana* years there are 5 × 367 = 1,835 sidereal days
- In a *yuga* the number of days when the moon is visible, or the number of risings of the moon, is the number of days in a *yuga* (1,830) minus the number of lunations (62) that is 1,768 days (*YJ*. 29). In other words, the number of days in a *yuga* when the moon is visible is the number of days in a *yuga* minus the number of new moon days.
- In a *yuga* the total number of times a *nakṣatra* will transit (say the meridian) or come round is the number of days in a *yuga* minus 21 or 1,830–21=1,809 (*YJ*.29). The interval between successive passages of the moon by a *nakṣatra*-sector (sidereal month) is about 27.32 days. Thus in a *yuga* (of 1,830 days) the moon will pass the same *nakṣatra*-sector 1830÷27.32=67 times. As there are 27 *nakṣatra*-sectors, the moon will pass by each 67×27=1809 times.
- In a *yuga* the sun will pass by 135 *nakṣatra*s (*YJ*.30a-c). In a year the sun traverses through 27 *nakṣatra*-sector and therefore, in a *yuga* of five years the sun will pass through (27×5=) 135 *nakṣatra*-sectors.
- In a *yuga* there are four *pāda parvan*s (*YJ*.30a-c). The word *pāda* means 31 and this verse means that there are 4×31=124 *parvan*s (joints of bright and dark fortnights) in a *yuga*. This is the definition of a *parvan*.

- A *yuga* comprises 61 *sāvana* months, 62 lunar (synodic) months, and 67 sidereal months (*YJ*.31). A *yuga* is 1830 *sāvana* days (*YJ*.28), therefore, a *sāvana* month (1,830 ÷ 61 =) 30 *sāvana* days. The number of synodic and sidereal months given in *Vedāṅga Jyotiṣa* implies that a synodic month is (1,830 ÷ 62 =) 29.52 *sāvana* days and a sidereal month is (1,830 ÷ 67 =) 27.31 *sāvana* days long. These lengths of the synodic and sidereal months are very close to the modern (mean) values.
- (Each year) the sun stays in each *nakṣatra* 13 and $^5/_9$ (*sāvana*) days (*RJ*.18; *YJ*.39). These verses unambiguously state that in *Vedāṅga Jyotiṣa* a *nakṣatra* is not considered to be a star or an asterism but a sector of the sky (see Chapter IV, Section 3) and the sun will transit a *nakṣatra*-sector in (365.24÷27) = 13.53 *sāvana* days (about 13 + $^5/_9$ *sāvana* days). The author(s) of *Vedāṅga Jyotiṣa* would have reasoned thus: in a *yuga* of 1,830 *sāvana* days (*YJ*.28), the sun traverses 135 *nakṣatra*(-sector)s (*YJ*.30a-c), therefore, each *nakṣatra*(-sector) is traversed in (1,830 ÷ 135 =) $13^5/_9$ *sāvana* days. Similarly, a *yuga* of 1,830 (*sāvana*) days has 67 sidereal months (*YJ*.31), that is, the moon traverses each *nakṣatra*-sector 67 times. So the moon traverses each *nakṣatra*-sector in (1,830 ÷ (27×67) =) $1^7/_{603}$ *sāvana* days. This interval is not given in *Vedāṅga Jyotiṣa*. These transit times indicate that a *nakṣatra*-sector is 13.33° wide. This is consistent with the analytically derived widths (12.65±3.25° and 11.38±5.07°, Chapter IV, Section 3.2). There is also here an implicit assumption that all *nakṣatra*-sectors are of equal width. This is the fundamental difference between *Vedāṅga Jyotiṣa* and the Saṃhitās and Brāhmaṇas. The stars and asterisms (or *nakṣatras*) of the earlier Vedic texts may have been used by the calendar-makers to indicate the position of a *nakṣatra*-sector, but the *nakṣatras* are never interpreted as stars in *Vedāṅga Jyotiṣa*. The division of the ecliptic into wide sectors to define the position of the sun and the moon is unique to the *Vedāṅga Jyotiṣa*. Pingree (1973) has misinterpreted *RJ*.18; he contends that the sun stays in each *nakṣatra* 13 and $^5/_9$ *sidereal* days.

The numbers given above must have been obtained by observation and not just by calculations.

2.6. NAKṢATRAS

Vedāṅga Jyotiṣa lists 27 *nakṣatra*s by their associated deities (*RJ*.25-28; *YJ*.32-35). This list starts with Kṛttikās like the lists of *nakṣatra*s in the Saṃhitās and Brāhmaṇas and is given in Table V.3 (see also Table IV.1). *Vedāṇga Jyotiṣa* prescribes that a sacrificer (*yajamāna*) substitute the name of the deity associated with his *nakṣatra*-name for his name during the performance (*saṅkalpa*) of the sacrifice (*yajña*). The text has a list of malign *nakṣatra*s (*YJ*.36; Table V.4.) and enjoins that the deities associated with these should not be invoked during sacrifices. This classification is not found in *Ārca-Jyotiṣa*. There is here an element of astrology, which generally is not encountered in *Vedāṅga Jyotiṣa* or the Vedic texts (apart from *AV*).

Vedāṅga Jyotiṣa also introduces a (so-called) *jāvādi* (*jau ādi*, beginning with *jau*, abbreviation for Aśvayuj) arrangement of *nakṣatra*s (*RJ*.14; *YJ*.18). In this arrangement the *nakṣatra*s are identified either by an abbreviated name or by an abbreviation of the name of the *devatā* of the *nakṣatra*. This arrangement is given in Table IV.3. The derivation of the *jāvādi* arrangement and its role in understanding the *nakṣatra*-sectors are described in Chapter IV

Table V.3: The *devatā* and the corresponding *nakṣatra* of the *Vedāṅga Jyotiṣa*

Devatā	*Nakṣatra*	*Devatā*	*Nakṣatra*
Agni	Kṛttikās	Mitra	Anurādhā
Prajāpati	Rohiṇī	Indra	Rohiṇī
Soma	Mṛgaśīrṣa	Nirṛti	Vicṛtau
Rudra	Ārdrā	Āpo	P. Aṣāḍhās
Aditi	Punarvasus	Viśedeva	U. Aṣāḍhās
Bṛhaspati	Puṣya	Viṣṇu	Śroṇā
Sarpā	Āśreṣās	Vasus	Śraviṣṭhās
Pitṛs	Maghās	Varuṇa	Śatabhiṣaj
Bhaga	P. Phālgunīs	Ajaikapād	P. Proṣṭhapadas
Aryaman	U. Phālgunīs	Ahirbudhnya	U. Proṣṭhapadas
Savitā	Hasta	Pūṣan	Revatī
Tvaṣṭā	Citrā	Aśvins	Aśvayujau
Vāyu	Svātī	Yama	Apabharaṇīs
Indrāgni	Viśākhās		

Table V.4: Classification of the malign *nakṣatra*

Ugra (Fierce)	*Krūra* (Cruel)
Ārdrā	Maghās
Citrā	Svātī
Viśākhās	Jyeṣṭhā
Śravaṇa	Mūla
Aśvins	Bharaṇīs

(Section 3.1). The *jāvādi* arrangement defines an ecliptic coordinate system in which the position of the sun and the moon are given by the (name of) the *nakṣatra*-sector and the *bhāṃśa* within this sector. The origin of this coordinate system is at new moon at the beginning of the *nakṣatra*-sector Śraviṣṭhās, around the winter solstice. *Vedāṅga Jyotiṣa* describes a number of algorithms to determine the position of the sun and moon during a *yuga*. These positions are given in terms of a *nakṣatra*-sector and the *bhāṃśa* within this sector. The coordinate system underpinning these positions is that defined by the *jāvādi* arrangement with origin described above. With this coordinate system and the origin, the algorithms of *Vedāṅga Jyotiṣa* can be interpreted unambiguously, without these, the algorithms are meaningless.

2.6.1. *Solar Nakṣatra-sectors*

In the Saṃhitās and the Brāhmaṇas the *nakṣatras* are stars and asterisms with which the moon conjoins every night. Thus in these texts the *nakṣatras* define the position of the moon (on the ecliptic). In these texts, the *nakṣatras* are never used to define the position of the sun on the ecliptic. This is not surprising as the position of the moon was defined by observations of the *nakṣatras* in the vicinity of the moon. However, during the day when the sun is visible, the *nakṣatras* are lost in the glare of the sun and at night, when the *nakṣatras* are visible, the sun is not. *Vedāṅga Jyotiṣa*, on the other hand, interprets *nakṣatras* as wide sectors of the ecliptic and the position of these sectors is defined with respect to a fixed origin. The position of the sun or moon can be defined or 'measured' in this frame of reference (or along the ecliptic) by timing the motion of these objects from the origin. Thus, constraints on the visibility of stars and asterisms no longer apply when the *nakṣatra*-sectors are used to measure or define the position of the sun and the moon on the ecliptic. *Vedāṅga Jyotiṣa* takes advantage of this coordinate system and gives algorithms to determine the position of the sun on the ecliptic during a *yuga*.

The *nakṣatra*-sector of the sun after elapse of *p parvans* in a *yuga* is obtained as follows:

In 1 *yuga* there are 124 *parvans*
and the sun traverses through 135 *nakṣatra*-sectors
therefore in 1 *parvan* the sun traverses through 135÷124 *nakṣatra*-sectors
$= 1 + {}^{11}/_{124}$ *nakṣatra*-sectors
that is, 1 complete *nakṣatra*-sectors plus
11th part of the next *nakṣatra*-sectors or 11 *bhāṃśa*
in *p parvan* the sun traverses through *p nakṣatra*-sectors + 11×*p* *bhāṃśa*

After *p parvans* the sun will have traversed *p* complete *nakṣatra*-sectors and 11×*p bhāṃśa*. This product is factored by 124 (or X where $11 \times p \equiv X$ modulo (124)) to allow for the complete number of *nakṣatra*-sectors. The complete number of *nakṣatra*-sectors is factored by 27, to allow for the complete number of *nakṣatra*-cycles, and the 'remainder' is the *nakṣatra*-sector of the sun as given in the *jāvādi* arrangement. Note that the complete *nakṣatra*-sectors and the complete *nakṣatra*-cycles are ignored as only the current *nakṣatra*-sector of the sun is required. The *nakṣatra*-sector of the sun at the start of each of the five years of a *yuga* is given in Table V.5. Dīkṣita (1896: 77-8) has given the *nakṣatra*(-sector)s of the sun at the start of

Table V.5: The *nakṣatra*-sectors of the sun at start of each year of a *yuga*

Year of the *yuga*	*parvan*		*Nakṣatra* number in the *jāvādi* arrangement	*Nakṣatra*-sector
1	0		0	Śraviṣṭhās
2	24	24×11 ≡ 16 mod(124)	16	Śravaṇa
3	48	48×11 ≡ 32 mod(124)	32 ≡ 5 mod (27)	Uttrāṣāḍhā
4	74	74×11 ≡ 70 mod(124)	70 ≡ 16 mod (27)	Śravaṇa
5	98	98×11 ≡ 86 mod(124)	86 ≡ 5 mod (27)	Uttrāṣāḍhā

each *parvan* of a *yuga*. Dīkṣita has assumed that a *yuga* starts after an elapse of one *parvan*, this is a mistake. A *yuga* starts at the beginning of the first *parvan* of the *nakṣatra*-sector Śraviṣṭhās and the sun is at the beginning of this sector at the start of a *yuga*, in agreement with *RJ*.5-6 and *YJ*.6-7.

The *nakṣatra*-sector of the sun at any time in a *yuga*, i.e. when the *parvan* and the *tithi* are known, can be obtained by correcting the *nakṣatra*-sector at the end of a *parvan* (obtained above) for the elapsed *tithi* (*YJ*.25).

In 1 *tithi* the sun traverses 9 *bhāṃśa* (see definition of *bhāṃśa*) and (following from above) after *p parvans* and *t tithis* the sun traverses

p nakṣatra-sectors + $(11\times p + 9\times t)$ *bhāṃśa*

$(11\times p + 9\times t) \equiv (q +)\ r$ modulo (124) or *q* complete *nakṣatra*-sectors plus *r bhāṃśa*

Thus at the end of *p parvans* and *t tithis* the sun will be in the *r*th *bhāṃśa* of $(p+q)$th *nakṣatra*-sector or *nakṣatra*-sector X where $(p+q) \equiv$ X modulo (27) to allow (eliminate) the complete number of *nakṣatra*-cycles.

This rule gives the *bhāṃśa* at the end of a given *parvan* and *tithi*. For a given (or the 'current') *tithi*, *bhāṃśa* (in a *nakṣatra*-sector) of the sun and the 'time' (on a scale of 124) of the day, *Vedāṅga Jyotiṣa* gives an algorithm (*YJ*.26) to obtain the *tithi* and part of the day (on a scale of 124) when the sun 'entered' the *nakṣatra*-sector.

Suppose at the end of a *parvan* the *tithi* is *t* and the sun is in *bhāṃśa* (*b*) of a *nakṣatra*-sector and part (*d*) (on a scale of 124) of a day

YJ.26 states that the *tithi* should be corrected by

$[b + (b\div 9)\times 2] \div 11$

to obtain the *tithi* when the sun entered the *nakṣatra*-sector

The logic behind this expression is not entirely clear. Kuppanna Sastry (1984) suspects that there might be a 'rule' lurking here that may have been lost due to the corruption of the text. That may be so, but the above expression reduces to $^{b}/_{9}$, which also follows from the definition of *bhāṃśa* (Section 1, above). That is, the *bhāṃśa b* traversed by the sun is equivalent to *b*/9 *tithi*. The 'current' *tithi* should be corrected by this amount to obtain the *tithi* when the sun entered the *nakṣatra*-sector. Although *YJ*.26 does not state so explicitly, it is implied that the 'current' time (*d*) should be corrected by the fraction of the *tithi* (after correcting the current *tithi*) converted to a fraction of a day (one *tithi* is equal to 122 parts of a day).

2.6.2. *Lunar Nakṣatra-sectors*

The position of the moon identified with respect to the background stars or *nakṣatra*s can be traced to *Ṛgveda Saṃhitā* (Chapter IV, Section 1). In the Saṃhitās and the Brāhmaṇas a grid of *nakṣatra*s is identified to determine the position of the moon along the ecliptic. *Vedāṅga Jyotiṣa* continues to define the position of the moon on the ecliptic but not with reference to the backdrop of stars. Instead, it defines the position of the moon in terms of the *nakṣatra*-sectors and the *bhāṃśa*s.

Two consecutive *nakṣatra*s (Table V.3 and Table IV.1) are separated by eleven *nakṣatra*-sectors in *jāvādi* arrangement. *Vedāṅga Jyotiṣa* uses this (in a rather obscure manner) to obtain the *nakṣatra* of the moon or *tithi-nakṣatra* at any given *tithi* and *bhāṃśa* of the moon (*YJ*.20). Thus, if *t* is the *tithi* and *b* is the *bhāṃśa* of the moon, then *YJ*.20 prescribes that the *tithi-nakṣatra* of the moon is X where

$$(11\times t + b) \equiv \text{X modulo } (27)$$

That is, the Xth *nakṣatra*-sector in the *jāvādi* arrangement is the *tithi-nakṣatra* of the moon. The factorization by 27 or modulo (27) is to eliminate the complete number of *nakṣatra*-cycles. The logic behind this expression is not obvious, and *Vedāṅga Jyotiṣa* appears to assume that the moon moves through one *nakṣatra*s-sector every *tithi*, which is incorrect. An accurate algorithm for the *tithi-nakṣatra* can be obtained as follows:

the moon passes the same *nakṣatra*-sector (or a sidereal month) in 27.32 days
or the moon crosses 27 *nakṣatra*-sector in 27.32 days or
$(27.32 \div 0.98) = 27.88$ *tithi*s

in 1 *tithi* the moon crosses 0.968 of a *nakṣatra*-sector
this is equivalent to 1 *nakṣatra*-sector less ≈4 *bhāṃśa*
suppose at the start of a *parvan* the moon is in *nakṣatra*-sector n and at *bhāṃśa* b
Then at *tithi* t the *nakṣatra*-sector or *tithi-nakṣatra* is X where
$(n + t) \equiv X$ modulo (27) #1
and *bhāṃśa* is $b + (124 - 4 \times t)$
(at #1 the *nakṣatra*-sector is obtained from Table V.3 and not the *jāvādi* arrangement)

This gives the accurate *nakṣatra*-sector or *tithi-nakṣatra* and *bhāṃśa* at any *tithi* in a *parvan*.

Vedāṅga Jyotiṣa ties the celestial position of the sun and the moon to terrestrial time through *kāla* (see definition above) and ultimately to mechanically measured time (by a water-clock). *Vedāṅga Jyotiṣa* obtains the time (in *kāla*) when the moon enters a *nakṣatra*-sector from the number of *bhāṃśa* traversed (*RJ*.11, *YJ*.19):

suppose the number of *bhāṃśa* traversed is b
$b \div 8 = q + r$
the time (in *kāla*) $= q \times 19 - r \times 73$

If the result is negative, then 603 (number of *kāla* in a *sāvana* day) is added to the result to obtain the time from the beginning of the previous day. This time is known as *parvan-bhā'dānakalā*. The logic behind this algorithm is not obvious, and the result (the time when the moon enters a *nakṣatra*-sector) can be obtained in a less opaque manner.

Suppose the time of the day (on a scale of 124 parts of a day) at the end of a *parvan* is d_p

this time in *kāla* is $t_p = d_p \times 603 \div 124$

suppose the *bhāṃśa* of the *nakṣatra*-sector traversed by the moon is b_p

the time taken to traverse these *bhāṃśas* is $t_n = b_p \times 610 \div 124$

the time (in *kāla*) when the moon enters a *nakṣatra*-sector or *parvan-bhā'dānakalā* is $t_p - t_n$

(603 is added if this number is negative and gives time on the previous day)

If the time of the start of a *nakṣatra*-sector when the moon enters a *tithi* after a *parvan* is required, the *parvan-bhā'dānakalā* is corrected by adding 7 times the *tithi* to the *parvan-bhā'dānakalā* (*RJ*.21; *YJ*.21). This is because a *sāvana* day is 603 *kāla* long and the moon transits a *nakṣatra*-sector in 610 *kāla*. Thus, a *nakṣatra*-sector will start 7 *kāla* later each day or each *tithi*. In the application of this algorithm, a *tithi* has been assumed equal to a *sāvana* day.

3. CONCLUSION

The versions of *Vedāṅga Jyotiṣa* that are available to us are redaction(s). The original text of Lagadha is not available, and it is impossible to tell if the present version is a complete or an accurate copy of the original. Lacunae described below indicate that the present version may indeed not be a complete copy of the original. In *Vedāṅga Jyotiṣa* a five-year intercalation period is explicitly stated and called a *yuga*, and indeed the purpose of *Vedāṅga Jyotiṣa* appears to be to determine various calendric parameters in this five-year period. However, the five years of a *yuga* are not named. In the Saṃhitās and the Brāhmaṇas a group of five years is often mentioned, and these are always associated with the calendar. The importance of this five-year period is recognized by naming these years individually (Table III.7). In these texts, these five years are collectively never called a *yuga* but, as has been shown in Chapter III, *MS*.I.10.8 identifies a five-year intercalation period and it is very likely that a five-year intercalation period is implied by the group of years given in the Saṃhitās and the Brāhmaṇas.

Only one name of the five years is common to the Saṃhitās, Brāhmaṇas and *Vedāṅga Jyotiṣa*: this is *saṃvatsara*. However, in *Vedāṅga Jyotiṣa*, *saṃvatsara* (*RJ*.1) simply means a year and not a particular year. The five years of a *yuga* with names similar to those given in the Saṃhitās and the Brāhmaṇas are given in the *Gārgī Saṃhitā*, a calendric text that post-dates *Vedāṅga Jyotiṣa*. The composer or the redactor of *Vedāṅga Jyotiṣa* may have left out the names of the five years because they may have considered these to be well known and obvious. However, this prompts a question. Why are these names included in the *Gārgī Saṃhitā*? It is possible that verses with the names of the five years of a *yuga* are missing from extant versions of *Vedāṅga Jyotiṣa*.

In the Saṃhitās and the Brāhmaṇas a calendar is described (see Chapter III, section 2.5) entirely in terms of the *ayanas*, with foci at the two solstices. In these texts, the sun starting from the winter solstice in the month of Māgha moves for six months and then stands still for a day before turning south (*KB*.xix.3). This day is called *viṣuvat*. The calendar of *Vedāṅga Jyotiṣa* also has two foci at the two solstices (*RJ*.6; *YJ*.7). However, *Vedāṅga Jyotiṣa* also introduces the equinox or the *viṣuva* and assumes that it is at the midpoint of each *ayana* of a *yuga*. However, nowhere in *Vedāṅga Jyotṣa* (or the Saṃhitās and the Brāhmaṇas) is it stated that day and night are of equal length on *viṣuva*. An algorithm is given in *Vedāṅga Jyotiṣa* to obtain the separation, in *parvans*, between the first and the *n*th *viṣuva* in a *yuga* (*RJ*.31; *YJ*.23).

a *yuga* of 5 years has 10 *viṣuva*
a *yuga* is divided into 124 *parvan*
interval between successive *viṣuva*
124 ÷ 10 *parvans* = 12 *parvans* and 6 *tithis*
(note: 1 *parvan* = 15 *tithi*)
separation between first *viṣuva* and the *n*th *viṣuva* in a *yuga*
(*n*–1) × (12 *parvans* + 6 *tithis*)
= 2(*n*–1) × 6 *parvans* + (*n*-1) × 6 *tithis*

It is not clear how this is of any value in describing the calendar of *Vedāṅga Jyotiṣa* or where and how this information was used in *Vedāṅga Jyotṣa*. In addition, this algorithm for the separation between equinoxes is inaccurate as the equinoxes are not midway between solstices (see Table III.1).

In *Vedāṅga Jyotiṣa* we witness a transition from a subjective and descriptive approach of the Saṃhitās and the Brāhmaṇas to a calendar to an analytical and quantitative approach. This can be seen in the treatment of intercalation. The Saṃhitās and the Brāhmaṇas stress the importance of the thirteenth month (Chapter III, Section 2.7) but do not give the reason for this 'extra' month. *Vedāṅga Jyotiṣa* points out succinctly and unambiguously that a synodic/lunar day (*tithi*) is shorter than the *sāvana* day and that extra (synodic) months (*adhimāsas*) are required to harmonize a synodic year with the seasons (*YJ*.37; see sub-section 1.5). *Vedāṅga Jyotiṣa* also gives an algorithm for intercalation that is similar to that given in *MS*.I.10.8. However, as shown above in Figure V.1, the proposed intercalation scheme is unsatisfactory. The calendar of *TS*.VII.4.8, (Chapter IV, Section 2.7) is self-correcting, but this is not mentioned in *Vedāṅga Jyotiṣa*, nor has any other procedure been mentioned for correcting the run-away divergence between the tropical and the synodic years that follows from the specified intercalation scheme. The purpose of *Vedāṅga Jyotiṣa* is to help determine the correct time for Vedic ceremonies and rituals. This clearly is not possible after a few years (from a fiducial year) with the five-year intercalation scheme given in *Vedāṅga Jyotiṣa*. The calendar-makers of *Vedāṅga Jyotiṣa* had a rather subtle scheme of intercalation that obviated the need for a correction to the scheme proposed in *YJ*.37 (Section 1.5).

The algorithms in *Vedāṅga Jyotiṣa* to determine the position of the sun and the moon within the *nakṣatra*-sectors are based on the assumption that a *yuga* is 1830 days long. Unlike the length of the tropical, synodic, or sidereal year (or the day and the month), this is not a naturally occurring number; it cannot be obtained from observations. *Vedāṅga Jyotiṣa* does not justify this length of the *yuga* nor does it give the source of this length of the *yuga*. It assumes that this length is known, and proceeds to give various algorithms based on it. An intercalation period of 1830 days follows from the 'mathematical' interpretation of *TS*.VII.4.8 and *PB*.V.9 (Chapter III, section 2.7). It is possible that the composer(s) of *Vedāṅga Jyotiṣa* had knowledge of this scheme and based the algorithms of *Vedāṅga Jyotiṣa* on this intercalation period.

In the calendar of *Vedāṅga Jyotiṣa* the New *Yuga* Day is the new moon day at the winter solstice. The Āryas must have known (from observations) that this can only occur every nineteen years (the Metonic cycle) but *Vedāṅga Jyotiṣa* does not refer to such a cycle. It is possible that *RJ*.5, and *YJ*.6 define the epoch of formulation of the calendar of *Vedāṅga Jyotiṣa* and *RJ*.32, and *YJ*.5, describe its practice. That is, the new *yuga* starts on the new moon day of the month of Māgha. This will make the New *Yuga* Day close to the winter solstice but not always on the winter solstice (see Section 1.5 above).

Allowing for these lacunae, the fundamental aspects of the calendar of *Vedāṅga Jyotiṣa* are very similar to the calendar that can be inferred from the Saṃhitās and the Brāhmaṇas.

4. THE TIME AND PLACE OF *VEDĀṄGA JYOTIṢA*

The location and the time of composition of either recension, of *Vedāṅga Jyotiṣa* are not known. Pingree (1973) has argued that much of mathematical astronomy of *Vedāṅga Jyotiṣa* was borrowed or copied from Mesopotamian sources during the Achaemenid occupation of northern parts of South Asia, 513-326 BCE (N.V.4). Pingree has contended that the borrowed elements were inserted into *Vedāṅga Jyotiṣa* without proper understanding or correction for the place and time of composition of *Vedāṅga Jyotiṣa*. A great deal of Pingree's argument is based on comparison of *Vedāṅga Jyotiṣa* with post-Vedic South Asian texts on mathematical astronomy. The post-Vedic texts include a number of Greek and Mesopotamian concepts of astronomy. This is not disputed. However, these texts are also a continuation of the *Vedāṅga Jyotiṣa* tradition, as it is unlikely that this tradition ended completely or that a new beginning of mathematical astronomy was made in South Asia influenced entirely by the Mesopotamian and Greek imports. It is absurd to claim that because calendric concepts in *Vedāṇga Jyotiṣa* and the post-Vedic astronomical texts are similar, *Vedāṅga Jyotiṣa* has also been influenced by Mesopotamian and Greek science.

Astronomical data given in *Vedāṅga Jyotiṣa* enable an objective determination of the place and time of composition of this text. *Vedāṅga Jyotiṣa* states (*RJ*.7; *YJ*.8) that the difference in the length of the longest and the shortest day is six *muhūrta*s or 4.8 hours (a *muhūrta* is 48 mins, see Chapter III, Section 2.1 and Section 1.2 above) in an *ayana*. This gives the ratio of the length of the longest to the shortest day to be 3:2. This ratio is also found in Mesopotamian astronomical texts (Pingree 1973). The value of the ratio of the longest to the shortest day depends on the latitude where the observation is made; a ratio of 3:2 is only possible at latitudes around 35° that is, only in the northern parts of South Asia. Pingree (1973) has argued that this ratio was copied from Mesopotamian sources into *Vedāṅga Jyotiṣa* without any correction for the change of location. However,

Gondhalekar (2008a) has shown that this ratio also excludes all cultural centres of the Achaemenid Empire (particularly Persepolis and Pasargadae) and most Mesopotamian cultural centres, leaving Babylon, Nineveh, and Nimrud, as possible Mesopotamian locations for determination of this ratio. It should be pointed out that the ratio 3:2, of the length of the longest to the shortest day in a year, could also have been determined in and around Takṣaśilā (modern-day Taxila, Punjab, Pakistan, latitude 33.75° north), and Shrinagar (Kashmir, India, latitude 34° north), both significant cultural centres in ancient South Asia. The narrowing down to just a couple of locations in West Asia where a ratio of 3:2 for the length of the longest to the shortest day could have been measured, has prompted Ôhashi (1993) and Gondhalekar (2008a) to re-examine this ratio. These authors have analysed in detail the relation given in *Vedāṅga Jyotiṣa* (*RJ*.22; *YJ*.40) for the change in the length of the day from winter solstice to the summer solstice (or vice versa). This relation can be expressed as:

$$\text{the length of daytime} = [12 + (^{2}/_{61})n] \; \textit{muhūrta}$$

where n is the number of days after (or before) the winter solstice. This is known as the 'linear zig-zag' function and shows that the Āryas considered the length of a day to change linearly from winter solstice to the summer solstice (and vice versa). The length (in hours) of days between the solstices, as given by this function, is plotted in Figure V.7 (reproduced from Gondhalekar 2008a). Plotted in this figure is also the computed length of daytime for three latitudes in the northern half of South Asia. The length of the day is defined as the interval between sunrise and sunset (N.V.3). That is, the interval between two times in a day when the zenith distance of the sun is 90.8° (or the altitude of the sun is -0.8°). In these computations, the effects of atmospheric refraction have been included. These data show that from (about 30 days after) winter solstice to spring equinox the zig-zag function matches the curve for 30° north and from the spring equinox to the summer solstice the zig-zag function does not match any of the three curves but comes close (for the first 60 days) to the curve for 25° north. For an error of 2 minutes (that is, the time between the first and the last contact of the solar disc with the horizon) in the observations of the sunrise and sunset times, the

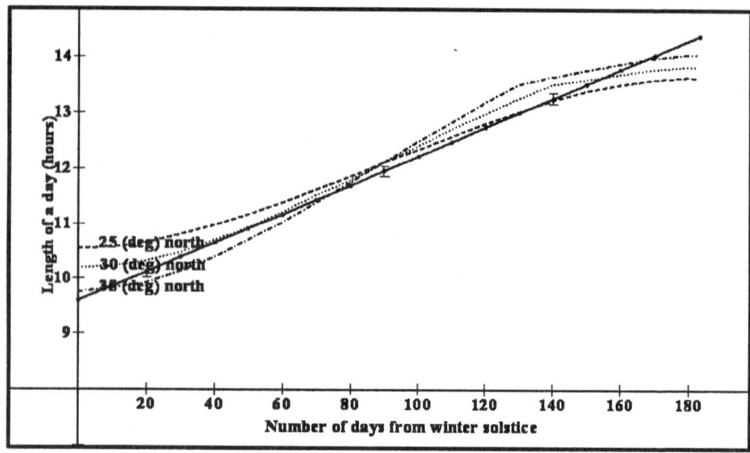

Figure V.7. The length of a day, from winter solstice to summer solstice. The linear zig-zag function is the dot-line. The error bars are for a 2 min. error in determining the time of sunrise and sunset (see text for details). Curves for the length of a day at latitudes of 25°, 30° and 35° north are also shown. Effects of atmospheric refraction have been included in these computations.

agreement between the zig-zag function and the curves for 25° and 30° is good at all times except close to the solstices. Compared to these two curves the match of the zig-zag function to the curve for 35° north is poor at all times. This suggests that the linear zig-zag function was most probably derived from observations made around spring equinox, at a location between latitudes 25° and 30° north. The curve was probably then extrapolated to the solstices, and the ratio of the length of the longest to the shortest day in a year was obtained from this extrapolated curve. It is entirely fortuitous that this ratio matches the value that could have been measured at 35° north. This discussion, of course, does not prove that this ratio was measured in South Asia, but it considerably weakens the case for a Mesopotamian origin for this ratio in *Vedāṅga Jyotiṣa.*

The ratio of 3:2 for the length of the longest to the shortest day is only found in the Mesopotamian astronomical texts produced after 700 BCE. Pingree (1973) asserts that since this ratio was copied into *Vedāṅga Jyotiṣa*, this text must have been composed after 700 BCE. His most likely date for the Ṛgvedic recension is the fifth or the fourth century BCE and the third or the fifth century CE for the Yajurvedic recension. Pingree does not provide independent corroborating evidence in support of his rather vague argument (N.V.4). Gondhalekar (2008a) has examined in detail the verses *RJ*.5-6 and *YJ*.6-7 in *Vedāṅga Jyotiṣa*. and argued that the Āryas would have determined the 'date' of the start of a new *yuga* by observations of the sun, moon, and *nakṣatras*. He has interpreted verses *RJ*.5-6 and *YJ*.6-7 as – the 'New *Yuga* Day' is the new moon day around the winter solstice when the *nakṣatra* Śraviṣṭhās rises heliacally (Section 1.5 above). Pingree and Morrissey (1989) have identified α *Delphini* as the *yogatārā* of *nakṣatra* Śraviṣṭhās. This star and therefore this *nakṣatra* rose and set heliacally within five days of the winter solstice between 1550 and 1150 BCE. The more usual interpretation of verses *RJ*.5-6 and *YJ*.6-7 is that the 'New *Yuga* Day' is the new moon day at the winter solstice when the moon and the sun conjoin the *nakṣatra* Śraviṣṭhās (Kuppanna Sastry 1984). That is, the ecliptic longitude of *nakṣatra* Śraviṣṭhās would have been 270° (the ecliptic longitude of the sun at the winter solstice) at the epoch of these verses. The J2000 ecliptic longitude of α *Delphini/nakṣatra* Śraviṣṭhās is 317.4° and would have been 270° in 1394 BCE. This epoch is consistent with the interval of 1550–1150 BCE obtained above for the heliacal rising and setting of *nakṣatra* Śraviṣṭhās/α *Delphini* around the winter solstice.

The analysis of the Vedic calendar presented here suggests that period from 1550 to 1150 BCE was a period of significant change in Vedic calendric science. The performance of the *cāturmāsya* sacrifices, by coincidence of the seasonal full moon and prescribed *nakṣatras* appears to have been specified around this epoch (Chapter IV, Section 5). This conclusion is based on the secure identification of *nakṣatra* Kṛittikās and is independent of the uncertainties of the data of Paitāmahasiddhānta (Chapter IV, Section 1). The implicit assumption of circumpolar (all stars of) *sapta ṛṣis*/Ursa Major suggests an epoch before 1300 BCE (Gondhalekar 2011). The *nakṣatra*-names of the Vedic months appear to have been formulated in this epoch (Chapter IV, Section 6). The cross-correlation of the (possible) *yogatārās* of the *nakṣatras* and the *nakṣatra*-sectors (Chapter IV, Section 7) suggests that the list of *nakṣatras* may have reached its final form around this epoch. This conclusion, of course, comes with a 'health warning' that is, the identification of *yogatārās* is based on data of Paitāmahasiddhānta, a post-Vedic text of uncertain provenance. However, the high correlation suggests that the identities of the *nakṣatra*s and their *yogatārā*s may have been transmitted down the ages with a high degree of fidelity. It is possible that, before this epoch, the Vedic year started in spring (*vasanta*) around the vernal equinox when the sun was close to the *nakṣatra* Kṛittikās and the spring (*vasanta*) full moon was in *nakṣatra Citrā*. Around 1300 BCE the start of the year or more accurately, the start of the *yuga* was fixed at new moon in *śiśira* and close to the winter solstice when the moon and the sun are close to *nakṣatra* Śraviṣṭhās. It is possible that this was also the epoch of start of the formulation of the ecliptic *nakṣatra*-sector coordinate system and mathematical astronomy in South Asia, although the texts available to us (*Vedāṅga Jyotiṣa*) may have been redacted at a later date.

The epoch suggested by four independent sets of parameters and their independent analysis is inconsistent with the chronology of the Vedic texts determined from the linguistic analysis (and the cultural and philosophical/theological context) of these texts (Witzel 1999c, 2001). This linguistic analysis suggests that the epoch of the Saṃhitās, Brāhmaṇas and Sūtras, that we have inherited, is the first half of the first millennium BCE, and that of *Vedāṅga Jyotiṣa* is the last half of the first millennium BCE. That is, these texts belong to the period after 1000 BCE. The late second millennium epoch is also inconsistent with reference to iron in *Atharvaveda Saṃhitā*, which places all post-Ṛgvedic texts after about 1200 BCE. The pre first millennium dates obtained from analysis of passages from the Vedic texts are usually dismissed as products of *'literal interpretation' of texts, or fanciful imaginations of the analysts* (Whitney 1895). Implicit in this dismissal is the belief that the editions of the texts that have come down to us are the first and the only editions of these texts. This is extremely unlikely if not impossible. These texts are a product of development over a long period and must have gone through numerous iterations before the redactions that we have inherited. The survival of earlier passages in texts that were redacted at a later date is possible and entirely likely. Arnold (1905) identified passages in *Ṛgveda Saṃhitā*, which he described as later additions to the original text. Arnold categorized these passages as 'popular' and these passages are in every *maṇḍala* of the Saṃhitā, this Saṃhitā is thus a mix of the old and new verses (Deshpande 1995). It is possible that passages from earlier sacrificial and sacerdotal texts are scattered in the editions of the Vedic texts that we have inherited. These passages may have been presented in the language and the context of the age of the redactors. The dates obtained above may tell us nothing about the epoch of the surviving edition of the Vedic texts but these dates are chronological markers in the history of the calendar of the Vedic Period. The similarity of the dates obtained from four independent sets of parameters, independently analysed, suggests that this was a significant epoch in the development of the Vedic calendar. At this epoch, the table of *nakṣatra*s was finalized and the *nakṣatra*s acquired a central role in the Vedic calendar; significant events and rituals of the Vedic year and *yuga* were identified by reference to the *nakṣatra*s and not just to the moon and seasons. By including the *nakṣatra*s (and synchronization of the synodic/lunar year with the seasons) in their calendar, the Āryas were able to reproduce more accurately the times of their rituals.

NOTES

1. According to the orthodox or traditionalist the true meaning of the Vedas, whose obscurity is admitted, can only be explained with the help of the Vedāṅgas along with other revealed and remembered texts and by reasoning. There are six Vedāṅgas (*aṅgas* – 'limbs' of the Veda) or auxiliary subjects, which in the course of time developed into separate and independent branches of study. The earliest reference to the Vedāṅgas is in *Muṇḍaka Upaniṣad* I.1.5. The six *aṅgas* are phonetics (Śikṣā), metrics (Chandas), grammar (Vyākaraṇa), etymology (Nirukta), ritual or religious practice (Kalpa) and astronomy (Jyotiṣa). Originating in the Vedic period, the Vedāṅgas were to promote a better understanding and recitation of the Vedic texts and their proper ritual employment. Jyotiṣa is composed in verse; the other five Vedāṅgas are composed in the *sūtra* style (short prose phases or sentences). This suggests that *Vedāṅga Jyotiṣa* is earlier than the other five Vedāṅgas.
2. There is also an *Atharvaṇa Jyautiṣa* very different in character from the Ṛgvedic and Yajurvedic recensions of *Vedāṅga Jyotiṣa*. This is a later text and whereas the two recensions of *Vedāṅga Jyotiṣa* deal with matters astronomical this text is concerned with astrological matters.
3. For computational purpose, sunrise and sunset are defined to occur when the geometric zenith distance of the sun is about 90.8° or the geometric centre of the Sun is 0.8° below the horizon. Before sunrise and after sunset there are periods when the Earth's surface does not receive direct sunlight but

sunlight scattered in the upper atmosphere reaches the Earth's surface; these periods are called twilight. The amount of light during twilight is determined by the state of the atmosphere generally and the local weather conditions particularly. Three definitions have evolved to identify the visibility during these periods:

- Civil twilight is defined to begin (in the morning) and to end (in the evening) when the centre of the sun is geometrically 6° below the horizon. During these periods and under good weather conditions, the illumination is sufficient to distinguish clearly objects on the ground. During these periods, the horizon can be clearly distinguished, and bright stars are visible under good atmospheric conditions and in the absence of moonlight or other illumination.
- Nautical twilight is defined to begin (in the morning) and to end (in the evening) when the centre of the sun is 12° below the horizon. During these periods, the horizon is indistinct and in the absence of any artificial illumination, only a general outline of objects on the ground can be distinguished.
- Astronomical twilight is defined to begin (in the morning) and to end (in the evening) when the centre of the sun is 18° below the horizon. Before this period in the morning and after this period in the evening, the sun does not illuminate the sky. Under good atmospheric and weather conditions, most stars are visible before the morning twilight and after the evening twilight.

Source: http://aa.usno.navy.mil/faq/docs/RST_defs.php

4. In the West, there is a long history of denying originality and creativity to Indians (and coloured people generally). One of the finest examples of this is *The History of British India* by James Mill, published in 1817 (Mill 1817). On the strength of this book, Mill was appointed an official of the East India Company and his book was required reading for all British government appointees in India. Mill never visited India, knew no Indian language, and distrusted anything written by Indians since 'they have a general disposition to deceit and perfidy'. Mill disputed and dismissed practically every claim ever made on behalf of Indian culture and chastised early British administrators (like William Jones) for their gullibility. Mill dismissed the Indian invention of the decimal system with place value and the role of zero in this system. He claimed that the invention of the numerals must have been very ancient, and the Indians must have borrowed it. Mill also ridiculed the attribution of arguments for a rotating earth and gravitational attraction to Āryabhaṭa and claimed that the Indians must have plagiarized these from the Europeans and claimed credit for Āryabhaṭa.

   In the late nineteenth century, William Dwight Whitney attempted to interpret *Vedāṅga Jyotiṣa*. He failed but lacked the intellectual integrity to admit it and dismissed *Vedāṇga Jyotiṣa* as 'mostly filled with unintelligible rubbish' (Whitney 1895). Whitney was a (American) Sanskrit scholar and seems to have been as well known for his contempt for Indians as for his scholarship. His principle contention seems to have been that the credit for astronomical observations 'belongs to the Greeks' and Greeks alone. The extension of this position; that rational and scientific thought emerged with the Greeks and only with the Greeks and is a uniquely European (and Christian) characteristic, is believed in the west even today (e.g. Wolpert 1992).

Pingree (1970-80) characterizes astronomy (and other scholarly disciplines) in India as 'repetitive' and lacking in 'innovation' and 'Indian astronomy was not completely static is due almost entirely to the repeated intrusions of new theories from the West'. In a series of papers he has repeatedly contended, without providing evidence of any substance, that mathematical astronomy of *Vedāṇga Jyotiṣa* and calendric science generally was imported into South Asia from Mesopotamian sources (e.g. Pingree 1996). He treats the entire Vedic literature as 'essentially a single body of more or less uniform material' (Pingree 1970-80) and believes that 'the dates of their composition lie roughly in the first half of the last millennium BCE' (Pingree 1996). He arrives at a 'jaw-dropping' conclusion, that since *MUL.APIN* was also written around 1000 BCE, the Indians must have borrowed almost all calendric concepts from the Mesopotamians (Pingree 1989b, 1996). The significant points in Pingree's contention that calendric science and mathematical astronomy were 'borrowed' by South Asians from Mesopotamian sources are that:

- The Vedic ritual year of 360 days divided into 12 months of 30 days each is mentioned in Mesopotamian texts.
- The oscillation of the rising-point of the sun along the eastern horizon between its extremities at the solstices is mentioned in Mesopotamian texts.
- A list of stars and constellations, headed by the Pleiades, to 'chart' the path of the moon is given in the Mesopotamian texts.
- The earliest version of *Vedāṇga Jyotiṣa*, (perhaps) the Ṛgvedic version, was transmitted to South Asia around fifth and fourth century BCE.

A number of early cultures, distributed widely in time and space, independently adopted an 'ideal' or quasi-synodic year of 360 days (N.III.9) the reason for doing so is obvious and was probably common to all these cultures. It is not difficult to observe the apparent annual motion of the sun and identify the two solstice points along the path of the sun and a number of early cultures had achieved this (N.III.6). The similarity in the descriptions proves nothing; if the people of Newgrange or the Incas had left us a description of the annual motion of the sun it would also have appeared very similar to that left by the Mesopotamians. The reason the Mesopotamian and the South Asian descriptions are similar is because they are describing the same phenomenon. Likewise, there are similarities in the Mesopotamian and South Asian lists of stars along the path of the moon, because both list-makers looked at the same sky, the same moon, and the same stars. The number of stars in the two lists is different (N.IV.1) because the ritual and calendric requirements of the two groups were different. The analogies between various cultures are not enough to prove actual historical exchange between them. The burden of proof always is with the one who proposes such an exchange and Pingree provides no proof. Pingree believes, wrongly, that 'Lagadha uses the list of twenty-seven *nakṣatras* to divide the ecliptic into twenty-seven equal arcs . . . in the same way as the Babylonian astronomers . . . had divided it into twelve equal arcs' (Pingree 1970-80). The *nakṣatra*-sectors of *Vedāṇga Jyotiṣa* are not simple divisions of the ecliptic into 27 sectors somehow positioned around the *nakṣatras* (as stars and asterisms). The 27 *nakṣatra*-sectors are evenly spaced equal sectors of the ecliptic, determined from the position of the new or the full moon (with respect to the *nakṣatras*) in a *yuga* with origin at *nakṣatra* Śraviṣṭhās (Chapter IV, Section 3.1). Each *nakṣatra*-sector may and probably was identified by the *nakṣatra* (as stars and asterisms) in that sector but the calendric parameters of *Vedāṇga Jyotiṣa* are determined with respect to the grid of the sectors and are independent of the *nakṣatras*. This is a fundamental departure from the star-lists of *MUL.APIN* and from the concept of *nakṣatras* (as stars and asterisms) in the earlier Vedic texts.

The start of a *yuga* when *nakṣatra* Śraviṣṭhās is at the winter solstice (*RJ*.5-6; *YJ*.6-7; Chapter V, Section 3) suggests a date of 1394 BCE (Kuppanna Sastry 1984). Pingree questions the relevance of this value 'to the date of the *Jyotiṣavedāṇga*'. He opines, 'We simply do not know where Lagadha would have placed the beginning of the equal *nakṣatra* Dhaniṣṭhā with respect to the fixed stars, nor do we know the accuracy with which he could have determined the sidereal longitude of the Sun at the winter solstice. Since a displacement of the beginning of the equal *nakṣatra* by some 10°, or an error of 10 days in computing the date of the winter solstice, or some combination of these two effects is all that is required to bring the date from the twelfth century to the fifth century BC, we should not lend much weight to this chronological argument' (Pingree 1973). These are astonishingly uninformed opinions. The error of 10° has not been independently justified and has been assumed to confirm Pingree's preconceptions about the chronology of *Vedāṇga Jyotiṣa*. Pingree has failed to understand the nature of the *nakṣatras*-sectors of *Vedāṇga Jyotiṣa* and he believes, wrongly, that Lagadha has assumed that 'the vernal equinox occurs when the sun is in *Kṛttikā*' (Pingree 1970-80). There is no evidence for this in *Vedāṇga Jyotiṣa* or the Saṃhitās and the Brāhmaṇas. There is actually no reference to (vernal) equinox in any Vedic text. A passing reference to equinox in *Vedāṇga Jyotiṣa* is of no calendric significance.

The *jāvādi* arrangement of *nakṣatras* (Chapter IV, Section 3.1) and verses *RJ*.5-6 and *YJ*.6-7 state that the origin of *nakṣatra*-sector Śraviṣṭhās is at the winter solstice. Pingree's comments about the sidereal longitude of the Sun at the winter solstice and the conclusions he draws are meaningless. Why would Lagadha have to 'determine' this sidereal longitude? And how would he have determined it? All Lagadha had to determine was the day of the winter solstice. In the Vedic texts, earlier then *Vedāṇga Jyotiṣa*, Āryas note that the seasonal year is between 364 days and 366 days long (*TS*.VII.1.10; Chapter III, Section 2.7). The *sattra* of *Gavām ayana* was performed to keep track of the number of days in a ritual year (Chapter III, Section 2.5) and the Āryas were aware that the seasonal year was longer than the ritual year by four to six days. The Āryas may have counted the number of days of a seasonal year from a fiducial day. The Vedic texts suggest that in the Middle to Late Vedic Period this fiducial day was the winter solstice (*KB* xix.3; *RJ*.5-6 and *YJ*.6-7). Could Lagadha have (as Pingree has suggested) believed or counted the winter solstice to be 375 days after the winter solstice of the previous year rather than 364 or 366 day? Once an approximate day of the winter solstice had been established, it would have been a simple observation (even for Indians!) to determine the day of the new moon. Admittedly, this day would not have been at the winter solstice most years. If, as is very likely, the start of the seasonal year was 'observed' over many years then a systematic error of +10 days (as Pingree has assumed) in determining the day of winter solstice is unlikely. Error of few days in the day of the winter solstice, is possible. Suppose this error is (improbably) ±5 days, i.e. five days either side of the winter solstice and not biased in one direction as assumed by Pingree. With an approximate day of the winter solstice, all Lagadha had to do to establish the day of the start of a *yuga* was to determine, visually, the day when the sun and the moon (i.e. new moon) 'are in the same region of the sky as *nakṣatra* Śraviṣṭhās', as stated in verses *RJ*.5-6 and *YJ*.6-7. That is, determine the day(s) of heliacal rising or setting of *nakṣatra* Śraviṣṭhās around the winter solstice. As shown in Chapter V, Section 2.5, the heliacal rising or setting of *nakṣatra* Śraviṣṭhās within ±5 days of the winter solstice is only possible from 1550 BCE to 1150 BCE if α *Delphini* is the *yogatārā* of *nakṣatra* Śraviṣṭhās.

VI

Epilogue

The development of the calendar of the Vedic Age can be divided into three wide overlapping periods, and these are summarized in Figure VI.1. The earliest period, the Ṛgvedic (*ṚV*) period, presupposes a seasonal calendar. The foundational element of a calendar, the season(s), had been identified before this period. In this period, a seasonal/tropical year had been identified and was used to define 'physical' time like the age of a person. During the later part of this period, the seasonal year was divided into six seasons. The division of a year into six seasons of equal duration is meteorologically unrealistic in South Asia but is required to maintain symmetry between various concepts (that become obvious in the post-Ṛgvedic texts) central to Vedic thought and rituals. This conclusion is based on the assumption that the climate of northern South Asia during the Vedic Age was similar to current climate of this region; at present, this can neither be proved nor disproved. In addition, during the *ṚV*-period a synodic (lunar) calendar had been formulated and the first step to synchronize the lunar and tropical/seasonal year had been taken by defining a ritual year of 360 days consisting of 12 months of 30 days each. Five days were added to this ritual year to harmonize it with the tropical year. There is evidence that the inadequacy of this synchronization had been appreciated and a more sophisticated synchronization involving the intercalation of a thirteenth month(s) was considered. It is also evident that, during this period, some stars or asterisms had been identified as chronological markers.

In the post *ṚV*-period, that is, the period defined by the 'canonical'

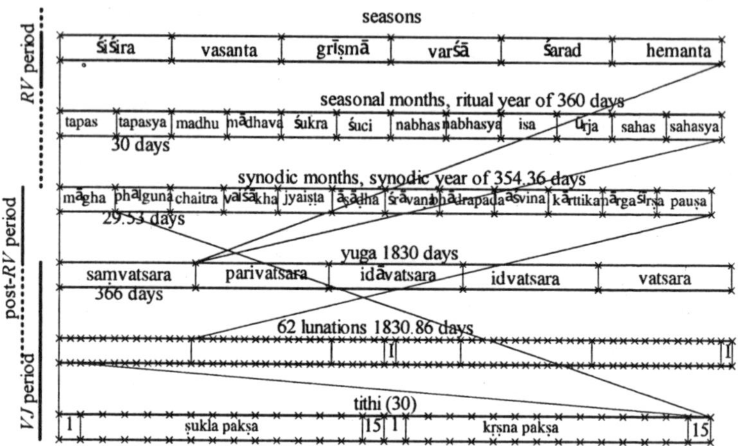

Figure VI.1: The fundamental layers of the Vedic Calendar. The 'seasonal layer' was identified before the *ṚV*-period, and it is possible that the division of the year into six seasons was made during the later part of this period. During the post *ṚV*-period, the seasonal layer had acquired a concrete form of two months per season. A stellar frame of reference (*nakṣatras*) and a synodic year of 12 months were established. The synchronization of the tropical and synodic years had been achieved. During the *VJ*-period, all layers of the previous periods were integrated into a stable calendar that could be projected into the future by mathematical methods. The 'I' in the fifth layer (from top) denotes the intercalated synodic months.

Saṃhitās, Brāhmaṇas, and Sūtras calendric concepts become more definite and transparent. A horizon reference or coordinate frame was identified to define the position of the heavenly bodies. The six (sometimes five) seasons were allocated two seasonal months that defined the weather and environment of the season. A number of arithmetic schemes to synchronize the seasonal and the ritual years were considered during this period. A perfect scheme (that is, a self-correcting scheme), to synchronize the synodic year with seasons was formulated. This scheme was based on the observations of the phases of the moon around (and soon after) winter solstice. This seems to have led to the identification of an 'intercalation period' of 1,830 *sāvana* days or 62 lunations. This in turn seems to have resulted in a scheme to harmonize the synodic/lunar years with seasons by the intercalation of two synodic months to 60 lunations.

The post-*Ṛ*V-period is characterized by the *nakṣatra*s moving to the centrestage. A grid of stellar chronological markers (*nakṣatra*s) along the path of the moon was established. The synodic year acquired a concrete form of 12 synodic months each identified by the *nakṣatra* near which the moon was full during the first year of the five-year intercalation period. The start of the seasons and the time of performance of various rituals were more accurately defined in terms of conjunction, in a season, of the moon (at a required phase) and a predefined *nakṣatra*. During this period, start of the year in both *vasanta* and *śiśira* can be identified. The two times suggest either a division of this period into two strata (earlier and later) or two calendars (possibly, agricultural and ecclesiastic) with different dates for the start of the year. The division of a day into smaller units (time-keeping) was introduced during this period, but the means of measuring this time cannot be identified. Whereas a calendar of the *Ṛ*V-period cannot be reconstructed, enough information is available in the post-*Ṛ*V texts to reconstruct a calendar of this period.

In the third, *Vedāṅga Jyotiṣa* (*VJ*) period, the calendric concepts of the earlier ages are expressed in mathematical form thus considerably enhancing the predictive capability of the Vedic calendar. The calendar of this period is based on an 'intercalation period' or a *yuga* of 1,830 *sāvana* days or (almost) 62 lunations. It is very likely that this period was obtained from the intercalation scheme developed from the interpretation of *TS*.VII.4.8, in the post-*Ṛ*V-period. Five synodic years were synchronized with seasons by intercalating two synodic months, one after 30 synodic months and another after 61 synodic months. A 'synodic *yuga*' was created of 62 synodic months or 1830.86 *sāvana* days. This procedure to synchronize the 'synodic *yuga*' with the seasons is remarkably similar to that described in *MS*.I.10.8. An accurate and self-correcting scheme to synchronize the synodic year with seasons was also developed during this period. This scheme was based on starting a *yuga* on the conjunction of the new moon with *nakṣatra* Śraviṣṭhās at (or around) winter solstice. This requirement intercalated two synodic months in a period of 62 lunations or one *yuga*. The residual error in this scheme was corrected every sixth (sometimes seventh) *yuga* as a synodic month was dropped in order to start the *yuga* on the new moon in *nakṣatra*-sector Śraviṣṭhās. This correction did not have to be made 'manually' as it happened 'automatically'.

In the calendar of the *VJ*-period, the *nakṣatras* were never interpreted as stars or asterisms but were reformulated as equal sectors of the ecliptic with the origin at new moon in the *nakṣatra*-sector Śraviṣṭhās at (or around) winter solstice. That is, a partial ecliptic coordinate system was developed. The position of the sun and the moon during a *yuga* was given within a *nakṣatra*-sector and was defined by the sector name and the separation from the leading edge of the sector. The position of the sun within a *nakṣatras*-sector is unique to this period and is not found in the earlier periods. The reformulation of *nakṣatras* as sectors of the ecliptic that are independent of stars and asterisms set the calendar of this period free of precession of the equinox and the calendar is stable over hundreds of years.

The calendar of the *VJ*-period has two fundamental requirements:

- An 'intercalation period' of 62 lunations or a *yuga* that starts in the season of *śiśira* at (or around) winter solstice when the new moon is in the *nakṣatra*-sector Śraviṣṭhās.
- An ecliptic coordinate system with origin at the start of the *nakṣatra*-sector Śraviṣṭhās.

A number of algorithms are given in *Vedāṅga Jyotiṣa* to determine parameters required for ritual observances. With these two requirements, the algorithms can be interpreted unambiguously – without these requirements, these algorithms would be meaningless.

Mechanical means (clepsydra) of measuring time (fractions of a day) were introduced during this period.

The epoch of individual periods of the Vedic calendar cannot be determined. However, three independent chronological markers can be identified from the analysis of the calendric passages in the Saṃhitās, Brāhmaṇas, and Sūtras. These are:

- The circumpolar (from Madhyadeśa, Delhi; 77°12'E; 28°35'N) *septa ṛṣis*/Ursa Major constellation.
- The unambiguous identification of the season, *nakṣatra* and the phase of the moon for the three *cāturmāsya* (seasonal) sacrifices.
- The identification of the *nakṣatra* at the winter solstice at the start of a *yuga*.

Analysis of the alignment of the seasonal full moon and the *nakṣatra*s for the *cāturmāsya* (seasonal) sacrifices suggest an epoch of 1400±500 BCE. This epoch is based on the secure identification of *nakṣatra* Kṛttikās with the Pleiades and is not affected by the uncertainties inherent in the analysis based on the available list of *yogatārā*s of the *nakṣatra*s. This epoch is in agreement with the period from 1550 to 1150 BCE suggested by the alignment of the sun, the moon, and the *nakṣatra* Śraviṣṭhās at the winter solstice at the start of a *yuga* of the calendar of *Vedāṅga Jyotiṣa*. The *sapta ṛṣi*s constellation can be seen to be circumpolar from Madhyadeśa (Delhi 77°12'E; 28°35'N) only before about 1000 BCE and certainly before 1500 BCE (Gondhalekar 2011). It is very likely that the sectors of the ecliptic of the *nakṣatra*-sector coordinate system were identified by the *nakṣatra*s (stars and asterisms) that were in use as calendric markers. Cross-correlation of the sectors and the *yogatārā*s of the *nakṣatra*s (stars and asterisms) suggests that the list of identifying *yogatārā*s may have been defined at 1300±300 BCE. This is consistent with the three chronological markers identified above. It is very likely this was also the epoch of the beginning of the formulation of *Vedāṅga Jyotiṣa*, although the available text may have been redacted at a later date.

Appendix

The earliest surviving collection of poems in any Indo-European language is in Vedic Sanskrit (N.I.1). The purpose of composition of these poems is not known but their metrical form suggests that at least some were for ritual recitation. Later these poems were gathered, edited, and arranged into ten books. This collection was called (and by which it is now known) the *Ṛgveda* (praise knowledge) *Saṃhitā* (placed together). Because some of these poems were recited during rituals, most modern translators believe that most if not all verses of *Ṛgveda Saṃhitā* refer in some way to the sacrificial rituals; for a critique of some of the available translations see Thomson (2004) and Thomson and Slocum (2009). Bearing this in mind, the translations of the verses given here should not be considered definitive, nor should the suggested interpretations be considered exclusive, as alternate ones are possible. With each verse, most available (in English) translations have been reproduced to point out the variations in the interpretations of the same verse. The author has translated the verses, for which there are no available English translations.

The Ṛgvedic verses given here are from *The Rigveda: Metrically Restored Text* (Van Nooten and Holland, 1994) and the text is copied from the online version of Karen Thomson and Jonathan Slocum (2009). The Ṛgvedic verses were composed over a considerable period. On linguistic grounds, Arnold (1905) identified passages that he considered later additions; he categorized these as 'popular'. He also categorized as 'popular' verses that in their meter, position, subject matter, etc. are not consistent with the broad plan of the

Saṃhitā. In this Appendix the verses categorized as 'popular', on linguistic grounds, are identified by 'PL'. This classification has been taken from the Metrically Restored Text. Out of the 74 *ṚV* verses in this Appendix, 29 have been categorized as 'popular' on linguistic grounds. In this Appendix there are no verses that can be categorized as 'popular' on non-linguistic grounds. The (E)arly, (M)iddle and (L)ate identifiers are chronological identifiers of Witzel (1999), see Chapter II, Table II.2.

CHAPTER I – BRICKS AND BOOKS

1. *ṚV*.I.133.1. PL

   1.133.01a *ubhé punāmi ródasī r̥téna*
   1.133.01b *drúho dahāmi sám mahīr anindrāḥ*
   1.133.01c *abhivlágya yátra hatā amítrā*
   1.133.01d *vailasthānám pári tr̥ḷhā áśeran*

   I purify both heaven and earth with truth, I burn up the mighty evil sprits that are opposed to Indra, in the place where the enemies, having been overpowered and slain, lay shattered around Vailasthāna. (Burrow 1963)

2. *ṚV*.I.133.3. PL

   1.133.03a *ávāsām maghavañ jahi*
   1.133.03b *śárdho yātumátīnām*
   1.133.03c *vailasthānaké armak*
   1.133.03d *mahāvailasthe armaké*

   Strike down, O Maghavan, the host of these sorceresses, in the ruined city of Vailasthāna, in the ruined city of Mahāvailasthāna. (Burrow 1963)

3. *ṚV*.I.117.21.

   1.117.21a *yávaṃ vŕ̥keṇa aśvinā vápantā*
   1.117.21b *íṣaṃ duhántā mánuṣāya dasrā*
   1.117.21c *abhí dásyum bákureṇā dhámantā*
   1.117.21d *urú jyótiś cakrathur āriyāya*

O Aśvins, miracle workers, you sow the barley with the plough, 'milking' strength for man.
(You have) blown away the Dasyu with the thunderbolt, you have created spacious light for the Ārya.

4. *Ṛ*V.IV.57.4. PL

4.057.08a *śunám̐ naḥ phālā vi kaṣantu bhūmim̐*
4.057.08b *śunám̐ kīnāśā abhí yantu vāhaíḥ*
4.057.08c *śunám parjányo mádhunā páyobhiḥ*
4.057.08d *śúnāsīrā śunám asmāsu dhattam*

In contentment may the ploughshare turn up the sod, in contentment the ploughman follows the oxen, celestial rain pours down honey and water. Ploughshare and plough, grant us us joy! (Panikkar 1979)

CHAPTER II – SEASONS

1. *Ṛ*V.I.95.3.

1.095.03a *trīṇi jānā pári bhūṣanti asya*
1.095.03b *samudrá ékam̐ diví ékam apsú*
1.095.03c *pūrvām ánu prá díśam pārthivānām*
1.095.03d *ṛtūn praśāsad ví dadhāv anuṣṭhú*

In his three places of birth they adore/honour (him), in the ocean, in the firmament (and) in the waters. Ruling in the eastern direction of earth's regions, he grants/establishes seasons in (their) order.

2. *Ṛ*V.VII.66.16. PL

7.066.16a *tác cákṣur deváhitam̐ śukrám uccárat*
7.066.16b *páśyema śarádaḥ śatám̐*
7.066.16c *jīvema śarádaḥ śatám*

(That) eye divinely placed, rising resplendent
May we behold (it) for a hundred autumns, may we live for a hundred autumns.

3. *Ṛ*V.X.161.4. PL (*AV*.XX.96.1)
10.161.04a *śatáṃ jīva śarádo várdhamānaḥ*
10.161.04b *śatáṃ hemantāñ chatám ū vasantān*
10.161.04c *śatám indrāgnī savitā bṛ́haspátiḥ*
10.161.04d *śatāyuṣā havíṣemám púnar duḥ*

Live prosperously, a hundred autumns, a hundred winters and through hundred springs. Indra-Agni, Savitṛ and Bṛhaspati, by the 'hundred-lives oblation/sacrifice' (give this person) again (a life of) a hundred (years)

4. *Ṛ*V.II.33.2.
2.033.02a *tvādattebhī rudara śáṃtamebhiḥ*
2.033.02b *śatáṃ hímā aśīya bheṣajébhiḥ*
2.033.02c *ví asmád dvéṣo vitaráṃ ví áṃho*
2.033.02d *ví ámīvāś cātayasvā víṣūcīḥ*

By the most salutary medicines given by thee, O Rudra, I would attain a hundred winters.
Drive far away from us hatred, away distress, away diseases in all directions. (Macdonell 1995)

By those most healing medicines that you give, Rudra, I would attain a hundred winters. Drive hatred far away from us, and anguish farther away; drive diseases away in all directions. (O'Flaherty 1981)

5. *Ṛ*V.V.14.4.
5.014.04a *agnír jātó arocata*
5.014.04b *ghnán dásyūñ jyótiṣā támaḥ*
5.014.04c *ávindad gā apáḥ súvaḥ*

Agni being born shone. By lustre (he) destroys Dasyu, (and) darkness.
(He has) found the cows, the water (and) the sun.

6. *Ṛ*V.VII.103.9. PL

7.103.09a *deváhitiṃ jugupur dvādaśásya*
7.103.09b *ṛtúṃ náro ná prá minanti eté*
7.103.09c *saṃvatsaré prāvṛ́ṣi āgatāyāṃ*
7.103.09d *taptā gharmā aśnuvate visargám*

They have followed the divine order of 12 months: these men (Maruts, the storm clouds) do not disobey the season. At the end of saṃvatsara when the rainy season comes, the heated gharma [in terms of the phenomenon, the warm atmosphere of the transitional period of summer and rainy season, in terms of the ritual, the heated milk-offering or the milkpot] obtains release. (Vajracharya 1997)

Three Seasons

7. *Ṛ*V.I.34.2. (M)

1.034.02a *tráyaḥ paváyo madhuvāhane ráthe*
1.034.02b *sómasya venām ánu víśva íd viduḥ*
1.034.02c *tráya skambhāsa skabhitāsa ārábhe*
1.034.02d *trír náktaṃ yāthás trír u aśvinā dívā*

Three fellies in (your) wealth bearing chariot. . . . Three pillars held/set for support, O Aśvin, thrice (you) journey by night and thrice by day.

8. *Ṛ*V.I.34.5. (M)

1.034.05a *trír no rayíṃ vahatam aśvinā yuvám*
1.034.05b *trír devátātā trír utāvataṃ dhíyaḥ*
1.034.05c *tríḥ saubhagatváṃ trír utá śrávāṃsi nas*
1.034.05d *triṣṭháṃ vāṃ sūre duhitā ruhad rátham*

Thrice, O Aśvin, you bring (us) wealth/riches, . . .
Thrice (you bring) us happiness and thrice fame, Sun's daughter has mounted (your) chariot of three seats.

9. *Ṛ*V.I.118.2. (E)

1.118.02a *trivandhuréṇa trivṛ́tā ráthena*
1.118.02b *tricakréṇa suvṛ́tā yātam arvāk*
1.118.02c *pínvataṃ gā jínvatam árvato no*
1.118.02d *vardháyatam aśvinā vīrám asmé*

Come herewards in your chariot, which has three seats,which rolls forward on three sides, which is fitted with three wheels, and which gracefully glides along. Fill our cows (with milk), and quicken our horses. Lead our brave offsprings to prosperity, O Aśvinā. (Velankar 1972)

10. *Ṛ*V.I.157.3. (E)

1.157.03a *arvāṅ tricakró madhuvāhano rátho*
1.157.03b *jīrāśuvo aśvínor yātu súṣṭutaḥ*
1.157.03c *trivandhuró maghávā viśvásaubhagaḥ*
1.157.03d *śáṃ na ā vakṣad dvipáde cátuṣpade*

Come to us praised/lauded three-wheeled, wealth laded chariot of Aśvin, drawn by swift horses.
With three seats, gracious, bestowing blessings and welfare on all. Increase (our) two-footed (sons) and four-footed (cattle).

11. *Ṛ*V.I.164.2. (E), PL

1.164.02a *saptá yuñjanti rátham ékacakram*
1.164.02b *éko áśvo vahati saptánāmā*
1.164.02c *trinābhi cakrám ajáram anarváṃ*
1.164.02d *yátremā víśvā bhúvanādhi tasthúḥ*

Seven yoke the one-wheeled chariot drawn by one horse with seven names. All these creatures rest on the ageless and unstoppable wheel with three naves. (O'Flaherty 1981)

Seven (priest-seers) yoke (employ in their sacrifice) the one-wheeled chariot (the sun and the year); one horse (the sun) with seven names draws it.

Triple-naves, ageless, unstoppable is the wheel (the year) on which all creatures stand. (Houben 2000a)

12. *Ṛ*V.I.164.48. (E), PL; (*AV*.X.8.4)

1.164.48a	*duvā́daśa pradháyaś cakrám ékaṃ*
1.164.48b	*trī́ṇi nábhyāni ká u tác ciketa*
1.164.48c	*tásmin sākáṃ · triśatā́ ná śaṅkávo*
1.164.48d	*arpitā́ḥ ṣaṣṭír ná calācalā́saḥ*

The felly-pieces are twelve, the wheel (the year) is one, the nave-pieces three; who has understood this? On it are placed, as it were, 360 pegs that do not wobble. (Houben 2000)

13. *Ṛ*V.IV.36.1. (E)

4.036.01a	*anaśvó jātó anabhīśúr ukthíyo*
4.036.01b	*ráthas tricakráḥ pári vartate rájaḥ*
4.036.01c	*mahát tád vo devíyasya pravā́canaṃ*
4.036.01d	*dyā́m ṛbhavaḥ pṛthivī́ṃ yác ca púṣyatha*

Born without horses, without a rein, (worthy) of holy song, the three-wheeled chariot traverses around the firmament, (that is the) great proclamation of that divine power. O Ṛbhus you/who nourish the heaven and earth.

FIVE SEASONS

14. *Ṛ*V.I.20.7. (L)

1.020.07a	*té no rátnāni dhattana*
1.020.07b	*trír ā́ sā́ptāni sunvaté*
1.020.07c	*ékam-ekaṃ suśastíbhiḥ*

You confer wealth, for offering of *Soma* sacrifice, three (times) seven, each (one-by-one) with well recited (hymn).

15. *Ṛ*V.I.164.12. (E), PL

1.164.12a	*páñcapādam pitáraṃ dvā́daśākṛtiṃ*
1.164.12b	*divá āhuḥ páre árdhe purīṣíṇam*
1.164.12c	*áthemé anyá úpare vicakṣaṇáṃ*
1.164.12d	*saptácakre ṣáḷara āhur árpitam*

The father (sun as year) with five feet and twelve shapes, the affluent, they say, is in the upper half of heaven. But these others say that the wide seeing one (the sun) is in the lower (half of the heaven), fixed in the seven-wheeled, six-spoked chariot. (Houben 2000a)

16. *Ṛ*V.I.164.13. (E), PL

1.164.13a	*páñcāre cakré parivártamāne*
1.164.13b	*tásminn ā́ tasthur bhúvanāni víśvā*
1.164.13c	*tásya nā́kṣas tapyate bhū́ribhāraḥ*
1.164.13d	*sanā́d evá ná śīryate sánābhiḥ*

On this five-spoked wheel as it revolves all creatures/worlds have their support.
Its axle, carrying a heavy load, does not get heated. Never at all does it wear out in its nave. (Houben 2000a)

17. *Ṛ*V.VII.87.4. (M)

7.087.04a	*uvā́ca me váruṇo médhirāya*
7.087.04b	*tríḥ saptá nā́ma ághniyā bibharti*
7.087.04c	*vidvā́n padásya gúhiyā ná vocad*
7.087.04d	*yugā́ya vípra úparāya śíkṣan*

The wise Varuṇa said to me; the cow bears thrice seven names. The wise, (who) know the secret of the place, speak to teach 'prince/student' for (future) age.

18. *Ṛ*V.X.90.15. (L), PL; (*AV*.XIX.6.15)

10.090.15a	*saptā́syāsan paridháyas*
10.090.15b	*tríḥ saptá samídhaḥ kṛtā́ḥ*
10.090.15c	*devā́ yád yajñám tanvānā́*
10.090.15d	*ábadhnan púruṣam paśúm*

Seven were his enclosing sticks; thrice seven were the faggots made, when the gods performing the sacrifice bound Puruṣa as the victim. (Macdonell 1995)

Seven sticks enclosed it like a fence. Thrice seven were the sticks of firewood. When the Gods bound Man as the offering for the sacrifice. (de Nocolás 1978)

SIX SEASONS

19. *Ṛ*V.VII.87.5. (M)
7.087.05a *tisró dyā́vo níhitā antár asmin*
7.087.05b *tisró bhū́mīr úparāḥ ṣáḍvidhānāḥ*
7.087.05c *gṛ́tso rā́jā váruṇaś cakra etáṃ*
7.087.05d *diví preṅkháṃ hiraṇyáyaṃ śubhé kám*

Three heavens were placed in this, three earths situated below, forming an order of six. Wise king Varuṇa, who has placed this bright/beautiful golden swing (sun) in the sky.

20. *Ṛ*V.X.90.6. PL
10.090.06a *yát púruṣeṇa havíṣā*
10.090.06b *devā́ yajñám átanvata*
10.090.06c *vasantó asyāsīd ā́jyaṃ*
10.090.06d *grīṣmá idhmáḥ śarád dhavíḥ*

When the gods performed a sacrifice with Puruṣa as an oblation, the spring was its melted butter, the summer its fuel, the autumn its oblation. (Macdonell 1995)

When the gods spread the sacrifice with the Man as the offering, spring was the clarified butter, summer the fuel, autumn the oblation. (O'Flaherty 1981)

21. *Ṛ*V.I.164.15. (E), PL
1.164.15a *sākaṃjā́nāṃ saptátham āhur ekajáṃ*
1.164.15b *ṣáḷ íd yamā́ ṛ́ṣayo devajā́ íti*
1.164.15c *téṣām iṣṭā́ni víhitāni dhāmaśá*
1.164.15d *sthātré rejante víkṛtāni rūpaśáḥ*

They say that besides those born in pairs there is a seventh born alone, while the six sets of twins are the sages born from the gods. The sacrifices for them are firmly set, but they change their forms and waver as he stands firm. (O'Flaherty 1981)

The seventh of those born together, they say, is born alone. Six form pairs (3×2)*. They are regarded as seers born from the

* should be (6×2).

gods. Those desired by these (the days and nights) are arranged in order; to the one who stands (on the chariot) those who are alternating their form (days and nights) are trembling. (Houben, 2000a)

22. *ṚV*.V.32.2.

5.032.02a	*tuvám útsām̐ ṛtúbhir badbadhānā́m̐*
5.032.02b	*áraṃha ū́dhaḥ párvatasya vajrin*
5.032.02c	*áhiṃ cid ugra práyutaṃ śáyānaṃ*
5.032.02d	*jaghanvā́m̐ indra táviṣīm adhatthāḥ*

(O) cloud of the mountain, you prepare the springs (that are) oppressed by the seasons.

Wielding a thunderbolt, (O) Indra, (you) kill even the powerful dragon. . . .

CHAPTER III – DAYS, MONTHS AND YEARS

1. *ṚV*.III.54.19.

3.054.19a	*devā́nāṃ dūtáḥ purudhá prásūto*
3.054.19b	*ánāgān no vocatu sarvátātā*
3.054.19c	*śṛṇótu naḥ pṛthivī́ dyaúr utā́paḥ*
3.054.19d	*sū́ryo nákṣatrair urú antárikṣam*

(That) messenger of gods, frequently born, pronounce us completely without sin.

May the earth, the sky, even water, the sun, the constellations/ stars (and) the vast firmament, hear us.

2. *ṚV*.I.31.14.

1.031.14a	*tuvám agna uruśáṃsāya vāgháte*
1.031.14b	*spārháṃ yád réknaḥ paramáṃ vanóṣi tát*
1.031.14c	*ādhrásya cit prámatir ucyase pitā́*
1.031.14d	*prá pā́kaṃ śā́ssi prá díśo vidúṣṭaraḥ*

You Agni desire the highest wealth/prize for the good (one with a good hymn) sacrificer. . . .

(You) who are all-wise, we call you father, (caring) even for the weak/poor, you teach (the cardinal points) of directions. . . .

3. *ṚV*.I.50.9.
   1.050.09a *áyukta saptá śundhyúvaḥ*
   1.050.09b *sū́ro ráthasya naptíyaḥ*
   1.050.09c *tā́bhir yāti sváyuktibhiḥ*

   Sūrya has yoked the pure, bright seven (the) daughters of the Sun's chariot.

   With this willing team, he moves on. (de Nicolás 1978)

   The sun has yoked the seven splendid daughters of the chariot; he goes with them, who yoke themselves. (O'Flaherty 1981)

4. *ṚV*.I.50.4.
   1.050.04a *taráṇir viśvádarśato*
   1.050.04b *jyotiṣkṛ́d asi sūriya*
   1.050.04c *víśvam ā́ bhāsi rocanám*

   Crossing space, you are the marker of light,seen by everyone, O sun. You illumine the whole, wide realm of space. (O'Flaherty 1981)

5. *ṚV*.II.38.4.
   2.038.04a *púnaḥ sám avyad vítataṃ váyantī*
   2.038.04b *madhyā́ kártor ní adhāc chákma dhī́raḥ*
   2.038.04c *út saṃhā́yāsthād ví ṛtū́m̐r adardhar*
   2.038.04d *arámatiḥ savitā́ devá ā́gāt*

   Weaving her extended web, She has once more spread it out (over the world). The wise artisan has laid down his tools in the midst of his work. The divine Savitṛ, having rested for a while, has arisen once more (and) properly distributed the seasons; the quick-witted god has now arrived (towards the evening, after his day's work). (Velankar 1972)

   She who weaves has rolled up again what was stretched out. The skilful worker has laid down the work half completed. He stirs and stands up; he has set apart the different times. With his thoughts gathered, the god Savitṛ has come. (O'Flaherty 1981)

6. *Ṛ*V.III.58.1.
3.058.01a *dhenúḥ pratnásya kā́miyaṃ dúhānā*
3.058.01b *antáḥ putráś carati dákṣiṇāyāḥ*
3.058.01c *ā́ dyotaníṃ vahati śubhráyāmā*
3.058.01d *uṣása stómo aśvínāv ajīgaḥ*

The cow grants the desire of ancient. The son from/of south walks/travels within (the firmament). She who has a radiant chariot (Uṣas) brings light/ splendour. The hymn of Uśas (drives/ drove/awakens/awoke) the Aśvins.

7. *Ṛ*V.IV.13.3.
4.013.03a *yáṃ sīm ákṛṇvan támase vipṛ́ce*
4.013.03b *dhruvákṣemā ánavasyanto ártham*
4.013.03c *táṃ sū́riyaṃ harítaḥ saptá yahvī́*
4.013.03d *spáśaṃ víśvasya jágato vahanti*

Master of unchallenged domains, they send forth the Sun on his way with never-failing precision. The Sun, dispeller of darkness, whose eye contemplates all things, is borne onwards by seven shining mares. (Panikkar 1979)

8. *Ṛ*V.IV.13.5. PL (*RV*.IV.14.5)
4.013.05a *ánāyato ánibaddhaḥ kathā́yáṃ*
4.013.05b *níaṅṅ uttānó áva padyate ná*
4.013.05c *káyā yāti svadháyā kó dadarśa*
4.013.05d *divá skambháḥ sámṛtaḥ pāti nā́kam*

How strange the Sun! Untethered, unsupported, he hangs in space. Why does he never fall? What inner power propels him? Who can observe it? He guards heaven's vault, the sky's pillar. (Panikkar 1979)

9. *Ṛ*V.III.32.9.

3.032.09a	*ádrogha satyáṃ táva tán mahitváṃ*
3.032.09b	*sadyó yáj jātó ápibo ha sómam*
3.032.09c	*ná dyā́va indra tavásas ta ójo*
3.032.09d	*nā́hā ná mā́sāḥ śarádo varanta*

O unmolested Indra, that greatness of yours is unfailing when as soon as you were born you drank *soma*. Neither the heavens nor the days, neither the months nor the year, may oppose the might of you, who are mighty, O Indra. (Velankar 1972)

10. *Ṛ*V.I.34.8.

1.034.08a	*trír aśvinā síndhubhiḥ saptámātṛbhis*
1.034.08b	*tráya āhāvā́s trēdhā́ havíṣ kṛtám*
1.034.08c	*tisráḥ pṛthivī́r upári pravā́ divó*
1.034.08d	*nā́kaṃ rakṣethe dyúbhir aktúbhir hitám*

With three rivers, seven mothers, triple vessels, the triple oblation is prepared by the Aśvins.

Passing/blowing in the sky above three earth(s), you protect the firmament with days and nights.

11. *Ṛ*V.I.113.3.

1.113.03a	*samānó ádhvā svásaror anantás*
1.113.03b	*tám anyā́nyā carato deváśiṣṭe*
1.113.03c	*ná methete ná tasthatuḥ suméke*
1.113.03d	*náktoṣā́sā sámanasā vírūpe*

Common is the path of the two sisters, unending. One after another they traverse it, instructed by the Gods. They quarrel not, they stay not, firmly established. Night and Dawn, of one mind, of different hues. (Thomas 1923)

12. *Ṛ*V.X.85.13. PL

10.085.13a	*sūryā́yā vahatúḥ prā́gāt*
10.085.13b	*savitā́ yám avā́sṛjat*
10.085.13c	*aghā́su hanyante gā́vo*
10.085.13d	*árjunyoḥ pári uhyate*

The wedding procession of Sūryā went forward as Savitṛ sent it off. When the sun is in *Aghā* they kill the cattle, and when it is in Arjunī she is brought home. (O'Flaherty 1981).

Sūryā's bridal procession went forth, which Savitaṛ sent out. Under the (constellation) Aghās the cattle are slaughtered (in honour of the bridegroom, when he comes to the bride's house); under the (constellation of the) two Arjunīs the departure (to the groom's house) takes place. (Edgerton 1965)

In AV.XIV.1.13. Aghā is replaced by Maghā and Arjunī by Phālgunī the post-*ṚV* names of these *nakṣatra*s.

13. *ṚV*.VI.59.6. (*VS*.XXXIII.93)
6.059.06a *índraagnī apā́d iyám*
6.059.06b *pū́rvā́gāt padvátībhiyaḥ*
6.059.06c *hitvī́ śíro jihváyā vā́vadac cárat*
6.059.06d *triṃśát padā́ ní akramīt*

Indra and Agni, this (lady) without feet (Uṣas – dawn) came before those with feet (animals and humans).
Leaving (her) head (?), speaking with (her) tongue (she) traversed down (her course) in thirty steps.

14. *ṚV*.X.189.3. PL
10.189.03a *triṃśád dhā́ma ví rājati*
10.189.03b *vā́k pataṃgā́ya dhīyate*
10.189.03c *práti vástor áha dyúbhiḥ*

(Sun) rules/governs/shines (over) thirty domain(s) (of the) day. The song of the morning is bestowed on the bird (sun) by the days.

15. *ṚV*.X.138.6.
10.138.06a *etā́ tiyā́ te śrútiyāni kévalā*
10.138.06b *yád éka ékam ákṛṇor ayajñám*
10.138.06c *māsā́ṃ vidhā́nam adadhā ádhi dyávi*
10.138.06d *tváyā víbhinnam bharati pradhím pitā́*

Only you have done those glorious deeds. . . .
You have placed over in the sky the regulator of the months, the Father conveys the orb (day) split in two.

16. *ṚV*.X.85.5. PL (*AV*.XIV.1.4)
    10.085.05a *yát tvā deva prapíbanti*
    10.085.05b *táta ā́ pyāyase púnaḥ*
    10.085.05c *vāyúḥ sómasya rakṣitā́*
    10.085.05d *sámānām mā́sa ā́kṛtiḥ*

    When Deva, they drink of thee then thou swellest out again
    Vāyu is Soma's protector, it is the moon that shapes the year. (Bose 1966)

    When they drink you, O god, then (as waxing Moon) you fill yourself up again. (The wind-god) Vāyu is protector of Soma. The Moon (also 'month') is the pattern of the year. (Edgerton 1965)

17. *ṚV*.V.78.8. PL
    5.078.08a *yáthā vā́to yáthā vánaṃ*
    5.078.08b *yáthā samudrá éjati*
    5.078.08c *evā́ tváṃ daśamāsiya*
    5.078.08d *sahā́vehi jarā́yuṇā*

    As the wind and the wood and the ocean stir, so let the 10-month-old child come down together with the afterbirth. (O'Flaherty 1981)

18. *ṚV*.V.78.9. PL
    5.078.09a *dáśa mā́sāñ chaśayānáḥ*
    5.078.09b *kumāró ádhi mātári*
    5.078.09c *niraítu jīvó ákṣato*
    5.078.09d *jīvó jī́vantiyā ádhi*

    When the boy has lain for ten months in the mother, let him come out alive and unharmed, alive from the living woman. (O'Flaherty 1981)

19. Ṛ*V*.V.45.7.

5.045.07a *ánūnod átra hástayato ádrir*
5.045.07b *ā́rcan yéna dáśa māsó návagvāḥ*
5.045.07c *ṛtáṃ yatī́ sarámā gā́ avindad*
5.045.07d *víśvāni satyā́ áṅgirāś cakāra*

Here, . . . with handheld stone, the Navagvā worshipped for ten months.

Sarmā going by the divine way found the cows. Aṅgiras. . . .

20. Ṛ*V*.X.2.7.

10.002.07a *yáṃ tvā dyā́vāpṛthivī́ yáṃ tvā ā́pas*
10.002.07b *tváṣṭā yáṃ tvā sujánimā jajā́na*
10.002.07c *pánthām ánu pravidvā́n pitṛyā́ṇaṃ*
10.002.07d *dyumád agne samidhānó ví bhāhi*

Who, you the heaven and earth, you the water, you the Tvaṣṭṛ, the creator of fair things, have created. You who know the path of the Pitṛ, O Agni, may you shine brightly. . . .

21. Ṛ*V*.I.164.44. PL

1.164.44a *tráyaḥ keśína ṛtuthā́ ví cakṣate*
1.164.44b *saṃvatsaré vapata éka eṣām*
1.164.44c *víśvam éko abhí caṣṭe śácībhir*
1.164.44d *dhrā́jir ékasya dadṛśe ná rūpám*

Three long-haired once show themselves in due season; during the year one of them (fire, with flames as hair) shaves (himself, strewing his hair/flames all around); one (the sun, with rays as hair) surveys everything by means of his powers; of one (the wind, (with lightning as hair) the rushing is seen but not his form. (Houben 2000a)

22. Ṛ*V*.VII.103.8. PL

7.103.08a *brāhmaṇā́saḥ somíno vā́cam akrata*
7.103.08b *bráhma kṛṇvántaḥ parivatsarī́ṇam*
7.103.08c *adhvaryávo gharmíṇaḥ siṣvidānā́*
7.103.08d *āvír bhavanti gúhiyā ná ké cit*

Brahmins with Soma raise their voices offering the prayer for the beginning of the year; the officiating priests come forth heated and sweating. None remain hidden. (O'Flaherty 1981)

Soma-pressing Brahmins have uttered their voice while offering the prayer associated with *parivatsara* [a synonym for *saṃvatsara*]. Adhvaryu priests, heated and sweating, have made their appearance. None of them are hiding (anymore). (Vajracharya 1997)

23. *ṚV*.X.190.2. PL
10.190.02a *samudrā́d arṇavā́d ádhi*
10.190.02b *saṃvatsaró ajāyata*
10.190.02c *ahorātrā́ṇi vidádhad*
10.190.02d *víśvasya miṣató vaśī́*

From the billowy ocean was born the year, that arranges days and nights, ruling over all that blinks its eyes. (O'Flaherty 1981)

24. *ṚV*.I.164.11. PL
1.164.11a *duvā́daśāraṃ nahí táj járāya*
1.164.11b *várvarti cakrám pári dyā́m ṛtásya*
1.164.11c *ā́ putrā́ agne mithunā́so átra*
1.164.11d *saptá śatā́ni viṃśatíś ca tasthuḥ*

With 12 spokes – for it does not become old – the wheel of the truthful order (the sun as year) turns on and on around the sky. Sons, in pairs, O Agni, 720 (360 days and 360 nights), fixed in the seven-wheeled, six-spoked chariot. (Houben 2000a)

25. *ṚV*.X.60.8. PL
10.060.08a *yáthā yugáṃ varatráyā*
10.060.08b *náhyanti dharúṇāya kám*
10.060.08c *evā́ dādhāra te máno*
10.060.08d *jīvā́tave ná mṛtyáve*
10.060.08e *átho ariṣṭátātaye*

Just as one binds a yoke with a cord (belt) to secure, so has he fastened your spirit to life, not to death but to keep safe.

26. *Ṛ*V.X.101.3. PL

10.101.03a *yunákta sī́rā ví yugā́ tanudhvaṃ*
10.101.03b *kṛté yónau · vapatehá bī́jam*
10.101.03c *girā́ ca śruṣṭíḥ sábharā ásan no*
10.101.03d *nédīya ít sṛṇíyaḥ pakvám éyāt*

Harness the plough and stretch the yoke on it, sow the seed in the prepared womb. And if the hearing of our song is weighty enough, then the ripe crop will come nearer to the scythes. (O'Flaherty 1981)

kṛté yónau – 'prepared field' would be a better interpretation then 'prepared womb'.

27. *Ṛ*V.X.72.1. PL

10.072.01a *devā́nāṃ nú vayáṃ jā́nā*
10.072.01b *prá vocāma vipanyáyā*
10.072.01c *ukthéṣu śasyámāneṣu*
10.072.01d *yáḥ páśyād úttare yugé*

Let us now speak with wonder of the birth of the gods – so that someone may see them when the hymns are chanted in this later age. (O'Flaherty 1981)

28. *Ṛ*V.I.92.11.

1.092.11a *viūrṇvatī́ divó ántām̐ abodhi*
1.092.11b *ápa svásāraṃ sanutár yuyoti*
1.092.11c *praminatī́ manuṣíyā yugā́ni*
1.092.11d *yóṣā jārásya cákṣasā ví bhāti*

She has awakened, uncovering the very edges of the sky; she pushed aside her sister. Shrinking human generations, the young woman shines under her lover's gaze. (O'Flaherty 1981)

29. *Ṛ*V.VI.15.8.

6.015.08a *tvā́ṃ dūtám agne amṛ́taṃ yugé-yuge*
6.015.08b *havyavā́haṃ dadhire pāyúm ī́ḍiyam*
6.015.08c *devā́saś ca mártiāsaśca jā́gṛviṃ*
6.015.08d *vibhúṃ viśpátiṃ námasā ní ṣedire*

O Agni, the immortal envoy, age-after-age you (have been) the bearer of oblation, . . .

Both gods and mortals have established you with honour, the mighty, vigilant chief of men.

30. *Ṛ*V.X.72.2. PL

10.072.02a	*bráhmaṇas pátir etā́*
10.072.02b	*sáṃ karmā́ra ivādhamat*
10.072.02c	*devā́nām pūrviyé yugé*
10.072.02d	*ásataḥ sád ajāyata*

The lord of sacred speech, like a smith, fanned them together. In the earlier age of the gods, existence was born from non-existence. (O'Flaherty 1981)

31. *Ṛ*V.I.158.6.

1.158.06a	*dīrghátamā māmateyó*
1.158.06b	*jujurvā́n daśamé yugé*
1.158.06c	*apā́m ártham yatī́nā́m*
1.158.06d	*brahmā́ bhavati sā́rathiḥ*

Dīrghatamā, the son of Mamatā, became old in the tenth *yuga*. Priests of the waters, he became helper/charioteer of the devotees.

32. *Ṛ*V.I.25.8.

1.025.08a	*véda māsó dhṛtávrato*
1.025.08b	*duvā́daśa prajā́vataḥ*
1.025.08c	*védā yá upajā́yate*

Dhṛtavrata (Varuṇa) knew the 12 productive months, he also knew about the thirteenth additional month. (Abhyankar 1993)

He, who keeps his vows, knows the twelve months that bestow children, he knows the month which is additional. (Pingree 1989)

He knows the twelve months with their progeny (the days) and also the month that is added. (Kane 1968)

Note: '*upajā̆yate*' has been interpreted as the 'thirteenth additional month', 'month which is additional' and 'the month that is added' in the three interpretations above. However, neither 'month' nor 'thirteen' can be construed from *upajā̆yate*. An appropriate interpretation of *upajā̆yate* is 'that which in addition' or 'that which has to be added' (Monier-William 1999). The verse *ṚV*.I.25.8 can therefore be translated as

Dhṛtavrata knew (has knowledge of) the twelve productive months, he also knew (has knowledge of) that which has to be added.

33. *MS*.I.10.8.
Ṛtuyājī vā anyaś, cāturmāsya-yājy anyo; yo ,vasanto 'bhūt, prāvṛḍ abhūñ, śarad abhūd' iti yajati, sa ṛtuyājy; atha yas trayodaśaṃ māsaṃ saṃpādayati, trayodaśaṃ māsam abhiyajate, sa cāturmāsya-yājy, ṛjūṃs trīn iṣṭvā, caturtham utsṛjeta; rjū dvau parā iṣṭvā tṛtīyam utsṛjeta; ye vai trayaḥ saṃvatsarās, teṣāṃ ṣaṭtriṃśat pūrṇamāsā; yau dvau, tayoś caturviṃśatis; tad ye 'mī ṣaṭtriṃśaty adhi tān asyāṃ caturviṃśatyām upasaṃpādayaty, eṣa vāva sa trsyodaśo māsas, tam evaitat saṃpādayati, tam abhiyajate.

It is another who celebrates the seasonal sacrifices, and it is another who brings the *cāturmāsya*-sacrifices. He who sacrifices whilst he thinks; spring has come, the raining season has come, autumn has come, is a person who celebrates the seasonal sacrifices. But he who causes a thirteenth month to arise and celebrates this thirteenth month with sacrifices, is a person who brings the *cāturmāsya*s; after having brought these offerings for three continuous [synodic-lunar] years he has to postpone the sacrifice a month at the time when the fourth course should begin, and having then brought these offerings for two other continuous [synodic-lunar] years he has to postpone the sacrifice a month at the time when the third course should begin. In the first mentioned three years there are 36 full-moon days, and in the last mentioned two years there are 24 full-moon days. When the sacrificer thus adds the *cāturmāsya*-sacrifices which are contained in the 24 full-moon days to the sacrifices belonging to the 36 full-moon days, there takes place what is called the 'thirteenth month', there the sacrificer gives rise to the 'thirteenth month', there he celebrates it. (Faddegon 1926)

34. *TS*.VII.4.8.
saṃvatsarā́yadīkṣiṣyámāṇā ekāṣṭak kā́yāṃ dīkṣeran
eṣā́ vái saṃvatsar ásya pátnī yád ekāṣṭakā́
etá syāṃ vā́ eṣá eṭáṁ rā́triṃ vasati
sākṣā́d evá saṃvatsrám ārábhya dīkṣante
ā́rtaṃ vā́ eté saṃvatsarásyābhí dīkṣante yá ekāṣṭakā́yāṃ
dīkṣante 'ntanāmānāv ṛtū́ bhavatas
vyàstaṃ vā́ eté saṃvatsarásyābhí dīkṣante yá ekāṣṭakā́yāṃ
dīkṣante 'ntanāmānāv ṛtū́ bhavataḥ phalgunīpūrṇamāsé
dīkṣeran múkhaṃ vā́ etát//

Those who are about to consecrate themselves for the year (rite) should consecrate themselves on the Ekastaka. The Ekastaka is the wife of the year; on that night he dwells with her; verily they consecrate themselves grasping openly the year. Those who consecrate themselves on the Ekastaka consecrate themselves in the troubled part of the year, for then are the two months called the end. Those who consecrate themselves on the Ekastaka consecrate themselves on the torn part of the year, for them are the two seasons called the end. They should consecrate themselves on the full moon in Phalguni. The full moon in Phalguni is the beginning of the year. (Keith 1916)

CHAPTER IV – *NAKṢATRA*

1. *ṚV*.X.156.4.
10.156.04a *ágne nákṣatram ajáram*
10.156.04b *ā́ sū́ryaṃ rohayo diví*
10.156.04c *dádhaj jyótir jánebhiyaḥ*

O Agni, you have (made) the Sun, the ageless constellation (*nakṣatra*), rise in the sky. Granting light to the people.

2. *Ṛ*V.VII.86.1.
7.086.01a *dhī́rā tú asya mahinā́ janū́ṃṣi*
7.086.01b *ví yás tastámbha ródasī cid urvī́*
7.086.01c *prá nā́kam ṛṣváṃ nunude bṛhántaṃ*
7.086.01d *dvitā́ nákṣatram paprá̄thac ca bhū́ma*

Intelligent indeed are the generations by the might of him who has propped asunder the two wide worlds. He has pushed away the high, lofty firmament and the day-star (sun) as well; and he spread out the earth. (Macdonell 1995)

The creatures are wise owing to the greatness of him, who has stayed apart even the vast heaven and earth. He drove up the high and great firmament as indeed the sun; and he broadened out the earth. (Velankar 1963)

The generations have become wise by the power of him who has propped apart the world-halves even though they are so vast. He has pushed away the dome of the sky to make it high and wide; he has set the sun on its double journey and spread out the earth. (O'Flaherty 1981)

3. *Ṛ*V.I.50.2 (*AV*.XII.2.17)
1.050.02a *ápa tyé tāyávo yathā*
1.050.02b *nákṣatrā yanti aktúbhiḥ*
1.050.02c *sū́rāya viśvácakṣase*

The constellations (*nakṣatras*), along with the night, steal away like thieves, making way for the sun who gazes on everyone. (O'Flaherty 1981)

4. *Ṛ*V.VIII.55.2.
8.055.02a *śatáṃ śvetā́sa ukṣáṇo*
8.055.02b *diví tā́ro ná rocante*
8.055.02c *mahnā́ dívaṃ ná tastabhuḥ*

Hundred stars in the sky shine like white bulls. By their greatness they produce setting of the day (they cause the day to set).

5. *Ṛ*V.I.24.10. PL

1.024.10a *amī́ yá ṛ́kṣā níhitāsa uccā́*
1.024.10b *náktaṃ dádṛśre kúha cid díveyuḥ*
1.024.10c *ádabdhāni váruṇasya vratā́ni*
1.024.10d *vicā́kaśac candrámā náktam eti*

Those bears/stars which are seen, set high in the firmament/ night, where are they all in the day? By the divine rite of Varuṇa, the arrival of the radiant moon by/at night is not impaired.

6. *Ṛ*V.V.85.3.

5.085.03a *nīcī́nabāraṃ váruṇaḥ kávandham*
5.085.03b *prá sasarja ródasī antárikṣam*
5.085.03c *téna víśvasya bhúvanasya rā́jā*
5.085.03d *yávaṃ ná vṛṣṭír ví unatti bhū́ma*

Varuṇa lets the huge cask, open downwards,
Flow through the heavens, the earth, the mid-air regions
For the waters of the universe to fall
In rain and moisten the young grain
(de Nicolás 1978)

Over the two world-halves and the realm of space between them Varuna has poured out the cask, turning its mouth downwards. With it the king of the whole universe waters the soil and the rain waters the grain. (O'Flaherty 1981)

Varuna has poured out the cask, with its rim turned downwards, over heaven, earth and the interspace.
Thereby the king of the whole world sprinkles the soil, as the rain (sprinkles) the barley. (Witzel 1995)

7. *Ṛ*V.VIII.72.8.

8.072.08a *ā́ daśábhir vivásvata*
8.072.08b *índraḥ kóśam acucyavīt*
8.072.08c *khédayā trivṛ́tā diváḥ*

With the ten (fingers) of Vivasvat, Indra has pulled up the heavenly bucket, with a threefold cord. (Witzel 1995)

8. *AV*.X.8.9.
tiryágbilaś camasá ūrdhvábudhnas tásmin yáśo níhitaṃ viśvárūpam/
tád āsata ṛṣayaḥ saptá sākáṃ yé asyá gopā́ maható babhūvúḥ

A bowl (camasa) with orifice downwards, bottom-side up – in it is deposited glory of all forms; there sit together the seven seers, who have become the keepers of it, the great one (Whitney 1962)

9. *Ṛ*V.IV.51.2.
4.051.02a *ásthur u citrā́ uṣásaḥ purástān*
4.051.02b *mitā́ iva sváravo adhvaréṣu*
4.051.02c *ví ū vrajásya támaso duvā́rā*
4.051.02d *uchántīr avrañ chúcayaḥ pavākā́ḥ*

The brilliant Dawns have stood in the east, like posts set up at sacrifices. Shining they have unclosed the two doors of the pen of darkness, bright and purifying. (Macdonell 1995)

The lovely Dawns(?) has stood up in the east like posts erected at the sacrifice. The pure and purifying Dawns(?) while shining, have thrown open the doors of the stalls of Darkness. (Velankar 1972)

10. *Ṛ*V.X.19.1. PL
10.019.01a *ní vartadhvam mā́nu gāta*
10.019.01b *asmā́n siṣakta revatīḥ*
10.019.01c *ágnīṣomā punarvasū*
10.019.01d *asmé dhārayataṃ rayím*

Come, do not turn (from) us, … wealthy ones. Agni and Soma, the two 'wealth restorers', maintain wealth for us/ make us wealthy.

11. *Ṛ*V.V.54.13.
5.054.13a *yuṣmā́dattasya maruto vicetaso*
5.054.13b *rāyáḥ siyāma rathíyo váyasvataḥ*
5.054.13c *ná yó yúchati tiṣíyo yáthā divó*
5.054.13d *asmé rāranta marutaḥ sahasríṇam*

Wise Maruts, may we be the carriers of treasurers bestowed on us by you. It should not depart (from us) just as Tiṣya (does not depart from) the sky. Maruts, we bow thousand times/we bestow thousand bows.

12. *ṚV*.X.61.7.

10.061.07a	*pitā́ yát svā́ṃ duhitáram adhiṣkán*
10.061.07b	*kṣmayā́ rétaḥ saṃjagmānó ní ṣiñcat*
10.061.07c	*suādhíyo 'janayan bráhma devā́*
10.061.07d	*vā́stoṣ pátiṃ vratapā́ṃ nír atakṣan*

When the father approached his own daughter, (his) semen poured down on the earth/ground.
. . . the gods fashioned Vastpati, the upholder of religious ordnances.

13. *ṚV*.X.85.2. PL (*AV*. XIV.1.2)

10.085.02a	*sómenādityā́ balínaḥ*
10.085.02b	*sómena pṛthivī́ mahī́*
10.085.02c	*átho nákṣatrāṇām eṣā́m*
10.085.02d	*upásthe sóma ā́hitaḥ*

By Soma the Ādityas possess strength, by Soma the earth is great; and Soma is placed in the lap of the constellations. (Edgerton 1965)

Through Soma the Ādityas are mighty; through Soma the earth is great. And in the lap of these constellations Soma has been set. (O'Flaherty 1981)

CHAPTER V – *VEDĀṄGA JYOTIṢA*

1. *AB*.vii.11 (xxxii.10).
yām paryastamayād adbhyudiyād iti sā tithiḥ

Where (the moon) sets and rises that is *tithi.** (Dīkṣita 1896)

* Translated from Marāṭhī

Tithi is that period of time during which the moon sets and rises (again). (Subbarayappa and Sarma 1985)

2. *AV*.XIX.53.3.
pūrṇḥ kumbho dhikāla āhitas tam vai paśyāmo bahudhā nu santaṃ
pūrṇḥ / kumbhaḥ / adhi / kāle / āhitaḥ /
taṃ / vai / paśyāmḥ / bahudhā / nu / santaṃ /

A full vessel is set upon time.
We indeed see it, being now manifoldly.
(Whitney, 1962)

pūrṇḥ / kumbhaḥ / adhikāle / āhitaḥ /
taṃ / vai / paśyāmḥ / bahudhā / nu / santaṃ /

A full vessel is set (up) with reference to (measurement of) time. We indeed see it existing in many different forms. (Narahari Achar 1998a)

3. *TB*.1.5.2.1.
yát púṇyaṃ nákṣatram/
tād báṭ kurvīia upavyuṣám/
yadā́ vái sū́rya udéti/
átha nákṣatraṃ ná eti/
yāvatị tátra sū́ryo gácchet/
yátra jaghanyàṃ páśyet/
tā́vati kurvīta yatkārī́ syāt/
puṇyāhá ha vái yajñéṣuṃ ca śatádyumnaṃ ca
mātsyó niravasāyayā́ñ cakāra//

The auspicious star, its position has to be determined at sunrise. But when the sun rises, the star would not be visible (on account of the brightness of the Sun). So, before the Sun rises, watch for the adjacent star. By performing the rite with due time adjustment, one would have performed the rite at the correct time. Thus it was that sage Mātsya performed sacrifice of Yajñeṣu and Śatadyumna to perfection. (Subbarayappa and Sarma 1985)

Glossary

Agnicayana – piling of the fire altar (Chapter III, Section 2.8).

Agnihotra – a daily (morning and evening) ritual involving 'attendance' of *agni* and an oblation of milk.

Agniṣṭoma – 'hymns in praise of *agni*'. The full sacrifice lasted for five days and was performed in spring each year. The principle participants in this sacrifice were the sacrificer and his wife and the priests stood on the side to instruct the sacrificer.

Agnyādhāna – a rite of establishing the sacrificial fire, this is a rite of the *iṣṭi* type. The rite lasts for two days and can be performed on a new or a full moon day.

Āraṇyaka – texts taught and studied in the solitude of a forest because their contents were considered too sacred, secret, esoteric, etc.

Arithmetic Calendar – a calendar with simple but fixed rules for synchronizing the synodic year with seasons. Such calendars are easy to use but are unstable and inaccurate over long periods.

Astronomical Calendar – (or Observation-based calendar) a calendar based on astronomical observations. In particular, the tropical year and the synodic year are synchronized by astronomical observations. Such calendars are stable and accurate over extended periods but a level of expertise in astronomical observations is demanded.

BCE – Before the Common (Current) Era, this is a secular alternative to BC (Before Christ). See CE below.

Brahmacārin – a student of the Veda.

Brāhmaṇa – utterance of a teacher or a theologian, later generalized to the collections of such utterances (Chapter I, Section 2.1).

Cāturmāsya – four-monthly or seasonal sacrifices. The three principal sacrifices are separated by four synodic months and the ceremonies are performed when the full moon conjoins the prescribed *nakṣatra*s (Chapter IV, Section 6). The sacrifices are of the *iṣṭi* type.

CE – Common (Current) Era, a secular alternative to AD (Anno Domini). CE like BCE is used to number years in the Gregorian calendar, the globally used calendar at present. The numbering system to which BCE and CE refer is based on the assumed year of the conception or birth of Jesus Christ; BCE denotes the years before the start of this epoch and CE denotes the years after this epoch. There is no year zero in this scheme, so 1 CE follows 1 BCE.

Darśapūrṇamāsa – *darśa* means the times when the moon is 'seen' only by the sun (also *amāvāsyā*) and *pūrṇamāsa* means the full moon. This is the new and the full moon rite and it is an archetype (*prakṛti*) of *iṣṭi* rites. Four principal priests (*Hotṛ*, *Adhavaryu*, *Āgnidhra* and *Brahman*) are required for the performance of this rite. The two half-monthly sacrifices were to be performed regularly for a period of 30 years from the time a householder established a sacrificial fire of his own (*agny-ādhāna*). The ceremony lasts for two consecutive days; first day for the preparations (*upvasatha*) and the second for the performance. The day of the performance can be either the last two or the first two days of the lunar month.

Daśahotṛ – the *mantra* of ten Hotṛ. A mystical *mantra* that identifies 10 parts of a body with ten offerings at a sacrifice.

Dvādaśāha – a *Soma* sacrifice that lasts for twelve days.

Iṣṭi – Oblation offered by the Adhvaryu in a standing position.

Jyotiṣṭoma – a *Soma* sacrifice performed just after the setting up of the (householder's) sacred fire in *vasanta*. The main oblation is *soma* juice but substitutes are allowed. The procedure is elaborate and lasts for five days and 16 priests are involved (four for each Veda). There is a hierarchy in each group of priests and the priestly fees are distributed accordingly. The *Jyotiṣṭoma* sacrifice is the model for all *Soma* sacrifices with one '*sutyā*' or pressing day and the *Jyotiṣṭomasāman* chanted at the end. There are four variations of this sacrifice:

- *Agniṣṭoma* – *Agniṣṭomasāman* is chanted at the end.
- *Ukthya* – *Ukthyasāman* is chanted at the end.
- *Ṣoḍaśin* – *Ṣoḍaśinsāman* is chanted at the end.
- *Atirātra* – *Atirātrasāman* is chanted at the end.

Each of the *sāman* is followed by the recitation of the relevant Śāstra (normally a group of *ṛcās*).

Kāmyeṣṭi – an optional rite performed out of desire of a special reward.

Maṇḍala – a book of *Ṛgveda Saṃhitā.*

Mantra – a Vedic verse.

Nirukta – the oldest known text on etymology, philology and semantics; author Yāska (fifth or sixth century BCE).

Paśubandha – an animal sacrifice.

Rājasūya – a royal consecration or a coronation.

Ṛc (plural *ṛcās*) – (sacred) verse of the *Ṛgveda Saṃhitā.*

Saṃhitā – (placed together), a continuous text compiled according to the rules of *sandhi* (euphonic combination).

Sahasradakṣiṇā – thousand offerings presented to the officiating priest.

Sautrāmaṇī – a rite for Indra (Sautrāma – a good protector). The principle characteristic of this rite is the offering of wine, which is not drunk by the officiating priest but is given to a non-participating brahmin.

Smṛti – A body of sacred knowledge and tradition remembered by human teachers.

Stellar Magnitude – a measure of star's brightness as seen by an observer on the earth. Magnitude value increases as the brightness of a star decreases. The unaided eye can detect a star as faint as magnitude 6.

Sūkta – a good recitation, also a Vedic hymn.

Sūtra – a thread. In theological and religious discourses, it means a 'short rule' or a theorem condensed into a few words. Many 'short rules' woven together to form a book of rules is also called a Sūtra (Chapter I, Section 2.1).

Śākhā – community or 'school' of the Vedic texts, traditions and interpretations.

Śrauta – a ceremony based on Śruti and conforming to the sacred tradition of the Veda. Śrauta sacrifice requires a minimum of four priests, these are:

- the Hotṛ who recites the verses (*ṛcās*) in praise of the gods and invites them to the sacrifice.
- the Udgātṛ who accompanies with songs the preparation and performance of the sacrifice especially the *Soma* sacrifice.
- the Adhvaryu who performs the acts of the sacrifice and mutters the prose prayers and formulae specific to a sacrifice.
- the Brahman whose duty is to protect the sacrifice against malign influences or mistakes in the acts or songs by pronouncing appropriate sacred words.

The Brahman has to be knowledgeable in all three Vedas (*Ṛg*, *Yajur* and Sāma) but the other three priests involved in the Śrauta sacrifice only have to know the Veda specific to their part in the ceremony. Thus, each Veda has its own Śrauta Sūtras and it does not include material pertinent to other Veda, e.g. a *sūtra* of *Ṛgveda* for most part excludes actions and recitations of the Udgātṛ, Adhvaryu and Brahman.

Śruti – that which has been heard. The sacred knowledge, the eternal and infallible truth that is embodied in the Veda.

Upanayanam – to lead near. The ceremony of initiation of a young man in which he is symbolically led to a teacher.

Upaniṣad – *upa-ni-sad*, 'to sit down near someone' and probably refers to the pupil sitting down near his teacher. These texts are anti-ritualistic and mark the culmination of the line of inquiry into the nature of ultimate reality. The Upaniṣads are considered Vedānta or the end of Veda because they came at the end of Vedic period and were taught towards the end of Vedic instruction.

Vidhi – rule or direction for performance of a rite, sacrifice, etc.

Vājapeya – 'a drink of strength or power', a form of *Soma* sacrifice performed by a ruler to gain a 'higher position'.

Bibliography

Abhyankar, K.D., 'Misidentification of some Indian Nakṣatras', *Indian Journal of History of Science*, vol. 26, no. 1, 1991, pp. 1-10.

———, 'A Search for the Earliest Vedic Calendar', *Indian Journal of History of Science*, vol. 28, no. 1, 1993, pp. 1-14.

Agrawal, D.P. and B.M. Pande, 'Ecology and Archaeology of Western India', *Proc. Workshop held at Physical Research Laboratory*, Ahmedabad, 1976.

Amaladass, A., 'Jesuits and Sanskrit Studies', *Jesuits in India: In Historical Perspective*, ed. T.R. De Souza and C.J. Borges, Goa: Xavier Centre of Historical Research, 1992, pp. 209-31.

Arnold, E.V., *Vedic Metre*, Cambridge: Cambridge University Press, 1905.

Ashfaque, S.M., 'Astronomy in the Indus Valley Civilization', *Centaurus*, vol. 21, no. 2, 1977, pp. 149-93.

Aveni, A.F. *Skywatchers*, Austin: University of Texas Press, 2001, pp. 110-13.

Balachandra Rao, S., *Indian Astronomy: An Introduction*, Hyderabad: University Press, 2000, pp. 2-10.

Bentaleb, I., C. Caratini, M. Fontugne, M.-T. Morzadec-Kerrfourn, J.-P. Pascal, C. Tissot, 'Monsoon Regime Variability during the Late Holocene in Southwestern India', *Third Millennium* BC *Climate Change and Old World Collapse*, ed. N. Dalfes, G. Kukla and H. Weiss, 1997, pp. 475-88.

Billard, R., *L'astronomie indienne*: *investigation des textes sanskrits et des données numériques*, Paris: École française d'Extrême-Orient, 1971.

Bhide, V.V., *The Chaturmasya Sacrifice*, Pune: University of Poona, 1979.

Bose, A.C., *Hymns from the Vedas*, London: Asia Publishing House, 1966.

Brereton, J.P., 'Edifying Puzzlement: Ṛgveda 10.129 and the Uses of Enigma', *Journal of the American Oriental Society*, vol. 119, 1999, pp. 248-60.

Bryant, E., *The Quest for the Origin of Vedic Culture: The Indo-Aryan Migration Debate*, Oxford: Oxford University Press, 2001.

Bryson, R.A., 'Proxy Indications of Holocene Winter Rains in Southwest Asia Compared with Simulated Rainfall', *Third Millennium BC Climate Change and Old World Collapse*, ed. N. Dalfes, G. Kukla and H. Weiss, 1997, pp. 465-73.

Burrow, T., 'On the Significance of the term arma-, armaka- in Early Sanskrit Literature', *Journal of History of India*, vol. 41, no. 121, 1963, pp. 159-66.

———, *The Sanskrit Language*, London, 1955, rpt. Delhi: Motilal Banarsidass, 2001.

Caland, W., *Pañcavimsa Brāhmaṇa – The Brāhmaṇa of Twenty Five Chapters*, Calcutta: Asiatic Society of Bengal, 1931.

Chakravarty, A.K., *Origin and Development of Indian Calendrical Science*, Calcutta: Indian Studies, Past & Present, 1975, pp. 13-15.

Colebrooke, H.T., 'On the Vedas or Sacred Writings of Hindus', *Asiatick Researches*, vol. 8, 1805.

Converse, H.S., 'The Agnicayana rite: Indigenous origin?', *History of Religion*, 14, 1974, pp. 81-95.

Deshpande, M.M., *The Indo-Aryans of Ancient South Asia: Language, Material Culture and Ethnicity*, ed. G. Erdosy, New York: Walter de Gruyter, 1995, pp. 67-84.

Dīkṣita, Ś.B., *Bhāratīya Jyotiṣāstra* (in Marāṭhī), Pune: Varadā Books, 1896.

Dumont, P.E., 'Indo-Aryan Names from Mitanni, Nuzi and Syrian Documents', *Journal of the American Oriental Society*, vol. 67, 1947, pp. 251-3.

———, 'Taittirīya Brāhmaṇa (A Brahmana of the Black Yajur Veda)', *Proceedings of the American Philosophical Society*, vol. 92, 1948, pp. 447-503.

———, 'The Special Kind of Agnicayana (or the Special Method of Building the Fire-altar) in the Taittirīya Brāhmaṇa', *Proceedings of the American Philosophical Society*, vol. 95, 1951, pp. 628-75.

———, 'The Iṣṭis to the Nakṣatras (or the oblations to the Lunar mansions) in the Taittirīya Brāhmaṇa', *Proceedings of the American Philosophical Society*, vol. 98, 1954, pp. 204-23.

———, 'The Kāmya Animal Sacrifice in the Taittirīya Brāhmaṇa', *Proceedings of the American Philosophical Society*, vol. 113, 1969, pp. 34-66.

Edgerton, F., *The Beginnings of Indian Philosophy*, London: George Allen & Unwin Ltd., 1965.

Eggeling, J., *The Śatapatha-Brāhmaṇa According to the Text of the Mādhyaṃdina School, Part I (Books I and II)*, Oxford: Clarendon Press, 1882.

———, *The Śatapatha-Brāhmaṇa According to the Text of the Mādhyaṃdina School, Part II (Books III and IV)*, Oxford: Clarendon Press, 1885a.

———, *The Śatapatha-Brāhmaṇa According to the Text of the Mādhyaṃdina School, Part IV (Books VIII, IX and X)*, Oxford: Clarendon Press, 1885b.

———, *The Śatapatha-Brāhmaṇa According to the Text of the Mādhyaṃdina School, Part III (Books V, VI, and VII)*, Oxford: Clarendon Press, 1894.

———, *The Śatapatha-Brāhmaṇa According to the Text of the Mādhyaṃdina School, Part V (Books XI, XII, XIII and XIV)*, Oxford: Clarendon Press, 1900.

Elizarenkova, T.Y., *Rigveda: Izbrannye Gimny*, Moscow: Nauka, 1972.

Faddegon, B., 'The Thirteenth Month in Ancient Hindu Chronology', *Acta Orientalia*, vol. IV, 1926, pp. 124-33.

Fagan, B., *The Long Summer*, London: Granta Books, 2004.

Falk, H., 'The Purpose of Ṛgvedic Ritual', *Inside the Texts, Beyond the Texts*, ed. M. Witzel, Cambridge: Harvard University Press, 1997.

Frazer, J.G., *The Golden Bough: A Study in Magic and Religion*, vol. I, London: Macmillan Press, 1933.

Fleitmann, D. et al., 'Holocene Forcing of the Indian Monsoon Recorded in a Stalagmite from Southern Oman', *Science*, vol. 3000, 2003, pp. 1737-9.

Geldner, K.F., *Der Rig-Veda*, aus dem Sanskrit ins Deutsche Übersetzt, Cambridge, Harvard Oriental Series, vols. 33-36, 1951-7.

Gonda, J., *Vedic Literature (A History of Indian Literature,* ed. J. Gonda), Wiesbaden: Otto Harrassowitz, 1975.

———, *Triads in the Veda*, Amesterdam: North-Holland Publishing Co., 1976.

———, *Prajāpati and the Year*, Leiden: North-Holland Publishing Co., 1984.

Gondhalekar, P., 'Vedāṅga Jyotiṣa: Where and When?', *Indian Journal of History of Science*, vol. 43, no. 3, 2008a, pp. 339-52.

———, 'Intercalation in the Vedic Texts', *Indian Journal of History of Science*, vol. 43, no. 4, 2008b, pp. 495-514.

———, 'The Vedic Nakṣatras: A Reappraisal', *Indian Journal of History of Science*, vol. 44, no. 4, 2009, pp. 479-96.

———, 'The Vedic *Nakaṣtra*-names of the months', *Indian Journal of History of Science*, vol. 45, no. 3, 2010, pp. 331-41.

———, 'Possible Chronological Markers in the Vedic Texts', *Indian Journal of History of Science*, vol. 46, no. 1, 2011, pp. 1-22.

'Gray, L.H., 'Calendar (Persian)', *Encyclopaedia of Religion and Ethics,* ed. J. Hastings, vol. III, Edinburgh, T.&T. Clark, 1908-21, pp. 128-31.

Griffith, R.T.H., *The Hymns of the Ṛgveda*: translated with popular commentary, Benares, 1889-92, ed. J.L. Shastri, Delhi: Motilal Banarsidass, 1973.

———, *The Text of the White Yajurveda* (1899), rev. and enlarged edn., Delhi: Munshiram Manoharlal, 1987.

Griswold, H.D., *The Religion of the Rigveda*, Delhi: Motilal Banarsidass, 1971.

Gupta, S.P., 'The Indus-Sarasvatī Civilization Beginnings and Developments', *The Aryan Debate*, ed. T.R. Trautmann, Delhi: Yoda Press, 2005.

Hillebrandt, A., *Vedic Mythology*, Breslau (1927). English edn., Delhi: Motilal Banarsidass, 1980.

Houben, J.E.M., 'The Ritual Pragmatics of a Vedic hymn: The "Riddle hymn" and the Pravargya Ritual', *Journal of the American Oriental Society*, vol. 120, 2000a, pp. 499-536.

———, 'On the Earliest Attestable Forms of the Pravargya Ritual: Ṛg-Vedic References to the Gharma-Pravargya, Especially in the Atri-family Book (Book 5)', *Indo-Iranian Journal*, vol. 43, 2000b, pp. 1-25.

Hunger, H. and D. Pingree, *MUL.APIN An Astronomical Compendium in Cuneiform*, Archiv für Orientforschung, vol. 24, Austria: F. Berger, 1989.

Jaggi, O.P., *Indian Astronomy and Mathematics*, New Delhi: Oxford University Press, 1986.

Jamison, S.W. and M. Witzel, Vedic Hinduism, Harvard, www.people.fas.harvard.edu/~witzel/vedica.pdf, 1992.

Kane, P.V., *History of Dharmaśāstra*, Pune: Bhandarkar Oriental Research Institute, 1968.

Kansara, N.M., 'Vedic Source of Vedāṅga Jyotiṣa', *Issues in Vedic Astronomy and Astrology*, ed. H. Pandya, S. Dixit and N.M. Kansara, Delhi: Rashtriya Veda Vidya Pratishthan in association with Motilal Banarsidass, 1992.

Kashikar, C.G., 'Agnicayana: Extension of Vedic Aryan Ritual', *Aṇnals of the Bhandarkar Oriental Research Institute*, vol. X, 1981, pp. 121-33.

Keay, J., *India: A History*, London: HarperCollins, 2000, p. 8.

Keith, A.B., *The Veda of the Black Yajur Veda School* entitled *Taittiriya Sanhita*, Cambridge: Harvard University Press, 1914.

———, *Rigveda Brahmanas*: *The Aitareya and Kauṣītaki Brāhmaṇas of the Rigveda*, London: (1920), Harvard Oriental Series, vol. 25, rpt. Delhi, Motilal Banarsidass, 1998.

Kenoyer, J.M., *Ancient Cities of the Indus Valley*, Karachi: Oxford University Press, 1998.

———, 'Cultures and Societies of the Indus Tradition', *Historical Roots in the Making of 'the Aryan'*, ed. R. Thapar, New Delhi: National Book Trust, 2006, pp. 21-49.

Kochhar, R., *The Vedic People*, New Delhi: Sangam Books, 2000.

Krick, H., *Das Ritual der Feuergründung*, Wien: Verlag der österreicheschen Akademie der Wissenschaften, 1982.

Kuiper, F.B.J., *Ancient Indian Cosmology*, Essays selected and introduced by John Irwin, New Delhi: Vikas, 1983, p. 52.

———, *Aryans in the Rigveds*, Amsterdam: Rodopi, 1991.

Kuppanna Sastry, T.S., *Vedāṅga Jyotiṣa of Lagadha*, critical ed. K.V. Sarma, New Delhi: Indian National Science Academy, 1984.

Lubotsky, A., *A Ṛgvedic Word Concordance*, New Haven: American Oriental Society, 1997.

Lal, B.B., 'It is Time to Rethink', *The Aryan Debate*, ed. T.R. Trautmann, Delhi: Yoda Press, 2005.

Macdonell, A.A., *Vedic Mythology*, Strasburg (1897), rpt. Delhi: Munshiram Manoharlal, 2000.

———, *A Vedic Reader for Students*, Delhi (1917), rpt. Low Price Publications, 1995.

Macdonell, A.A. and A.B. Keith, *Vedic Index of Names and Subjects*, London: John Murray, 1912.

Mahadevan, I., 'Dravidian Parallels in Proto-Indian Script', *Journal of Tamil Studies*, vol. 2, no. 1 (1970), pp. 157-276.

Marshack, A., *The Roots of Civilization: The Cognitive Beginning of Man's First Art, Symbol and Notation*, London: Weidenfeld and Nicolson, 1972, p. 43.

McIntosh, J.R., *The Ancient Indus Valley: New Perspectives*, Santa Barbara: ABC-CLIO, Inc., 2008.

Mill, J., *The History of British India*, London (1817), abridged rpt. Chicago: University of Chicago Press, 1975.

Monier-Williama, M., *A Sanskrit-English Dictionary*, Oxford (1899), rpt. Delhi: Motilal Banarsidass, 1999.

Müller, F.M., *History of Ancient Sanskrit Literature*, London: Williams and Norgate, 1859.

———, *Rig-Veda Samhitā: The Sacred Hymns of Brāhmans*, vol. IV, London: Oxford University Press, 1892.

Narahari Achar, B.N., 'On the Meaning of AV.XIX.53.3. Measurement of Time?', *Electronic Journal of Vedic Studies*, vol. 4. no. 2, 1998a, pp. 21-6.

———, 'Enigma of the Five Year Yuga of Vedāṅga Jyotiṣa', *Indian Journal of History of Science*, vol. 33, no. 2, 1998b, pp. 101-9.

———, 'On the Vedic Origins of the Ancient Mathematical Astronomy in India', *Journal of Studies of Ancient India*, vol. I, 1998c, pp. 95-108.

———, 'Searching for Nakṣatrās in the Ṛgveda', *Electronic Journal of Vedic Studies*, vol. 6, no. 2, 2000a.

———, 'On the Caitrādi Scheme', *Indian Journal of History of Science*, vol. 35, no. 4, 2000b, pp. 295-310.

de Nicolás, A.T., *Meditations through the Ṛgveda*, Boulder & London: Shambhala, 1978.

O'Flaherty, W. D., *Hindu Myths*, London: Penguin Books, 1975.

———, *The Rig Veda: An Anthology*, London: Penguin Books, 1981.

Ôhashi, Y. '*Development of Astronomical Observations in Vedic and post-Vedic India*', *Indian Journal of History of Science*, vol. 28, no. 3, 1993, pp. 185-251.

Oldenberg, H., *The Gṛhya Sūtras*, Part I, Oxford (1886), rpt. Delhi: Motilal Banarsidass, 1964.

———, *Prolegomena on Metre and Textual History of the Ṛgveda*, Berlin (1888), tr. V.G. Paranjape and M.A. Mehendale, Delhi: Motilal Banarsidass, 2005.

———, *The Religion of the Veda*, Kiel (1894), tr. Shridhar B. Shrotri, Delhi: Motilal Banarsidass, 1988.

Panikkar, R., *The Vedic Experience Mantramañjarī*, London: Longman & Todd, 1979.

Parpola, A. et al., *Further Progression the Indus Script Decipherment*, Copenhagen: Scandinavian Institute of Asian Studies, Special Publication no. 3, 1970.

Parpola, A., 'From Archaeology to a Stratigraphy of Vedic Syncretism', *The Vedas: Texts, Language, and Ritual,* Proc. Third International Vedic Workshop, Leiden, 2002, ed. A. Griffiths and J.E.M. Houben, Groningen: Egbert Forsten, 2004, pp. 521-55.

———, *Study of the Indus Script*, Special Lecture, 50th International Conference of Eastern Studies, Tokyo, 2005, pp. 28-66.

Pingree, D., 'The Mesopotamian Origin of Early Indian Mathematical Astronomy', *Journal of History of Astronomy*, vol. IV, 1973, pp. 1-12.

———, 'History of Mathematical Astronomy in India', *Dictionary of Scientific Biography*, ed. C. Coulston, New York: Charles Scribner's Sons, 1970-80, pp. 533-633.

———, 'Jyotiḥśāstra, Astral and Mathematical Literature', *A History of Indian Literature*, Wiesbaden: Otto Harrassowitz, 1981.

———, 'MUL.APIN and Vedic Astronomy', *Studies in Honor of Åka Sjöberg*, Philadelphia: University of Pennsylvania Museum Publication, 1989b, pp. 439-45.

———, 'Astronomy in India', *Astronomy before the Telescope*, ed. C. Walker, London: British Museum, 1996.

Pingree, D. and P. Morrissey, 'On the Identification of the Yogatārā of the Indian Nakṣatra', *Journal of History of Astronomy*, vol. XX, 1989a, pp. 99-119.

Possehl, G.L., *A Short History of Archaeological Discoveries at Harappa*, Harappa Excavations 1986-1990: A Multidisciplinary Approach to the Third Millennium Urbanism, ed. Richard H. Meadow, Madison: Prehistory Press, 1991.

———, *The Indus Civilization: A Contemporary Perspective*, Walnut Creek: AltaMira Press, 2002.

von Rad, U. et al., 'A 5000-yr Record of Climate Change in Varved Sediments from the Oxygen Minimum Zone off Pakistan, North-eastern Arabian Sea', *Quaternary Research*, vol. 51, 1999, pp. 39-53.

Raghavan, V., *Rtu in Sanskrit Literature*, Delhi: Shri Lal Bahadur Shastri Kendriya Sanskrit Vidyapeetha, 1972.

Ranade, H.G., *Katyayana Śrautra Sūtra*, Pune: Ranade, 1978.

———, *Lātyāyana Śrautra Sūtra*, Delhi: Motilal Banarsidass, 1998.

Ratnagar, S., 'The End of the Harappan Civilization', *The Aryan Debate*, ed. T.R. Trautmann, Delhi: Yoda Press, 2005.

Renou, L., 'Vedic Ṛtu', *Indian Culture*, vols. 14-15, 1948-49, pp. 21-6.

———, Études védiques et pàninéennes, 17 fasc. Paris, E. de Broccard, 1955-69.

Richards, E.G., *Mapping Time: The Calendar and its History*, Oxford: Oxford University Press, 1998, p. 3.

Saha, M.N. and N.C. Lahiri, *History of the Calendar*, Report of the Calendar Reform Committee, Part C, New Delhi: Council of Scientific and Industrial Research, 1992.

Sarkar, A. et al., 'High Resolution Holocene Monsoon Record from the Eastern Arabian Sea', *Earth and Planetary Science Letters*, vol. 177, 2000, pp. 209-18.

Sarasvati Amma, T.A., *Geometry in Ancient and Medieval India*, Delhi: Motilal Banarsidass, 1979.

Sarup, L., *The Nighaṇṭu and the Nirukta*, Delhi: Motilal Banarsidass, 1920-7.

Sen, C., *A Dictionary of the Vedic Rituals*, Delhi: Concept, 1978.

Seidenberg, A., 'The Origin of Mathematics', *Archive for History of Exact Science*, vol. 18, 1978, pp. 301-42.

Sewell, R. and Ś.B. Dīkṣita, *The Indian Calendar*, London: S. Sonnenschein, 1896.

Shamasastry, R., *Vedangajyautisha* (Sanskrit Commentary and English translation), Mysore: Asst. Supt., Govt. Branch Press, 1936.

Staal, F., 'The Ignorant Brahmin of the Agnicayana', *Annals of the Bhandarkar Oriental Research Institute*, 1978, p. 345.

———, *Agni: The Vedic Ritual of the Fire Altar*, Berkeley: Asian Humanities Press, 1983.

———, 'From Prāṅmukham to Sarvetomukham: A Thread through Śrauta Maze', *The Vedas: Texts, Language, and Ritual*, Proc. Third International Vedic Workshop, Leiden, 2002, ed. A. Griffiths and J.E.M. Houben, Groningen: Egbert Forsten, 2004.

Stuhrmann, R., 'Ṛgvedisch púr', *Electronic Journal of Vedic Studies*, vol. 15, no. 1, 2008, pp. 1-42

Subbarayappa, B.V. and K.V. Sarma, *Indian Astronomy: A Source Book*, Bombay: Nehru Centre, 1985.

Talageri, S.G., *The Ṛgveda: A Historic Analysis*, New Delhi: Aditya Prakashan, 2000.

Thibaut, G., 'On the Śulvasutras', *Journal of the Royal Asiatic Society of Bengal*, vol. 44, 1875, pp. 227-75.

———, 'Contribution to the Explanation of Jyotisha-Vedánga', *Journal of the Royal Asiatic Society of Bengal*, 46, 1877, pp. 411-37.

———, Grundriss der indo-arischen philologie und altertumskunde, begründet von G. Bühllr, fortgesetzt von F. Kielhorn. III. Band, 9. Heft. Astronomie, Astrologie und Mathematik, 1899, p. 10.

Thieme, P., 'The 'Aryan' Gods of the Mittanni Treaties', *Journal of the American Oriental Society*, vol. 80, 1960, pp. 301-17.

Thomas, E.J., *Vedic Hymns*, London: John Murray, 1923.

Thomson, K., 'Sacred Mysteries: Why the Rigveda has Resisted Decipherment', *Times Literary Supplement*, 26 March 2004, pp. 14-15.

Thomson, K. and J. Slocum, www.utexas.edu/cola/centers/lrc/RV7/, 2009

Tilak, B.G., *The Orion*, Poona: Tilak Bros., 1893, p. 31.

———, *Vedic Chronology and Vedāṅga Jyotiṣa*, Poona: Tilak Bros., 1925, p. 45.

Trautmann, T.R. (ed.), *The Aryan Debate*, New Delhi: Oxford University Press, 2005.

Vajracharya, G.V., 'The Adaptation of Monsoonal Culture by Rgvedic Aryans: A Further Study of the Frog Hymn', *Electronic Journal of Vedic Studies*, 3, 1997, pp. 3-16.

Velankar, H.D., *Ṛgveda Maṇḍala VII*, Bombay: Bharatiya Vidya Bhavan, 1963.

———, *Ṛksūktaśatī*, Bombay: Bharatiya Vidya Bhavan, 1972.

Vogel, C von., 'Die Jahreszeiten im Spiegel der altinindischen Literatur', *Zeitschrift der Deutschen Morgenländischen Gesellschaft,* vol. 121, 1971, pp. 285-310.

Wallis, H.W., *The Cosmology of the Rigveda*, Delhi: Cosmo Publications, 1999.

Weber, A., *Die vedischen Nachrichten von den naxatra (Mondstationen).* II., Abhandlungen der Königlichen Akademie der Wissenschaften zu Berlin, 1861, pp. 267-400.

———, *Uber den Vedakalender, namens Jyitisham*, Abhandlungen Der Königlichen Akademie der Wissenschaften zu Berlin, 1862, pp. 1-130.

Weiss, H. Courty et al., *Science,* vol. 261, 1993, pp. 995-1004.

Wheeler, R.E.M., 'Harappa 1946: The Defences and Cemetery R37', *Ancient India*, vol. 3, 1947, pp. 58-130.

Whitney, W.D., *Oriental and Linguistic Studies*, New York: Scribner, Armstrong Company, 1874.

———, 'On a Recent Attempt by Jacobi and Tilak to Determine on Astronomical Evidence the Date of the earliest Vedic Period 4000 BC', *Indian Antiquary*, vol. 24, 1895, pp. 361-67.

———, *Atharva-Veda-Saṃhitā*, Cambridge (1905), rpt. Delhi, Motilal Banarsidass, 1962, pp. 987-91.

Williams, S.W., http://www.math.buffalo.edu/mad/Ancient-Africa/lebombo.html, 2005.

Wilson, H.H., *Ṛgveda Saṃhitā: A Collection of Ancient Hindu Hymns*, 4 vols. London, (1850-88); ed. and rev. Ravi Prakash Arya and K.L. Joshi, Delhi: Parimal Publications, 1997.

Winternitz, M., *A History of Indian Literature*, Weinberge (1907), tr. V. Srinivasa Sarma, Delhi: Motilal Banarsidass, 1981.

Witzel, M., 'On the Localisation of Vedic Texts and Schools', *India and the Ancient World: History, Trade and Culture before AD 650*, ed. Gilbert Pollet, Leuven: Orientalia Lovaniensia Analecta 25, 1987, pp. 173-214.

———, 'Early Indian History: Linguistic and Textual Parameters', *The Indo-Aryans of Ancient South Asia*, ed. George M. Erdosy, Berlin: Walter de Gruyter, 1995, pp. 85-125.

———, 'Early Sanskritization: Origin and Development of the Kuru State', *Electronic Journal of Vedic Studies*, vol. 1, no. 4, 1995a, pp. 1-26.

———, 'Looking for the Heavenly Casket', *Electronic Journal of Vedic Studies*, vol. 1, no. 2, 1995.

———, 'The Development of the Vedic Canon and its Schools: The Social and Political Milieu', *Inside the Texts, Beyond the Texts*, Opera Minora 2, ed. M. Witzel, Cambridge: Harvard University, Dept. of Sanskrit and Indian Studies, 1997, pp. 257-400.

———, 'Aryan and non-Aryan Names in Vedic India: Data for the Linguistic Situation, *c.* 1900-500 BC', *Proc. International Seminar on Aryan and non-Aryan in South Asia: Evidence, Interpretation and Ideology*, Opera Minora 3, ed. J. Bronkhorst and M.M. Deshpande, Cambridge: Harvard University, Dept. of Sanskrit and Indian Studies, 1999a, pp. 337-60.

———, 'Early Sources for South Asian Substrate Languages', *Mother Tongue* (Special Issue), 1999b, p. 8.

———, 'Autochthonous Aryans? The Evidence from Old Indian and Iranian Texts', *Electronic Journal of Vedic Studies*, vol. 7, no. 3, 2001.

———, 'Vala and Iwato, the Myth of the Hidden Sun in India, Japan, and Beyond', *Electronic Journal of Vedic Studies*, vol. 12, 2005, pp. 1-69.

Witzel, M. and T. Goto, *Rig-Veda: Das heilige Wissen. Erster and Zweiter Leiderkris*, Frankfurt am Main: Verlag der Weltreligionen, 2007.

Wolpert, L. *The Unnatural Nature of Science,* London: Faber and Faber, 1992, pp. 45-7.

Zimmer, H., *Altindischens Leben die Cultur der Vedischen Arier Nach den Samhita dargestellt*, Berlin: Weidmann, 1879.

Index of Passages from the Vedic Texts

Passages in the Appendix are preceded by 'A'

AB.i.7 (ii.1), 90
AB.i.30(v.4), 66
AB.ii.17 (vii.7), 104
AB.iii.33 (xiii.9), 143
AB.iii.41 (xiv.3), 98
AB.iii.44 (xiv.6), 90
AB. iv.6 (xvi. 6), 97
AB.iv.12 (xvii.6), 107
AB.iv.15 (xviii.1), 97
AB.iv.18 (xviii.4), 107
AB.iv.26 (xix.4), 76, 107, 169, 176
AB.iv.29 (xx.1), 97
AB.v.30 (xxv.5), 104
AB.vii.11 (xxxii.10), 119, 199, A271
AB.vii.13 (xxxiii.1), 98
AB.viii.19 (xxxix.5), 98
AB.viii.28 (xl.5), 91, 168

AV.II.28.4, 103
AV.III.10.5, 104
AV.V.6.4, 100
AV.V.25.10, 98
AV.VI.30.1, 58
AV.VIII.2.22, 77
AV.VIII.9.16-18, 80
AV.X.8.5, 114
AV.X.8.7, 90
AV.X.8.9, A270
AV.XI.3.7, 55
AV.XII.2.25, 91
AV.XIV.1.13, 140
AV.XIX.6.10, 65
AV.XIX.7.2-5, 141
AV.XIX.8.2, 145
AV. XIX.53.3, 203, A272
AV. XIX.53, 54, 203

ĀGS.II.3.1 175
ĀGS.III.5.2, 175

KB.iii.1, 119
KB.iii.2, 104
KB.iii.4, 74
KB.iv.4, 180
KB.v.1, 145, 180
KB.v.8, 101
KB.vi.15, 126
KB.vii.6, 89
KB.xi.7, 104
KB.xix.2, 175
KB.xix.3, 98, 101, 102, 107, 120, 121, 176, 228

KB.xix.8, 107
KB.xxvi.1, 98

KS.28.6, 134
KS.39.13, 141
KS.40.4, 143

KŚS 5.1.1, 169
KŚS 5.3.1, 169
KŚS 5.6.1, 169, 171
KŚS 5.11.2, 170

LŚS.IV.8.1-7, 124
LŚS.IV.8.8-21, 124
LŚS.IX.12.13, 175
LŚS X.13.4-16, 106

MS.I.10.8, 119, 120, 124, 125, 172, 206, 207, 214, 219, 227, 229, 242, A266
MS.II.8.13, 143
MS.II.13.20, 141
MS.III.12, 100

PB.V.9, 120, 122, 124, 176, 180, 229
PB.V.10, 124
PB.IV.6.17, 102
PB.VI.2.2, 66
PB.VI.10.4, 107
PB.IX.3.5, 6, 104
PB.XVII.13.17, 104, 110
PB.XXII.3, 97
PB.XXIII.23, 145
PB.XXIV.20, 92, 93, 104
PB.XXV.10.1, 58

RJ.1, 2, 3, 197
RJ.4, 200
RJ.5, 157, 161
RJ.5-6, 208, 209, 210, 213, 224, 233, 238, 239
RJ.6, 228
RJ.7, 230
RJ.8, 217
RJ.9, 138, 147
RJ.10, 201
RJ.11, 226
RJ.12, 215
RJ.14, 155, 221
RJ.16, 17, 203
RJ.18, 155, 165, 220
RJ.20, 205
RJ.21, 227
RJ.22, 231
RJ.23, 202
RJ.25-28, 142, 155, 221
RJ.30, 197
RJ.31, 228
RJ.32, 199, 216, 217, 229

ṚV.I.15, 63
ṚV.I.20.7, 65, A253
ṚV.I.23.15, 36, 65
ṚV.I.24.10, A269
ṚV.I.25.8, 111, 112, 135, A265
ṚV.I.25.36
ṚV.I.31.14, A256
ṚV.I.34.2, 65, A251
ṚV.I.34.5, 65, A251
ṚV. I.34.8, A259
ṚV.I.50.2, A268
ṚV.I.50.4, A257
ṚV.I.50.9, A257
ṚV.I.92.11, 109, A264
ṚV.I.95.3, A249
ṚV.I.112.11, 37
ṚV.I.113.3, A259
ṚV.I.117.21, 36, A248
ṚV.I.118.2, 65, A252
ṚV.I.133.1, 59, A248
ṚV.I.133.3, 59, A248
ṚV.I.157.3, 65, A252
ṚV.I.158.6, 109, A265
ṚV.I.164.2, 64, 65, A252

ṚV.I.164.11, A263
ṚV.I.164.12, 65, 67, A253
ṚV.I.164.13, 65, A254
ṚV.I.164.15, 65, 114, A255
ṚV.I.164.44, A262
ṚV.I.164.48, 65, A253
ṚV.I.179, 43
ṚV.II.33.2, A250
ṚV.II.36, 63, 76, 99
ṚV.II.37, 63
ṚV.II.38.4, A257
ṚV.III.3, 101
ṚV.III.7.7, 79
ṚV.III.32.9, A259
ṚV.III.33.5, 94
ṚV.III.53.8, 94
ṚV.III.54.19, A256
ṚV.III.58.1, A258
ṚV.IV.13.3, A258
ṚV.IV.13.5, A258
ṚV.IV.36.1, 65, A253
ṚV.IV.51.2, 140, A270
ṚV.IV.57.4, 36, A249
ṚV.V.14.4, 63, A250
ṚV.V.32.2, A256
ṚV.V.45.7, A262
ṚV.V.51.14, 140
ṚV.V.51.17, 144
ṚV.V.54.13, A270
ṚV.V.62.7, 36
ṚV.V.78.8, 9, 98
ṚV.V.78.8, A261
ṚV.V.78.9, A261
ṚV.V.83, 43, 69
ṚV.V.83.4, 69
ṚV.V.83.8, 69
ṚV.V.85.3, 36, A269
ṚV.VI.15.8, 109, A264
ṚV.VI.59.6, 94, A260
ṚV.VII.18.5-10, 42
ṚV.VII.19.2-4, 42
ṚV.VII.33.3-6, 42
ṚV.VII.66.16, A249
ṚV.VII.83.4, 69
ṚV.VII.83.8, 69
ṚV.VII.86.1, A268
ṚV.VII.87.4, 65, A254
ṚV.VII.87.5, 65, A255
ṚV.VII.95.2, 58
ṚV.VII.101, 69
ṚV.VII.103, 69, 134
ṚV.VII.103.8, A262
ṚV.VII.103.9, 64, 111, A251
ṚV.VIII.2.3, 36
ṚV.VIII.55.2, A268
ṚV.VIII.68.14, 65
ṚV.VIII.72.8, A269
ṚV.VIII.93.13, 140
ṚV.IX.72.9, 140
ṚV.IX.86.29, 79
ṚV.X.2.7, A262
ṚV.X.13.3, 79
ṚV.X.14.2, 36
ṚV.X.19.1, 140, A270
ṚV.X.27.8, 36
ṚV.X.48.7, 36
ṚV.X.60.8, 109, A263
ṚV.X.61.5-7, 143
ṚV.X.61.7, 143, A271
ṚV.X.72.1, 109, A264
ṚV.X.72.2, 109, A265
ṚV.X.85.2, A271
ṚV.X.85.5, 98, A261
ṚV.X.85.13, 140, A259
ṚV. X.90.6, 65, 69, A255
ṚV.X.90.7, 65
ṚV.X.90.15, 65, A254
ṚV.X.95, 43
ṚV.IX.97.54, 125
ṚV.X.101.3, 109, A264
ṚV.X.124.1, 79
ṚV.X.129, 36
ṚV.X.131.2, 36
ṚV.X.138.6, 98, A260

ṚV.X.156.4, A267
ṚV.X.161.4, A250
ṚV.X.184.3, 98
ṚV.X.189.3, 94, A260
ṚV.X.190.2, 133, A263

ŚB.I.3.4.7, 107
ŚB.I.3.5.11, 66
ŚB.I.3.5.16, 63
ŚB.I.5.3.9-13, 107
ŚB.I.5.4.1, 46
ŚB.I.6.4.18-20, 91, 168
ŚB.I.7.2.26, 92
ŚB.I.8.3.17, 94
ŚB.II.1.1.12, 104
ŚB.II.1.2.4, 139
ŚB.II.1.2.2, 148
ŚB.II.1.2.4, 139
ŚB.II.1.2.18, 137
ŚB.II.1.3.1, 74, 75, 102
ŚB.II.1.3.3, 101
ŚB.II.2.3.7, 77
ŚB.II.2.6-10, 143
ŚB.II.3.2.5, 94
ŚB.II.6.1.1, 102
ŚB. II.6.3.13, 170
ŚB.II.6.4.2-4, 65
ŚB.III.2.2.4, 104
ŚB.IV.2.1.18, 57
ŚB.IV.3.1.14-20, 76, 99
ŚB.IV.5.2.4, 134
ŚB.V.4.1.3-7, 89
ŚB.V.5.2.5, 62
ŚB.VI.2.2.18, 168, 180
ŚB.VI.6.4.3, 104
ŚB.VI.7.1.18, 104
ŚB.VI.7.1.19, 79
ŚB.VI.7.3.9, 103
ŚB.VII.2.2.14, 128
ŚB.VII.2.4.26, 109
ŚB.VII.4.2.31, 103
ŚB.VII.4.11, 102
ŚB.VII.5.1.37, 90
ŚB.VIII.2.1.3, 62
ŚB.VIII.3.2.6, 77, 104, 129
ŚB.VIII.5.2.10, 63
ŚB.VIII.7.1.1, 62
ŚB.VIII.7.1.3, 62, 128
ŚB.IX.1.1.43, 117
ŚB.IX.1.4.1, 119
ŚB.IX.1.5.1, 119
ŚB.IX.3.3.18, 117
ŚB.X.4.2.18, 94
ŚB.X.4.2.20, 96
ŚB.X.4.2.23-25, 96
ŚB.X.4.3.8, 117, 128
ŚB.X.4.3.19, 117
ŚB.X.5.4.5, 117, 145
ŚB.XI.1.1.7, 175
ŚB.XI.1.6.2, 134
ŚB. XII.1.2.1-22, 92, 93, 104
ŚB.XII.1.3.9, 107
ŚB.XII.2.3.6, 104
ŚB.XII.3.2.5, 96
ŚB. XII.8.2.35, 63

SGS.I.26, 145

TB.1.1.2.1, 146
TB.1.1.2.6, 81
TB.1.1.2.8, 168
TB.1.5.1, 145
TB.1.5.2.1, 168, 216, A272
TB.2.1.4, 90
TB.2.6.19, 76
TB.2.8.6, 103
TB.2.8.7, 92
TB.3.1.1-2, 145, 153
TB.3.1.2, 148
TB.3.1.4, 142, 143
TB.3.1.5, 145, 154
TB.3.4.4, 19, 147
TB.3.7.11, 103
TB.3.8.1, 115

TB.3.8.3, 100
TB 3.8.10, 66
TB.3.10.1, 74, 95, 99, 100, 109, 110, 129, 130
TB.3.10.4, 104, 107, 110
TB.3.11.10, 107
TB.3.12.8.7, 8, 200

TS.I.4.14, 99, 100
TS.II.5.8.3, 104
TS.II.6.1.1, 74, 75
TS.III.4.7, 90
TS.III.5.1, 99
TS.IV.4.5.1, 143
TS.IV.4.10, 141, 145
TS.IV.4.11, 76
TS.IV.11.1, 99
TS.VI.5.3, 81, 100, 102
TS.V.2.5, 90, 128
TS.V.4.12, 66
TS.V.7.6, 107
TS.VI.4.10.2, 3, 90
TS.VI.5.3, 81, 100, 102
TS.VII.1.10, 111, 136, 239
TS.VII.2.3, 89
TS.VII.4.8, 108, 120, 121, 122, 124, 167, 168, 175, 176, 180, 199, 206, 217, 219, 229, 243, A267
TS.VII.5.6, 98, 99
TS.VII.5.7, 98
TS.VII.5.7.1, 124
TS.VII.5.25, 139

VS.IV.25, 89
VS.VII.30, 99, 100
VS.IX.7, 145
VS.XXII.30, 100
VS.XXII.31, 100
VS.XXIV.11, 76
VS.XXIV.20, 76
VS.XXIV.36, 139
VS.XXVII.45, 104, 110
VS.XXX.10, 147
VS.XXXV.15, 103
VS.XXXVI.24, 103

YJ.1, 2, 197
YJ.3, 197
YJ.5, 199, 216, 217, 229
YJ.6, 157, 161, 146, 216, 229
YJ.6-7, 208, 209, 210, 213, 224, 233, 238, 239
YJ.7, 228
YJ.8, 230
YJ.9, 217
YJ.10, 206, 218
YJ.11, 206, 207
YJ.12, 204
YJ.12b, 30d, 39d, 203
YJ.13, 200
YJ.15, 201
YJ.18, 155, 221
YJ.19, 226
YJ.20, 225
YJ.21, 227
YJ.22, 205
YJ.23, 228
YJ.24, 203
YJ.25, 224
YJ.26, 224, 225
YJ.27, 215
YJ.28, 199, 205, 206, 218, 220
YJ.29, 219
YJ.30a-c, 219, 220
YJ.31, 220
YJ.32-35, 155, 221
YJ.36, 221
YJ.37, 207, 213, 229
YJ.38, 203
YJ.39, 155, 165, 220
YJ.40, 231
YJ.41, 202
YJ.43, 197

Index

acitaḥ, 125
adhvaryu, 47, 79, 90, 102, 106, 129, 130, 263
Agastya, 43
Aghā, 140
agnicayana, 23, 81
agricultural societies, 51
aha, 17, 92, 103
ahargaṇa, 131
āhavanīya, 125, 126, 128, 129
ahorātra, 17, 92, 96, 133
Aitareya Brāhmaṇa, 11, 47, 49, 52, 79, 199
akṣara, 202, 203
altitude, 85, 86, 209, 231
amānta month, 99
amāvāsyā, 91, 94, 199, 274
aṃśa, 200
aṅgas, 235
Anukramaṇī, 42
Āpastamba Śrauta Sūtras, 49
Āraṇyakas, 39
Archaeological Survey of India (ASI), 26
Arjunī, 140
Ārca-Jyotiṣa, 11, 195, 196, 213, 221
Arundhatī, 144
Āryan debate, 36
Ārya, 23, 24, 33, 34, 36, 42
Āśvalāyana Śrauta Sūtras, 11, 38, 49
Atharvaveda Saṃhitā, 11, 38, 45, 49, 55, 65, 77, 80, 140, 234
atmospheric refraction, 231
autumnal equinox, 81, 86, 145
avaka, 129
ayana, 76, 101, 121, 197, 217, 228, 230
azimuth, 85, 86, 87

Bactria-Margiana Archaeological Complex (BMAC), 34
battle of ten kings, 42
Baudhāyana Śrauta Sūtras, 38, 49, 52
bhāga, 204
bhāṃśa, 158, 162, 200
Bhandarkar Devadatta Ramkrishna, 26
Brāhmaṇas, 39, 46
Bṛhaddevatā, 42

calendar stick, 83, 106, 133

cardinal day, 14, 133; directions, 74, 79, 131; point, 85, 89
Cāturhotra Cayana, 126, 132
cāturmāsya sacrifices, 49, 65, 66, 72, 119, 169, 173, 183, 189, 233, 244
cāturmāsya-yājins, 119
celestial coordinate system, 86, 193
celestial equator, 86, 193
circumpolar, 140, 190, 191, 233, 244
Citra, 108, 140, 165, 167, 168
clepsydra, 165, 197, 202
clock arithmetic, 192
Colebrooke Henry Thomas, 53, 196
Cunningham, Alexander 26

dakṣiṇāgni, 125
Dakṣiṇāyana, 101, 102
Darśapūrṇamāsa, 44, 99, 118, 272
decans, 137
declination, 193
deva-nakṣatras, 145, 153
Devāyana, 76, 101, 146
Dravidian language family, 26

ecliptic coordinate system, 192, 222, 244
ecliptic latitude, 162, 183, 193, 194
ecliptic longitude, 162, 171, 177, 192, 193, 194
ecliptic, 19, 81, 86, 144, 155, 168
Ellis, Francis Whyte 26
equatorial coordinate system, 86, 192

firmament, 22, 85, 89, 126, 139, 190
First Point of Aries, 162, 193
five seasons, 66, 74, 77, 82

gārhapatya, 125
Gavām ayana, 17, 92, 97, 101, 120, 169
Ghaggar-Hakra, 27, 30, 59
Gharma-Pravargya ritual, 67, 115
Gṛhya Sūtras, 39, 50
grīṣma, 62, 102, 128, 170
o
Harappa, 26, 32
heliacal rising, 86, 168, 181, 209, 216
heliacal setting, 86, 168, 181, 209, 216
hemanta, 62, 75, 102, 129, 170
hiraṇyagarbha, 42
Hittite-Mitanni treaty, 33
horizon reference system, 85
hotṛ, 49, 105, 126
Hsiu (*Xiù*), 137
Hsüang Tsang, 26
hunter-gatherer societies, 14, 27, 62, 82, 83

ideal year, 18, 135
Indo-European language, 25, 33, 59, 247
Indus Civilization, 26, 28, 29, 33, 138
Indus-Sarasvati Civilization, 29
intercalation, 18, 21, 100, 110, 114, 119, 123, 126, 174, 180, 200, 206, 208, 229, 242
interregnum, 121, 124
Iṣa, 76, 128

Jaiminīya Brāhmaṇa, 11, 52
Janaka, 48
jāvādi arrangement, 155, 157, 161, 167, 221
Jhukar Phase, 32
Jones William, 25, 53, 196, 236
Julian days, 163, 193

kāla, 198, 202
kālavidhānaśāstra, 197
Kāṇva Saṃhitā, 44
Kassites, 33
kāṣṭā, 203
Kathāsaritsāgra, 54
Kātyāyana Śrauta Sūtra, 11, 49
Kātyāyana, 42, 54
Kauṣītaki Brāhmaṇa, 11, 47, 56
Kṛṣṇa (Black) *Yajurveda*, 39, 44
kṛṣna pakṣa, 130, 199
Kṛttikās, 138, 143, 146, 149, 153, 165, 167, 171, 190, 245
kṣudra-cayaya, 125
kṣudra-muhūrta, 96, 131

lokampṛṇā bricks, 128
Lopāmudrā, 43
lunar year, 19, 108, 112
lunation, 84, 93, 98, 113

Madhava, 76, 128
Madhu, 76, 128, 167
Mādhyaṃdina Saṃhitā, 45
Maghās, 140, 157
mahā-cayaya, 125, 126, 132
Mahāvrata, 101, 105, 107, 124
Mahidāsa Aitareya, 47
maṇḍala, 40
mantra, 39, 273
māsa, 17, 98, 103
Marsden William, 25
Marshall, John, 27
mātrā, 203
Mātsya, 216
Mehrgarh, 29
Mesopotamian astronomy, 147, 189
Metonic cycle, 113, 135
mīmāṁsakas, 120
mīn, 138
Modular arithmetic, 192
Mohenjo-daro, 26
Month(s), 13, 17, 84, 90, 98, 99, 103, 104, 174, 198; thirteenth, 80, 100, 114, 120, 130, 135; twelve, 66, 111, 128, 205
mṛga, 143
Mṛgaśīrṣa, 138, 143
mṛgavyādha, 143
muhūrta, 94, 96, 128, 197, 203
MUL.APIN, 190, 237
Müller, F. Max 53, 198

Nabha, 76, 128
Nabhasya, 76, 128
Nāciketa Cayaṇa, 125, 131
nāḍikā, 198, 203
nāḍīmaṇḍala, 204
nakṣatra, 18, 93, 100, 145, 155, 161, 198, 221, 243
nakṣatradarśa, 19
nakṣatra-sector, 22, 138, 156, 158, 162, 198
nakṣatravidyā, 147
Nebra sky disc, 134
Newgrange, 133

pāda parvans, 219
pāda, 42, 96, 198
Padapāṭha, 41
Paippalāda, 45
Paitāmahasiddhānta, 148
pala, 203
Palaeoclimatic Data, 70
Pāṇini, 53, 175
parivatsara, 104
Parvan, 200
parvan-bhā'ḍānakalā, 226
parvan-rāśi, 200
Phalgunīs, 108, 140, 145, 168, 169, 180
Pitrāyana, 101, 146
Pleiades, 134, 138, 144, 167, 171, 191, 245

Prajāpati, 16, 79, 94, 96, 126, 143
Prātiśākhya, 41
prāvṛṣ, 63, 64, 103
precession of the equinox, 165, 173, 180, 193, 209, 215, 245
proper motion, 177, 193
psudo-*yogatārā*s, 184
Punarvasū, 140
Punjab Phase, 32
pūrṇamāsī, 94
pūrṇimānta month, 99
Purūravas, 43
Puruṣa, 43, 65, 126
Puruṣṇī-Ravi, river, 42
Puṣya, 140

radial motion, 194
Rakhal Das Banerji, 27
Rangpur Phase, 32
ṛc (*ṛcā*s), 38
Revatī, 140
Ṛgveda Saṃhitā (*Ṛgveda*), 11, 40, 51, 53
right ascension, 148, 192
ritual year, 17, 66, 96, 98, 104, 107, 111, 120, 124, 199
ṛkṣa, 139
Roth, Fr. Heinrich, 25
ṛta, 16
ṛtu, 17, 62, 63, 205, 207
Ṛtuśeṣa, 202, 207
ṛtu-yājins, 119
ṛtvij, 62

Śabara, 121
ṣaḍaha, 97, 105
Saha, 76, 129
Sahasya, 76, 129
Sahni, Rai Bahadur Daya Ram, 27
Śākalya, 40
Sākamedha, 169, 170, 171
śākhā, 40
Sāmaveda Saṃhitā, 39, 45
saṃvatsara, 16, 103, 111, 126, 134
sāṃvatsarik sattra, 104
Śāṇḍilya, 48, 80, 125
sapta ṛṣi, 138, 140, 143, 190, 245
Sapta Sindhu, 35
śarad, 61, 63, 81, 103
Sarasvatī, 27, 29, 31, 58
Sarvānukramaṇī, 42
Śatapatha Brāhmaṇa, 11, 48, 52, 56, 77, 96, 117, 125
Śaunaka, 41, 45
sāvana, 198
Sāvitra Cayana, 74, 94, 129
Sāyaṇa, 120, 121
seasonal calendar, 61, 82, 146
seasons, 13, 21, 61, 63, 81, 82, 84, 99, 102, 103, 111, 128, 144, 169, 197, 202, 205, 215, 240; three, 64, 68, 73, 74; five, 66, 69, 74, 78; six, 66, 69, 76, 79; seven, 72, 80
Shastri, Pattabhiraman, 26
Shastri, Shankara, 26
Sidereal month, 202, 219;
śiśira, 75, 102, 107, 167, 170, 234
Skanda, 143, 167, 216
smṛti, 39, 273
spring equinox, 87, 108, 193
Śrauta Sūtras, 39, 49
Śuci, 76, 128
Sudās, 42
Śukla (White or Clear) *Yajurveda*, 44, 48
śukla pakṣa, 103, 199
Śukra, 76, 128
sūkta, 40, 276
Śulbasūtra, 50, 128, 134
summer solstice, 81, 86, 101
śunāsīrīya, 101, 169

synodic month, 15, 89, 98, 99, 145, 162, 198; period, 133, 176; year, 19, 112, 120
śyāma ayas-iron, 55

Taittirīya Brāhmaṇa, 11, 47, 74, 94, 96, 99, 125; *Saṃhitā*, 11, 44, 112, 125
tally stick, 106, 133
Tapas, 76, 107, 129, 209
Tapasya, 76, 107, 129
tāra, 139
Tiṣya, 140
Tithi, 199
tithi-nakṣatra, 225
tropical year, 100, 103, 110, 112, 121, 205, 241
Tṛtsus, 42
Tura Kāvaṣeya, 127

udgātṛ, 48, 106, 276
Upaniṣads, 39, 56, 235, 277
Ursa Major (Big Bear), 20, 138, 233, 244
Urvaśī, 43
Ussher, James, 53
Uttarāyana, 101

Vaiśvadeva, 169, 170
Vājasaneyī Saṃhitā, 11, 44, 127
varṣa, 77, 104
varṣā, 63, 75, 77, 84, 102, 128
Varuṇapraghāsa, 170
vasanta, 62, 64, 81, 102, 108, 128, 167, 170
Vasiṣṭha, 41, 43, 144
Vedāṅga Jyotiṣa, 11, 16, 19, 50, 108, 138, 155, 161, 195, 244
vedāṅgas, 235
Vedic calendar, 116, 124, 233, 243; corpus, 37, 38; hour, 96; minute, 96; months, 98, 121, 174; Sanskrit, 59, 245; week, 98
vernal equinox, 77, 81, 86, 145, 149, 234
Viṣuvat, 97, 101, 105, 107, 120, 228
Viśvāmitra, 41

Wheeler, Mortimer, 33
winter solstice, 81, 87, 101, 106, 122, 133, 157, 167, 199, 209, 231, 244

Xavier, Francis, 25

Yājnavalkya, 48, 56
Yājuṣa-Jyotiṣa, 11, 195
yajusmati bricks, 128
yama-nakṣatras, 145, 153
yogatārā, 20, 148, 153, 171, 177, 181, 184, 194, 234
yuga, 108, 116, 146, 197, 199, 216, 244
Yajurveda, 11, 40, 44

zenith distance, 86, 231, 235
zenith, 85, 89
zodiac, 137